Selected Papers on Design of Algorithms

Selected Papers on Design of Algorithms

DONALD E. KNUTH

CSLI PUBLICATIONS

STANFORD

21 20 19 18 17   6 5 4 3 2

Library of Congress Cataloging-in-Publication Data

```
Knuth, Donald Ervin, 1938-
   Selected papers on design of algorithms / Donald E. Knuth.
   xvi,453 p. 23 cm. -- (CSLI lecture notes ; no. 191)
   Includes bibliographical references and index.
   ISBN 978-1-57586-583-6 (cloth : alk. paper) --
   ISBN 978-1-57586-582-9 (pbk. : alk. paper)
1. Algorithms.  2. Computer Algorithms.  3. Computer programming.
I. Title.  II. Series.
 QA9.58.K6525  2009
 518'.1--dc22                              2009027824
                                                  CIP
```

Internet page
   http://www-cs-faculty.stanford.edu/~knuth/da.html
contains further information and links to related books.

to Robert W Floyd (1936–2001)
my partner in augrime

Contents

Preface

Algorithms are the threads that tie together most of the subfields of computer science. This book is a collection of technical papers in which I've tried to introduce or make improvements to algorithms for a wide variety of intriguing tasks that have come to my attention during the past fifty years. I'm grateful for this opportunity to put the materials into a consistent format, and to correct errors in the original publications that have come to my attention. If any of this work deserves to be remembered, it is now in the form that I most wish people to remember it.

Something magically beautiful happens when a sequence of commands and decisions is able to marshal a collection of data into organized patterns or to discover hidden structure. Every once in awhile something "clicks"; an elegant procedure suddenly emerges by which small steps combine to fulfill a large objective — aha! On several occasions I've been lucky enough to encounter some of these unexpected nuggets of algorithmic paydirt, and I hope that I can communicate the essence of what makes them tick so that readers will be able to share my pleasure.

Since many different kinds of algorithms are considered here, I've organized this book by topic instead of by chronology. I doubt if anybody will actually read this book straight through from Chapter 1 to Chapter 28; yet I've tried to arrange the chapters so that such a journey will not seem like a completely random walk. Algorithms that deal with similar topics or use similar paradigms are therefore grouped together.

Chapter 1, however, is not about any particular algorithm. It is a eulogy to Bob Floyd, my long-time colleague to whom this book is dedicated, because he helped significantly to inspire so many of the ideas that are described in the other chapters. I know of no better way to begin a book about the design of algorithms than to describe Floyd's life and work.

Bob and I jointly wrote Chapter 2, which discusses questions about networks for sorting that interested us deeply during the years when we first began to work together. Chapter 3 is about a different approach to sorting, based on C. A. R. Hoare's "quicksort" but with a different way to access the data and control the activities.

The focus shifts from sorting to searching in Chapter 4, which presents an efficient way to find the best possible binary tree that will support searches for a given word, when the frequencies of each word to be looked up are known. It illustrates an important case where the general techniques of "dynamic programming" can be made to run even faster than usual.

Optimum binary trees also define optimum codes. The subject of Chapter 5 is the task of maintaining a binary tree that remains optimum as the frequency data changes. Interesting choices of data structures and tactics are involved here, with some subtle surprises.

Chapter 6 deals with another type of data rearrangement, a problem that has turned out to be much more significant than I ever suspected when I worked on it briefly in 1978. It's a question about rearranging objects on a line, when some pairs of neighboring objects are able to change places while others are not. In this case the algorithm itself is not especially important; but the associated theory, which ties this problem to the notion of "topological sorting," is useful in many models of physical and computational processes.

Topological sorting plays a role also in Chapter 7, which efficiently finds all ways to rearrange a given word so that a given ordering relation will hold between the letters in various positions. The algorithm works only when the ordering relation is "hierarchical" (nested), but many important orderings do fall into this special class.

Words of ordered literals arise as clauses in Boolean algebra. Chapter 8 shows that any set of nested clauses can be tested for satisfiability in linear time, even though closely related problems are NP-complete.

The important problem of finding all appearances of a given word or phrase in a long text is tackled in Chapter 9. The appealing algorithms discussed here can justly be called "instructive," since they have found their way into many textbooks of computer science.

The theory of automata is the link between Chapter 9 and Chapter 10: Chapter 9's algorithms were inspired by theoretical results, while Chapter 10 — which is another chapter written jointly with Bob Floyd — develops another aspect of the theory. Here we consider (and often surmount) the limitations of abstract finite-state computing devices that are able to do only the operations of addition, subtraction, and comparison.

Chapter 11, similarly, discusses subtle algorithms for arithmetic, but in this case the focus is on a proof of correctness. I discuss four lines of code from the program for TEX, which appear quite simple and innocuous, yet for which the standard methods of formal verification do not readily apply.

Chapter 12 discusses simplified techniques for proving the correctness of algorithms that pass messages between concurrent processes. An early note about the subtleties of distributed control and "critical sections" appears in Chapter 13. Then Chapter 14 focuses on another aspect of asynchronous computation, namely the problem of optimally prefetching data during inactive parts of a process so that it is present later when needed.

The center of attention in the next few chapters is optimization of combinatorial structures. Although most combinatorial problems are too huge to solve optimally, two general techniques are known to lead to efficient algorithms. The first of these, which finds optimum strings in context-free languages, is explored in Chapter 15. The second, which is based on the theory of network flows and on the more general theory of matroids, is illustrated in Chapters 16 and 17. Chapter 18 goes further to a study of "interval" patterns, not yet well understood, that go even beyond matroids.

The scene changes from combinatorics and optimization to algebra and theorem proving in Chapter 19, which discusses a technique for machine-assisted manipulation of abstract axioms via term rewriting. However, this chapter also has ties to Chapter 18, since both of them rely on a fundamental "diamond lemma" (also called the principle of *confluence*).

Chapter 20 is about efficient computations in another kind of algebraic system, when we want to work with finite groups that are generated by given permutations.

The remaining chapters deal with algorithms that have a distinctly numerical flavor. Discrete mathematics gives way to continuous constructions in Chapter 21, which goes beyond familiar algorithms of random number generation to tackle a problem of "random function generation." Chapter 22 shows how to deal with linear dependencies without amplifying errors; Chapter 23 shows how to stabilize floating-point computations. Chapter 24 is a foray into cryptography, considering the recovery of data that has been hidden by truncation. Chapter 25 computes the famous Bernoulli numbers in a novel way.

Three of my very first papers — written when I was just beginning to "learn the ropes" — have been saved for the end. They are largely concerned with getting the most I could out of the small and slow machines that were available to me at the time.

Chapter 28 is probably the most interesting of these early works, to me at least, because it contains the first research paper that I dared to submit to a major technical journal. I must confess that I can't remember ever encountering anyone who claimed to have read that paper carefully, except for the referees who studied it before publication; in essence, I had solved a problem that was no longer relevant in 1961, due to rapid changes in technology, namely the optimum placement of instructions and data on a rotating drum. Still, after rereading that paper today, I'm glad to have written it, because it demonstrates a way to blend manual and automated methods that I've used with success many times since: A human being does the creative part, a machine does the drudgework and checks for mistakes. This paper also includes an intriguing integer programming problem that still serves as a useful benchmark for IP solvers.

I've been blessed with the opportunity to carry out this work in collaboration with many outstanding coauthors, in addition to Bob Floyd: Anatoly V. Anisimov, Peter B. Bendix, Thomas J. Buckholtz, David R. Fuchs, Huang Bing-Chao, James H. Morris, Jr., Vaughan R. Pratt, and John F. Reiser.

The original papers were converted to electronic form with the help of Max Etchemendy, Tony Gee, and Lauri Kanerva. Then I applied spit and polish and handed the text to the incredible proofreading team of Dan Eilers, Ashutosh Mehra, Mark Ward, and Udo Wermuth. I've used John Hobby's marvelous METAPOST to redraw all of the illustrations. Emma Pease facilitated many aspects of the production. As usual, Dikran Karagueuzian has masterminded the whole project.

I have included an Addendum at the close of most chapters, commenting on subsequent developments. And I have tried to put all of the bibliographies into a consistent format, so that they will be as useful as possible to everyone who pursues the subject. I have also seized this opportunity to improve the wording of the chapters that were originally written before my publishers' copy editors taught me a thing or two about writing.

This book is number seven in a series of eight that Stanford's Center for the Study of Language and Information (CSLI) plans to publish, containing archival forms of the papers I have written. The first volume, *Literate Programming*, appeared in 1992; the second, *Selected Papers on Computer Science*, appeared in 1996; the third, *Digital Typography*, in 1999; the fourth, *Selected Papers on Analysis of Algorithms*, in 2000; the fifth, *Selected Papers on Computer Languages*, in 2003; and the sixth, *Selected Papers on Discrete Mathematics*, also in 2003. Still in preparation is the final volume, to be entitled *Selected Papers on Fun and Games.*

Donald E. Knuth
Stanford, California
June 2009

Acknowledgments

"Robert W Floyd, in memoriam" originally appeared in *SIGACT News* **34**, number 4 (December 2003), pp. 3–13. (New York: Association for Computing Machinery, Inc.) Copyright presently held by the author.

"The Bose–Nelson sorting problem" originally appeared in *A Survey of Combinatorial Theory*, essays in honor of R. C. Bose, edited by J. N. Srivastava (1973), pp. 163–172. Copyright ©1973 by North-Holland Publishing Company. Reprinted by permission of Elsevier Science.

"A one-way, stackless quicksort algorithm" originally appeared in *BIT* **26** (1986), pp. 127–130. Copyright ©1986 by Nordisk Tidskrift for Informationsbehandling. Reprinted by permission.

"Optimum binary search trees" originally appeared in *Acta Informatica* **1** (1971), pp. 14–25 and p. 270. Copyright ©1971 by Springer-Verlag GmbH & Co. KG. Reprinted by permission.

"Dynamic Huffman coding" originally appeared in *Journal of Algorithms* **6** (1985), pp. 163–180. Copyright ©1985 by Academic Press. Reprinted by permission of Elsevier Science.

"Inhomogeneous sorting" originally appeared in *International Journal of Computer and Information Sciences* **8** (1979), pp. 255–260. Copyright ©1979 by Plenum Publishing Corporation. Reprinted by permission of Springer-Verlag GmbH & Co. KG.

"Lexicographic permutations with restrictions" originally appeared in *Discrete Applied Mathematics* **1** (1979), pp. 117–125. Copyright ©1979 by North-Holland Publishing Company. Reprinted by permission of Elsevier Science.

"Nested satisfiability" originally appeared in *Acta Informatica* **28** (1990), pp. 1–6. Copyright ©1990 by Springer-Verlag GmbH & Co. KG. Reprinted by permission.

"Fast pattern matching in strings" originally appeared in *SIAM Journal on Computing* **6** (1977), pp. 323–350. Copyright ©1977 by The Society for Industrial and Applied Mathematics. Reprinted by permission.

"Addition machines" originally appeared in *SIAM Journal on Computing* **19** (1990), pp. 329–340. Copyright ©1990 by The Society for Industrial and Applied Mathematics. Reprinted by permission.

"A simple program whose proof isn't" originally appeared in *Beauty Is Our Business: A Birthday Salute to Edsger W. Dijkstra*, edited by W. H. J. Feijen et al. (1990), pp. 233–242. Copyright ©1990 by Springer-Verlag GmbH & Co. KG. Reprinted by permission.

"Verification of link-level protocols" originally appeared in *BIT* **21** (1981), pp. 31–36. Copyright ©1981 by Nordisk Tidskrift for Informationsbehandling. Reprinted by permission.

"Additional comments on a problem in concurrent programming control" originally appeared in *Communications of the ACM* **9** (May and December

1966), pp. 321–322 and p. 878. (New York: Association for Computing Machinery, Inc.) Copyright presently held by the author.

"Optimal prepaging and font caching" originally appeared in *ACM Transactions on Programming Languages and Systems* **7** (1985), pp. 62–79. Copyright ©1985 by the Association for Computing Machinery, Inc. Reprinted by permission.

"A generalization of Dijkstra's algorithm" originally appeared in *Information Processing Letters* **6** (1977), pp. 1–5. Copyright ©1977 by North-Holland Publishing Company. Reprinted by permission of Elsevier Science.

"Two-way rounding" originally appeared in *SIAM Journal on Discrete Mathematics* **8** (1995), pp. 281–290. Copyright ©1995 by The Society for Industrial and Applied Mathematics. Reprinted by permission.

"Matroid partitioning" originally appeared in Stanford Computer Science Report 342 (Stanford, California, 1973), 12 pp. Copyright by the author.

"Irredundant intervals" originally appeared in *ACM Journal of Experimental Algorithmics* **1** (1996), article 1, 19 pp. Copyright ©1996 by the Association for Computing Machinery, Inc. Reprinted by permission.

"Simple word problems in universal algebras" originally appeared in *Computational Problems in Abstract Algebra*, edited by John Leech (1970), pp. 263–297. Copyright ©1970 by Pergamon Press. Reprinted by permission of Elsevier Science.

"Efficient representation of perm groups" originally appeared in *Combinatorica* **11** (1991), pp. 33–43. Copyright ©1991 by Akadémiai Kiadó, Budapest. Reprinted by permission of Springer-Verlag GmbH & Co. KG.

"An algorithm for Brownian zeroes" originally appeared in *Computing* **33** (1984), pp. 89–94. Copyright ©1984 by Springer-Verlag GmbH & Co. KG. Reprinted by permission.

"Semi-optimal bases for linear dependencies" originally appeared in *Linear and Multilinear Algebra* **17** (1985), pp. 1–4. Copyright ©1985 by Gordon and Breach Science Publishers, Inc. Reprinted by permission of Taylor & Francis Group.

"Evading the drift in floating-point addition" originally appeared in *Information Processing Letters* **3** (1975), pp. 84–87 and p. 164. Copyright ©1975 by North-Holland Publishing Company. Reprinted by permission of Elsevier Science.

"Deciphering a linear congruential encryption" originally appeared in *IEEE Transactions on Information Theory* **IT-31** (1985), pp. 49–52. Copyright ©1985 by The Institute of Electrical and Electronics Engineers, Inc. Reprinted by permission.

"Computation of tangent, Euler, and Bernoulli numbers" originally appeared in *Mathematics of Computation* **21** (1967), pp. 663–688. Copyright ©1967 by the American Mathematical Society. Reprinted by permission.

Chapter 1

Robert W Floyd, in Memoriam

[Originally published in SIGACT News **34**, *4 (December 2003), 3–13, and also in IEEE Annals of the History of Computing* **26**, *2 (April–June 2004), 75–83.]*

Nobody has influenced my scientific life more than Bob Floyd. Indeed, were it not for him, I might well have never become a computer scientist. In this note I'll try to explain some of the reasons behind these statements, and to capture some of the spirit of old-time computer science.

Instead of attempting to reconstruct the past using only incidents that I think I remember, I will quote extensively from actual documents that were written at the time things happened. The remarks below are extracted from a one-hour keynote speech I gave to the Stanford Computer Forum on 20 March 2002; many further details, including images of the original documents, can be seen in a video recording of that lecture, which has been permanently archived on the Internet by Stanford's Center for Professional Development [`scpd.stanford.edu`]. As in that lecture, I won't attempt to give a traditional biography, with balanced accounts of Bob's childhood, education, family life, career, and outside interests; I believe that the intriguing task of preparing such an account will be undertaken before long by professional historians who are much more qualified than I. My aim here is rather to present a personal perspective.

My first encounter with Floyd's work goes back to 1962, when I was asked by *Computing Reviews* to assess his article "A descriptive language for symbol manipulation" [*Journal of the Association for Computing Machinery* **8** (1961), 579–584]. At that time I was studying mathematics as a second-year grad student at Caltech; he was working as a programmer-analyst at Armour Research Foundation in Chicago. Since I had recently completed a compiler for a subset of ALGOL, and had read the writeups and source listings of several other compilers, I was immediately impressed by what he had written [see *Computing Reviews*

3 (1962), 148, review #2140]: "This paper is a significant step forward in the field of automatic programming. Over the past few years, simple algorithms for analyzing arithmetic expressions have been discovered independently by many people. But conventional methods for explaining such algorithms obscured the essential facts. Floyd has developed a new notation which lets the trees be distinguished from the forest, and which admirably points out what is really going on in a translation process. An algebraic compiler can be described very precisely and compactly in this notation, and one can design such a compiler in Floyd's form in a few hours." In essence, Bob had introduced the notion of *productions* as an organizing principle for programming, anticipating to a certain extent the future development of so-called expert systems.

My own work on programming was merely a sideline by which I could pay for my college education and prepare to start a family. During the summer of 1962 I wrote a FORTRAN compiler for a small UNIVAC computer; this work had almost no connection with what I viewed as my future career as a teacher of mathematics, except that I did spend one fascinating day studying the efficiency of "linear probing" (the simple hash table algorithm by which my compiler maintained its dictionary of symbols). I had never heard of "computer science." My whole attitude changed, however, when I met Bob for the first time in person at the ACM conference in Syracuse at the end of that summer.

We became fast friends, perhaps because we had both learned programming in the late 1950s by sitting at the consoles of IBM 650 computers. Bob showed me some work he had been doing about mathematical techniques for verifying that a program is correct — a completely unheard-of idea in those days as far as I knew. The accepted methodology for program construction was quite the opposite: People would write code and make test runs, then find bugs and make patches, then find more bugs and make more patches, and so on until not being able to discover any further errors, yet always living in dread for fear that a new case would turn up on the next day and lead to a new type of failure. We never realized that there might be a way to construct a rigorous proof of validity; at least, I'm sure that such thoughts never crossed my own mind when I was writing programs, even though I was doing nothing but proofs when I was in a classroom. I considered programming to be an entirely different category of human activity. The early treatises of Goldstine and von Neumann, which provided a glimpse of mathematical program development, had long been forgotten. I was also unaware of John McCarthy's paper, "Towards a mathematical science of computation," presented at the IFIP Congress in Munich that summer, nor did I associate McCarthy-style recursive functions with "real" programming at that time. But Bob taught me how to wear my programmer's cap and my mathematician's cloak at the same time.

Computer programs were traditionally "explained" by drawing flowcharts to illustrate the possible sequences of basic steps. Bob's proof method was based on decorating each branch in the flowchart with an invariant assertion such as "$R \geq Y > 0$, $X \geq 0$, $Q \geq 0$, $X = R + QY$," which captures the essential relations that hold between variables at that point of the calculation. If we can show that the assertions immediately following each step are consequences of the assertions immediately preceding, we can be sure that the assertions at the end of the program will hold if the appropriate assertions were true at the beginning. Of course this is a simple principle, once it has been formulated; but it dramatically opened my eyes. When Bob published it later ["Assigning meanings to programs," *Proceedings of Symposia in Applied Mathematics* **19** (1967), 19–32], he gave credit to unpublished ideas of Alan Perlis and Saul Gorn, but I'm sure that he had developed everything independently. His paper presented a formal grammar for flowcharts together with rigorous methods for verifying the effects of basic actions like assignments and tests; thus it was a direct precursor of the "preconditions" and "postconditions" subsequently developed by Tony Hoare.

We began writing letters back and forth. In one letter, for example, I mentioned among other things that I'd been trying without success to find a systematic way to decide whether a given context-free grammar is ambiguous, even in the simple case

$$\langle A \rangle ::= \langle x \rangle \mid \langle A \rangle \langle x \rangle$$

where $\langle x \rangle$ is a finite set of strings. He replied on 16 October 1963 — using the stationery of his current employers, Computer Associates of Wakefield, Massachusetts — as follows: "I applaud your results on TTL's [whatever those were ... I've forgotten]; but see Greibach, 'The Undecidability of the Ambiguity Problem for Minimal Linear Grammars', Information and Control, June 1963, pg. 119. The paper by Landweber in the same issue is also interesting." Then he proceeded to present an algorithm that solves the simple case I had mentioned. We both learned later that he had thereby rediscovered a method of Sardinas and Patterson that was well known in coding theory [August A. Sardinas and George W. Patterson, "A necessary and sufficient condition for unique decomposition of coded messages," *Convention Record of the I.R.E., 1953 National Convention*, Part 8: Information Theory (1953), 104–108].

Near the end of 1963, Bob came to visit me at Caltech, bringing fresh Maine lobsters with him on the plane. We spent several days hiking in Joshua Tree National Monument, talking about algorithms and languages as we went. (He loved the outdoors, and we hiked together in Coe State Park several years later.) At the time I was getting ready to draft the chapter on sorting for *The Art of Computer Programming* (*TAOCP*). Soon afterwards I had occasion to travel to Boston, where I visited him at his beautiful new home in Topsfield, Massachusetts. We talked about some new ideas in sorting that I had just learned, in particular the "heapsort" algorithm of J. W. J. Williams [soon to appear in *Communications of the ACM* **7** (1964), 347–348; I had been the referee]. Bob responded by introducing an improvement to the initialization phase of that procedure [*Communications of the ACM* **7** (1964), 701]. I also introduced him at that time to the notion of sorting networks, namely to the methods of sorting that had been introduced by R. C. Bose and R. J. Nelson ["A sorting problem," *Journal of the Association for Computing Machinery* **9** (1962), 282–296]. Shortly after my visit, Bob wrote me a letter dated 5 February 1964, which began by discussing the expected length of the longest decreasing subsequence of a random permutation. Then he said, "About the sorting system you showed us, I find that any sorting procedure by interchanges of adjacent lines will correctly sort all inputs if it will correctly sort an input which is in reverse (decreasing) order." This elegant result, and his lemma that proved it, eventually became exercise 5.3.4–36 of *TAOCP*. He ended his letter by saying, "Was this the theorem you wanted me to find? Ask me another."

Bose and Nelson had conjectured that the sorting networks they constructed were optimal, having the fewest possible comparators. Support for their conjecture was obtained by Thomas N. Hibbard ["A simple sorting algorithm," *Journal of the Association for Computing Machinery* **10** (1963), 142–150]. But Bob found improved methods soon after he had learned of the problem, first reducing the number of comparison-exchange modules needed to sort 21 elements from 118 to 117, and then (on 27 March) showing that 9 elements can be sorted with only 25 modules instead of 27.

This startling breakthrough was the beginning of an exciting exchange of letters. On 4 April I wrote back, "Dear Bob, I was quite impressed, indeed awe-struck, etc., by the example you sent me last week showing an improvement of two steps in the fixed-comparison sort for nine elements. ... Therefore I lost another day from book-writing as I pondered this question anew. Here are the few results I obtained today; I hope you check them for accuracy and I also hope you are inspired to find bigger and better improvements. ... I think I can break that $n^{\log_2 3} - n$ barrier, in the following algorithm for $n = 16$. (This is a little startling since powers of two are the best cases for the Bose–Nelson sort.) ..." Bob replied, on 10 April: "Very pretty! Now generalize to three-dimensional diagrams. ..." And we continued to exchange dozens of letters with respect to this question of efficient networks for sorting. Whenever I had mail from Bob, I'd learn that he had gotten ahead in our friendly competition; then it was my turn to put *TAOCP* on hold for another day, trying to trump his latest discovery. Our informal game of pure-research-at-a-distance gave us a taste of the thrills that mathematicians of old must have felt in similar circumstances, as when Leibniz corresponded with the Bernoullis or when Euler and Goldbach exchanged letters. However, by the time we finally got around to publishing the fruits of this work ["The Bose–Nelson sorting problem," in *A Survey of Combinatorial Theory*, edited by J. N. Srivastava (Amsterdam: North-Holland, 1973), 163–172], we learned that Kenneth E. Batcher had blown our main theorem away by finding a much better construction.

All of this was incidental to our main research, which at the time was focused on the translation of artificial languages like ALGOL into machine language. Indeed, all computer science research in those days was pretty much carved up into three parts: either (1) numerical analysis or (2) artificial intelligence or (3) programming languages. In 1963 Bob had written a masterful paper, "Syntactic analysis and operator precedence" [*Journal of the Association for Computing Machinery* **10** (1963), 316–333], in which he launched an important new way to approach the parsing problem, the first syntax-directed algorithm of practical importance. And he followed that up in 1964 with an even more wonderful work, "The syntax of programming languages — A survey" [*IEEE Transactions on Electronic Computers* **EC-13** (1964), 346–353], probably the best paper ever written about that subject. In this survey he masterfully brought order out of the chaos of various approaches that people had been using in the input phases of compilers, but even more important was his introduction of a completely new algorithm with a brand new control structure. He presented this novel method "... with a metaphor. Suppose a man is assigned the goal of analyzing a sentence in a PSL [phrase-structure language, aka context-free language] of known grammar. He has the power to hire subordinates, assign them tasks, and fire them if they fail; they in turn have the same power. ... Each man will be told only once 'try to find a G' where G is a symbol of the language, and may thereafter be repeatedly told 'try again' if the particular instance of a G which he finds proves unsatisfactory to his superiors."

I think the algorithm he presented in this paper can be justly regarded as the birth of what we now call object-oriented programming.

At the end of 1964 I had nearly finished drafting Chapter 10 of *TAOCP*, the chapter on syntax analysis, and I wrote Bob a long letter attempting to explain a general approach that had emerged from this work (now known as LR(k) parsing). “I must apologize for the complexity of my construction (indeed, it is too complicated to put in my book), but this seems to be inherent in the problem. I know of at least three Ph.D. theses which were entirely concerned with only the most simple cases of parts of this problem! As I go further into chapter 10 I become more convinced that only five really worthwhile papers on scanning techniques have ever been written, and you were the author of all five of them!”

Bob became an Associate Professor of Computer Science at the Carnegie Institute of Technology in the fall of 1965, introducing among other things a course on “the great algorithms,” and supervising the Ph.D. theses of Zohar Manna (1968), Jay Earley (1968), and Jim King (1969). He also wrote another major paper at this time, “Nondeterministic algorithms” [*Journal of the Association for Computing Machinery* **14** (1967), 636–644], setting out the general principles of exhaustive search in a novel and perspicuous way that has led to many practical implementations. I found in my files a letter that I'd written to Myrtle Kellington in June 1967, urging her to have the illustrations in this paper typeset by the printer instead of using the much cheaper alternative of “Leroy lettering.” I argued that “Floyd's article, perhaps more than any other article I have ever seen, is based almost entirely on illustrations coordinated with text. ... Saying he should prepare his own diagrams is, in this case, like telling our authors never to use any mathematical formulas unless they submit hand-lettered copy. ... Professor Floyd has been one of our best and most faithful referees in the ACM publications for many years, and he has volunteered much of his valuable time to this often thankless job. Now he is an associate editor of JACM. We certainly owe him a decent treatment of his article.” I'm happy to report that she agreed, even though she had received my letter more than two weeks after the publication deadline.

Meanwhile my publishers and I had asked Bob for a detailed critique of *TAOCP* Volume 1, which was being prepared for the press in 1967. Needless to say, his comments proved to be invaluable to me, although I didn't agree with every single one of his remarks. Here are some excerpts from what he wrote: “Chapter I: Overall opinion. I like the chapter, but I think it could be improved by chopping most of the humor and anecdotes, retaining the historical material. ... The system of rating problems underestimates their difficulty for, say, college seniors, and designates too many with the '▸'. The author's personal notes of advice, etc., are often valuable; at times, though, the non-technical material gets a little thick. The technical content meets very high standards of scholarship, and is a credit to the author." Then he gave hundreds of detailed suggestions for improvements to the text.

Our correspondence was not entirely technical. On 22 February 1967 I wrote, "Bob, I have the feeling that this is going to be a somewhat extraordinary letter. During the last year or so I have been getting lots of offers of employment from other colleges. I think I told you that I plan (and have planned for a long time) to decide on a permanent home where I will want to live the rest of my life, and to move there after the year I spend in Princeton working for the government. (Namely, to move in September 1969.) Due to the present supply and demand in computer science, I am fortunate enough to be able to pick just about any place I want to go; but there are several good places and it's quite a dilemma to decide what I should do. I believe the four places that are now uppermost in my mind are Stanford, Cornell, Harvard, and Caltech (in that order). I expect to take about a year before I make up my mind, with Jill's help. It occurs to me that I would very much like to be located at the same place you are, if this is feasible; at any rate your plans have a non-trivial place in the non-linear function I am trying to optimize! ... So I would like to explore some of these possibilities with you. ..." Bob responded with his own perspective on the current state of affairs at various universities; the bottom line of his reply was, "I'd say if you want to make the move, I don't have any plans that would conflict, and I will be very tempted to go to Stanford myself; I probably will go, in fact."

I received and accepted Stanford's offer a year later, and George Forsythe (the chair of Stanford's department) asked me in March of 1968 to write a letter of recommendation on Bob's behalf. I replied that "I don't know anyone I could recommend more highly. He is the most gifted man in his 'age bracket' that I have ever met. Several of his published papers have been significant mileposts in the development of computer science, notably his introduction of precedence grammars, tree-sort algorithms, and methods of 'assigning meanings to programs.' I have also had the pleasure of carrying on frequent correspondence with him for five years, so I am quite familiar with his unpublished work too; this correspondence covers a wide variety of topics, for example, graph theory, word problems in semigroups, mathematical notations, algorithms for syntactic analysis, theorems about languages, algorithms for manipulating data structures, optimal sorting schemes, etc., etc. While I was editing the ACM *Communications* and *Journal*, I asked him to serve as referee for several papers, and it was not uncommon for him to submit a four- or five-page review containing important suggestions for improvements. He also has a good record of working with students at Carnegie Tech on both the undergraduate and graduate level: He has supervised some excellent theses and he has kept several student projects going. He is one of the rare people who have considerable experience and expertise both in writing computer programs and in developing useful theories related to programming. ... He is a true Computer Scientist! His special talents seem to be (a) the ability to devise ingenious algorithms and combinatorial constructions; (b) the ability to develop nontrivial new theories which are of both practical and mathematical interest; (c) the ability to organize a large body of loosely connected material and to perceive the important ideas; (d) a talent for good exposition and for finding just the right words to express an idea. His only fault known to me is that he is sometimes a little too sensitive (too much the perfectionist); for example, although he has lived in the East nearly all his life, he has already decided that no California wine is worth drinking except B. V. Beaujolais. ... One further remark is perhaps necessary, considering contemporary 'standards of society'. Floyd has never gone through the formalities of obtaining a Ph.D. degree. I believe this was due primarily to the fact that he entered graduate school at the University of Chicago when he was only 16 or 17 years old, as part of an experimental accelerated education program; this was not a mature enough age to do graduate work. [Bob was born 8 June 1936, and he began graduate school after receiving a B.A. degree in 1953 at age 17 — about five years earlier than usual for American students at the time.] Certainly he has written at least a dozen papers by now each of which is superior to any Ph.D. thesis I have ever seen in computer science, so the mere fact that he has never formally received the degree should be quite irrelevant."

(Bob used to say that he was planning to get a Ph.D. by the "green stamp method," namely by saving envelopes addressed to him as 'Dr. Floyd'. After collecting 500 such letters, he mused, a university somewhere in Arizona would probably grant him a degree.)

To my delight, Bob did receive and accept an offer from Stanford, and he arrived during the summer of 1968.

While I'm quoting from letters of recommendation, I might as well continue with two more that I was asked to write later. The first of these was addressed to the American Academy of Arts and Sciences on

12 February 1974: "... it is difficult to rank [computer scientists] with respect to mathematicians, physicists, etc., since computer science is so young. A mathematician like Lehmer or Polya has been producing high quality work consistently for 50 years or more, and this is much more than a computer scientist could do ... perhaps it's too easy [for computer scientists like myself] to become a fellow. On the other hand there are outstanding mathematicians like Bellman and Thompson whose work spans only 20 years or so, and that is closer to what a leading computer scientist (say 15 years) would have to his credit. I will list the two candidates who are generally recognized as leading pioneers and whose names are 'household words', and whose contributions span a broad range of topics as well as a long span of time (consistent quality): 1. Edsger Dijkstra, who is responsible for more landmark contributions in non-theoretical aspects of computer science than any other man. Namely, the current revolution in programming methodology, the fundamental principles of synchronizing cooperating processes, the method for implementing recursive processes, as well as important algorithms. Such aspects of computer science are the hardest in which to make fundamental breakthroughs. He is also an able theoretician. 2. Robert Floyd, who is responsible for many of the leading theoretical ideas of computer science as well as chairman of what I think is the leading academic department (mine! but I'm trying to be unbiased). His work is cited more than twice as much as any other person's in the series of books I am writing (summarizing what is known about programming). His fundamental contributions to the syntax and semantics of programming languages, to the study of computational complexity, and to proofs of program correctness, have been a great influence for many years, and he has also invented dozens of important techniques which are now in common use."

Second, to the John Simon Guggenheim Memorial Foundation on 3 December 1975: "Professor Floyd is one of the most outstanding computer scientists in the world; in my mind he is one of the top three. During his career he has been a leading light in the development of many of the key concepts of our discipline: (a) A production language to describe algebraic parsing techniques (now called Floyd productions). (b) The notion of precedence grammars. (c) Semantics of programming languages. (d) Automatic techniques for constructing computer programs and proving their correctness. Each of these contributions has essentially created an important subfield of research later pursued by hundreds of people. Besides this he has invented many algorithms of practical importance (e.g. to find all shortest paths in a network, to sort numbers into order, to find the median of a set of data), and he has several landmark technical papers which show that certain problems are intrinsically hard. As an example of this, I can cite his recent discovery of the fastest possible way to add numbers: This result meant that he had to invent an addition procedure which was faster than any others heretofore known, and to prove that no faster method will ever be possible. Both of these were nontrivial innovations. In my recent book which reviews what is known about sorting and searching, his work is cited 20 times, far more than any other person. His published papers show an amazing breadth, especially when one realizes that so many of them have been pioneering ventures that became milestones in computer science. I am sure that whatever he proposes to do during his sabbatical year will be of great future value to science, and so I heartily recommend support by the Guggenheim Foundation."

Bob was elected to the American Academy in 1974 and awarded a Guggenheim Fellowship for 1976–1977.

Upon Bob's arrival at Stanford he immediately became a popular teacher. Indeed, students frequently rated his problem-solving seminar, CS204, as the best course of their entire college career. He also was dedicated to teaching our introductory programming course, CS106; we often talked about examples that might help to introduce basic concepts. He was promoted to Full Professor at Stanford in 1970, one of extremely few people to achieve this rank after having served only five years as Associate Professor (three of which were at Carnegie).

One of the greatest honors available to mathematicians and computer scientists in those days was to be invited to give a plenary lecture at an international congress. Floyd was one of only eight people in computer or system science to receive such an honor from the International Congress of Mathematicians held at Nice in 1970; the eight invitees were Eilenberg, Floyd, Knuth, and Winograd (USA); Schützenberger (France); Lavrov, Lupanov, and Kolmogorov (USSR). A year later, Bob was the only speaker to be invited to the IFIP Congress in Ljubljana by two *different* technical area committees.

I can't help mentioning also the fact that he helped me shape up my writing style. For example, I was on leave of absence at the University of Oslo when I received the following life-changing letter from his hand:

September 21, 1972

Prof. Donald Knuth

Dear Don:

> Please quit using so many exclamation points! I can't stand it!! I'm going out of my mind!!! (Don't get alarmed, I'm only kidding!!!!)
>
> Sincerely yours (!!!!!),
> Robert W. Floyd

Of course I took this advice to heart — and wrote the following reply in April of 1973, after learning that Bob was our dean's choice to succeed George Forsythe as department chair: "Bob, Congratulations, to you and to the Dean for his fine decision. I hope you will accept, since I think you will be able to accomplish important things for our department. Please be our Chairman. Sincerely, Don. P.S. If I were allowed to use exclamation points I would be more emphatic."

Bob served as department chair for three years, devoting considerable energy to the task. Above all he worked tirelessly with architects on the design of a new home for the department (Margaret Jacks Hall), in which our entire faculty would be together for the first time. For example, in December 1975 I sent him the following memo: "Thanks for all the fine work you've evidently done together with the architects for the new building. This will have a lasting payoff and we all owe you a huge debt of gratitude." And on 27 August 1976: "(Thank you)^n for the outstanding services you gave our department as its chairperson these last years. You focused on and solved the critical problems facing us, and the present and future strength of our department is due in large measure to your efforts; the effects will last for a long time. Not having the energy to be chairman myself, I am doubly grateful that you sacrificed so much of your time to these important tasks." Finally in November 1978, when he was enjoying a well-deserved second year of sabbatical at MIT, I sent the following message: "Dear Bob, The faculty had its first chance to walk through Margaret Jacks Hall yesterday. The building has taken shape nicely. The roof is finished, so the winter rains (if we get any) won't be a problem for the workmen doing the finishing. The carpentry work is super quality, and the spaces are nice to walk through and be in. So I'm writing to say 'thanks for all the energy you contributed towards the success of the project'. Thanks from all of us! Best regards, Don. P.S. Am enjoying the telephone poker game that Rivest told me about." [Okay, I let an exclamation point creep in, but that one was legitimate. Note also that the article "Mental poker" by Adi Shamir, Ronald L. Rivest, and Leonard M. Adleman in *The Mathematical Gardner*, edited by David A. Klarner (Wadsworth, 1981), 37–43, credits Bob with proposing the problem of playing a fair game of poker without cards.]

This picture of Bob Floyd was taken by a Stanford student and posted on our department's photo board about 1972. To simulate gray scale with binary pixels, I've rendered it here using the Floyd–Steinberg "error diffusion" algorithm, implementing that algorithm exactly as suggested in the famous article that Bob published with Louis Steinberg in 1976 (diffusing errors from bottom to top and right to left); the resolution is 600 dots per inch. *Caution:* Software for printing might have mangled the bits that you are now seeing. Furthermore, the physical model in Floyd and Steinberg's article matches the technology of inkjet printers better than that of laser-jet printers or offset printers, so you may not be seeing this image at its best. Error diffusion does, however, give beautiful results when it has been tuned to work with a typical inkjet device.

Bob received the ultimate honor in our field, the ACM Turing Award, at the end of 1978 — for "helping to found the following important subfields of computer science: the theory of parsing, the semantics of programming languages, automatic program verification, automatic program synthesis, and the analysis of algorithms. Your work has had a clear influence on methodologies for the creation of efficient and reliable software." I was surprised to notice, when rereading his Turing lecture "The paradigms of programming" [*Communications of the ACM* **22** (1979), 455–460, with a nice picture on page 455], that he had recommended already in 1978 many of the things that I've been promoting since 1984 under the banner of "literate programming." [See page 458 of his Turing lecture; and see Donald E. Knuth, "Literate programming," *The Computer Journal* **27** (1984), 97–111.]

At the time Bob was receiving this award, and for many years afterwards, I was immersed in problems of digital typography. Thus I wasn't able to collaborate very much with him in the latter part of his career, although we did have fun with one project ["Addition machines," *SIAM Journal on Computing* **19** (1990), 329–340]. He was destined to be disappointed that his dream of a new, near-ideal programming language called Chiron was never to be realized — possibly because he was reluctant to make the sorts of compromises that he saw me making with respect to TEX. Chiron was "an attempt to provide a programming environment in which, to the largest extent possible, one designs a program by designing the process which the program carries out." His plans for the Chiron compiler included novel methods of compile-time error recovery and type matching that have never been published.

I know that when he retired in 1994, the publication of his book *The Language of Machines* with Richard Beigel brought him enormous satisfaction, especially when he received a copy of the fine translation of that book into German.

Then, alas, a rare ailment called Pick's disease began to attack his mind and his body. His scientific life came sadly to a premature end. Yet dozens of the things he did in his heyday will surely live forever.

I'd like to close with a few anecdotes. First, people often assume that my books are in error or incomplete when I refer to Bob as "Robert W Floyd," since the indexes to my books give a full middle name for almost everybody else. The truth is that he was indeed born with another middle name, but he had it legally changed to "W" — just as President Truman's middle name was simply "S". Bob liked to point out that "W." is a valid abbreviation for "W".

Second, he loved his BMW, which was nicknamed Tarzan. He told me that it would be nice to own two of them, so that both cars could have license plate holders that said "my other car is a BMW."

Third, he had a strong social conscience. For example, he spent a significant amount of time and energy to help free Fernando Flores from prison in Chile. Flores was a former vice-rector of the Pontifical Catholic University of Chile who had developed a computer-based information system for the entire Chilean economy and had become a cabinet minister in the government of Salvador Allende. He came to Stanford as a research associate in 1976 largely because of Bob's efforts, after having been held without charges for three years by the military junta headed by Augusto Pinochet.

Fourth, Bob was a connoisseur of fine food. Some of the most delicious meals I've ever experienced were eaten in his presence, either as a guest in his house or in a restaurant of his choice. I particularly remember an unforgettable duckling with flaming cherry sauce, consumed during an ACM conference in Denver.

Fifth, I remember 1 May 1970, the day after Nixon invaded Cambodia. Tension was high on campus; Bob and I decided that such escalation by our president was the last straw, and we could no longer do "business as usual." So we joined the students and picketed Pine Hall (Stanford's Computation Center), preventing anyone from going inside to get work done that day. (I must admit, however, that we sat there talking about sorting algorithms the whole time.)

Finally, one last quotation — this one from the future instead of the past. The seventh volume of my collected works, to be entitled *Selected Papers on Design of Algorithms*, is scheduled to be published about two years from now.* For a long time I've planned to dedicate this book to Bob Floyd; indeed, the dedication page is the only page of the book that has been typeset so far, and it has been in my computer for several years. That page currently reads as follows, using an old word for algorithm that the Oxford English Dictionary traces back to Chaucer and even earlier: "To Robert W Floyd (1936–2001) / my partner in augrime."

I'm grateful to Richard Beigel, Christiane Floyd, Hal Gabow, Greg Gibbons, Gio Wiederhold, and Voy Wiederhold for their help in preparing these reminiscences.

---

* Well . . . , that was the schedule.

Publications of Robert W Floyd

(with B. Ebstein) "A formal representation of the interference between several pulse trains," *Proceedings of the Fourth Conference on Radio Interference Reduction and Electronic Compatibility* (Chicago: 1958), 180–192.

"Remarks on a recent paper," *Communications of the ACM* **2**, 6 (June 1959), 21.

"An algorithm defining ALGOL assignment statements," *Communications of the ACM* **3** (1960), 170–171, 346.

(with B. Kallick, C. J. Moore, and E. S. Schwartz) *Advanced Studies of Computer Programming*, ARF Project E121 (Chicago, Illinois: Armour Research Foundation of Illinois Institute of Technology, 15 July 1960), vi + 152 pages. [A description and user manual for the MOBIDIC Program Debugging System, including detailed flowcharts and program listings.]

"A note on rational approximation," *Mathematics of Computation* **14** (1960), 72–73.

"Algorithm 18: Rational interpolation by continued fractions," *Communications of the ACM* **3** (1960), 508.

"An algorithm for coding efficient arithmetic operations," *Communications of the ACM* **4** (1961), 42–51.

"A descriptive language for symbol manipulation," *Journal of the Association for Computing Machinery* **8** (1961), 579–584.

"A note on mathematical induction on phrase structure grammars," *Information and Control* **4** (1961), 353–358.

"Algorithm 96: Ancestor," *Communications of the ACM* **5** (1962), 344–345.

"Algorithm 97: Shortest path," *Communications of the ACM* **5** (1962), 345.

"Algorithm 113: Treesort," *Communications of the ACM* **5** (1962), 434.

"On the nonexistence of a phrase structure grammar for ALGOL 60," *Communications of the ACM* **5** (1962), 483–484.

"On ambiguity in phrase structure languages," *Communications of the ACM* **5** (1962), 526, 534.

"Syntactic analysis and operator precedence," *Journal of the Association for Computing Machinery* **10** (1963), 316–333.

"Bounded context syntactic analysis," *Communications of the ACM* **7** (1964), 62–67.

"The syntax of programming languages — A survey," *IEEE Transactions on Electronic Computers* **EC-13** (1964), 346–353. Reprinted in Saul Rosen, *Programming Languages and Systems* (New York: McGraw–Hill, 1967), 342–358.

"Algorithm 245: Treesort 3," *Communications of the ACM* **7** (1964), 701.

New Proofs of Old Theorems in Logic and Formal Linguistics (Pittsburgh, Pennsylvania: Carnegie Institute of Technology, November 1966), ii + 13 pages.

(with Donald E. Knuth) "Advanced problem H-94," *Fibonacci Quarterly* **4** (1966), 258.

"Assigning meanings to programs," *Proceedings of Symposia in Applied Mathematics* **19** (Providence, Rhode Island: American Mathematical Society, 1967), 19–32.

(with D. E. Knuth) "Improved constructions for the Bose–Nelson sorting problem," *Notices of the American Mathematical Society* **14** (February 1967), 283, Abstract 67T-228.

"The verifying compiler," *Computer Science Research Review* (Pittsburgh, Pennsylvania: Carnegie–Mellon University, December 1967), 18–19.

"Nondeterministic algorithms," *Journal of the Association for Computing Machinery* **14** (1967), 636–644.

"A machine-oriented recognition algorithm for context-free languages," *ACM SIGPLAN Notices* **4**, 5 (May 1969), 28–29.

(with D. E. Knuth) "Notes on avoiding 'go to' statements," *Information Processing Letters* **1** (1971), 23–31, 177. Reprinted in *Writings of the Revolution*, edited by E. Yourdon (New York: Yourdon Press, 1982), 153–162. Reprinted with corrections as Chapter 23 of Knuth's *Selected Papers on Computer Languages*, CSLI Lecture Notes 139 (Stanford, California: Center for the Study of Language and Information, 2003), 495–505.

"Toward interactive design of correct programs," *Information Processing 71*, Proceedings of IFIP Congress 1971, **1** (Amsterdam: North-Holland, 1972), 7–10. Reprinted in *Readings in Artificial Intelligence and Software Engineering*, edited by Charles Rich and Richard C. Waters (Los Altos, California: Morgan Kaufman, 1986), 331–334.

(with James C. King) "An interpretation-oriented theorem prover over integers," *Journal of Computer and System Sciences* **6** (1972), 305–323.

"Permuting information in idealized two-level storage," in *Complexity of Computer Computations*, edited by Raymond E. Miller and James W. Thatcher (New York: Plenum, 1972), 105–109.

(with Donald E. Knuth) "The Bose–Nelson sorting problem," in *A Survey of Combinatorial Theory*, edited by Jagdish N. Srivastava (Amsterdam: North-Holland, 1973), 163–172. [Reprinted as Chapter 2 of the present volume.]

(with Manuel Blum, Vaughan Pratt, Ronald L. Rivest, and Robert E. Tarjan) "Time bounds for selection," *Journal of Computer and System Sciences* **7** (1973), 448–461.

(with Alan J. Smith) "A linear time two tape merge," *Information Processing Letters* **2** (1973), 123–125.

(with Ronald L. Rivest) "Expected time bounds for selection," *Communications of the ACM* **18** (1975), 165–172.

(with Ronald L. Rivest) "Algorithm 489: The algorithm SELECT—for finding the ith smallest of n elements," *Communications of the ACM* **18** (1975), 173.

"The exact time required to perform generalized addition," *16th Annual Symposium on Foundations of Computer Science* (IEEE Computer Society, 1975), 3–5.

(with Louis Steinberg) "An adaptive algorithm for spatial greyscale," *Proceedings of the Society for Information Display* **17** (1976), 75–77. An earlier version appeared in *SID 75 Digest* (1975), 36–37.

(with Larry Carter, John Gill, George Markowsky, and Mark Wegman) "Exact and approximate membership testers," *10th Annual ACM Symposium on Theory of Computing* (1978), 59–65.

"The paradigms of programming," *Communications of the ACM* **22** (1979), 455–460. Reprinted in *ACM Turing Award Lectures: The First Twenty Years* (New York: ACM Press, 1987), 131–142. Russian translation in *Lektsii Laureatov Premii T'i͡uringa* (Moscow: Mir, 1993), 159–173.

(with Jeffrey D. Ullman) "The compilation of regular expressions into integrated circuits," *Journal of the Association for Computing Machinery* **29** (1982), 603–622.

"Riding the tiger," *Las Vegas Backgammon Magazine* **10**, 1 (April/May 1982), 34–35.

(with Jon Bentley) "Programming pearls: A sample of brilliance," *Communications of the ACM* **30** (1987), 754–757.

"What else Pythagoras could have done," *American Mathematical Monthly* **96** (1989), 67.

(with Donald E. Knuth) "Addition machines," *SIAM Journal on Computing* **19** (1990), 329–340. [Reprinted as Chapter 10 of the present volume.]

(with Richard Beigel) *The Language of Machines* (New York: Computer Science Press, 1994), xvii + 706 pages. French translation by Daniel Krob, *Le Langage des Machines* (Paris: International Thomson, 1995), xvii + 594 pages. German translation by Philip Zeitz and Carsten Grefe, *Die Sprache der Maschinen* (Bonn: International Thomson, 1996), xxvii + 652 pages.

(Floyd also wrote many unpublished notes on a wide variety of subjects, often revising and polishing them into gems of exposition that I hope will some day appear on the Internet for all to enjoy. More than 17 boxes of his papers are on deposit in the Stanford University Archives, with catalog number SC 625.)

Chapter 2

The Bose–Nelson Sorting Problem

[Written with Robert W. Floyd. Originally published in A Survey of Combinatorial Theory, edited by Jagdish N. Srivastava, a festschrift for R. C. Bose (Amsterdam: North-Holland, 1973), 163–172.]

A typical "sorting network" for four numbers is illustrated in Fig. 1; the network involves five "comparators," shown as directed wires connecting two lines. Four numbers are input at the left, and as they move towards the right each comparator causes an interchange of two numbers if necessary so that the larger number appears at the point of the arrow. At the right of the network the numbers have been sorted into nondecreasing order from top to bottom; it is easy to verify that this will be the case no matter what numbers are input, since the first four comparators select the smallest and the largest elements, and the final comparator ranks the middle two.

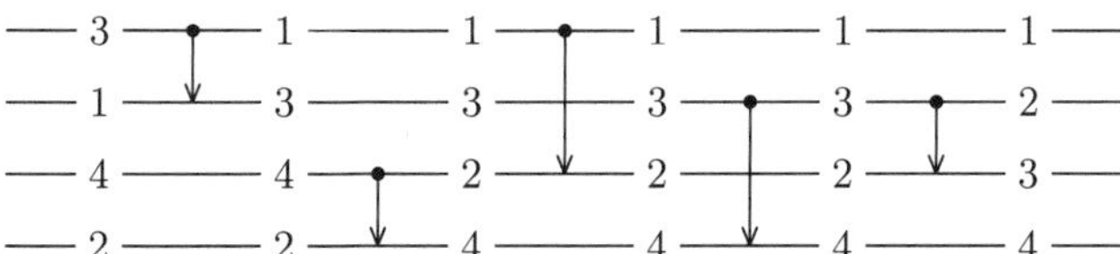

FIGURE 1. A sorting network.

Sorting networks were originally constructed prior to 1957 by R. J. Nelson, who developed special networks for eight or fewer elements. He also showed that n more comparators will always suffice to extend a network for n elements to a network for $n+1$ (see O'Connor and Nelson [8]).

In 1960–1961, he and R. C. Bose constructed n-element sorting networks that were considerably more economical as $n \to \infty$ (see Bose and Nelson [3]). The *Bose–Nelson sorting problem* is the problem of determining the minimum number, $\hat{S}(n)$, of comparators needed in an n-element sorting network.

TABLE 1. Bounds for $\hat{S}(n)$

Approximate date		$n = 1$	2	3	4	5	6	7	8	9	10	11	12	13	14	15	16	Asymptotic
1954	Nelson	0	1	3	5	9	12	18	19	27	36	46	57	69	72	86	101	$\frac{1}{2}n^2$
1960	Bose and Nelson	0	1	3	5	9	12	16	19	27	32	38	42	50	55	61	65	$n^{1.585}$
1964	Knuth and Floyd	0	1	3	5	9	12	16	19	25	31	37	41	49	54	60	64	$n^{1+c/\sqrt{\log n}}$
1964	Batcher	0	1	3	5	9	12	16	19	26	31	37	41	48	53	59	63	$\frac{1}{4}n(\log_2 n)^2$
1969	Best upper bounds known	0	1	3	5	9	12	16	19	25	29	35	39	46	51	56	60	$\frac{1}{4}n(\log_2 n)^2$
1971	Best lower bounds known	0	1	3	5	9	12	16	19	23	27	31	35	39	43	47	51	$n \log_2 n$

Bose and Nelson gave an upper bound for $\hat{S}(n)$, and conjectured that their method actually gave $\hat{S}(n)$ exactly; but subsequent constructions have shown that their upper bound can be improved for all $n > 8$ (see Knuth and Floyd [6], Batcher [2]). In this paper we develop a few aspects of the theory, and prove that Bose and Nelson's conjecture was correct for $n \le 8$.

Table 1 outlines some of the early work on the Bose–Nelson sorting problem and summarizes its current status; see Knuth [7] for further details of recent constructions, due to M. W. Green, A. Waksman, and G. Shapiro. The upper bounds listed for $n \le 12$ are probably exact.

In order to study the problem in detail, it is convenient to introduce a few notational conventions. Let $x = \langle x_1, \ldots, x_n \rangle$ and $y = \langle y_1, \ldots, y_n \rangle$ be sequences of n real numbers; x is said to be *sorted* if $x_1 \le x_2 \le \cdots \le x_n$. We define two operators on such sequences, the *exchange* operation (ij) and the *comparator* operation $[ij]$, for $1 \le i, j \le n$ and $i \ne j$, as follows:

$$x(ij) = y, \text{ where } y_i = x_j,\ y_j = x_i, \text{ and } y_k = x_k \text{ for } i \ne k \ne j; \tag{1}$$

$$x[ij] = \begin{cases} x, & \text{if } x_i \le x_j; \\ x(ij), & \text{if } x_i > x_j. \end{cases} \tag{2}$$

Thus $x[ij] = y$ if and only if $y_i = \min(x_i, x_j)$, $y_j = \max(x_i, x_j)$, and $y_k = x_k$ for $i \ne k \ne j$. It is clear that, when the indices i, j, k, l are distinct, we have (see Fig. 2)

$$(ij)[ij] = [ij], \qquad [ij](ij) = [ji]; \tag{3}$$

$$(ij)[jk] = [ik](ij), \qquad (ij)[kj] = [ki](ij), \qquad (ij)[kl] = [kl](ij). \tag{4}$$

FIGURE 2. A proof that $(ij)[jk] = [ik](ij)$.

A *comparator network* α is a sequence of zero or more exchange and/or comparator operations; a *sorting network* α is a comparator network such that $x\alpha$ is sorted for all x. We write $\alpha\beta$ for the network consisting of α followed by β; and we say that

$$\alpha \subseteq \beta \quad \text{if} \quad U\alpha \subseteq U\beta, \qquad \alpha \equiv \beta \quad \text{if} \quad U\alpha = U\beta, \tag{5}$$

where U is the set of all sequences $\langle x_1, \ldots, x_n \rangle$. Figure 1 illustrates the sorting network [12][34][13][24][23]. Clearly $\alpha \subseteq \beta$ implies that $\alpha\gamma \subseteq \beta\gamma$.

Furthermore, if β is a sorting network and $\alpha \subseteq \beta$ we must have $\alpha \equiv \beta$; in fact, $x\alpha = x\beta$ for all x in this case, since $x\alpha$ must be sorted.

Sorting networks can also be interpreted in a more general way, if we allow m numbers to be contained in each line for some fixed $m \geq 1$. If $x_1, \ldots, x_n$ are *multisets* (i.e., sets with the possibility of repeated elements), containing m elements each, we can redefine the comparator $[ij]$ to be the operation of replacing x_i and x_j by the smallest and largest m elements, respectively, of the original $2m$ elements in x_i and x_j. See Fig. 3, which illustrates the case $m = 2$. Our first result gives a basic property of this general interpretation.

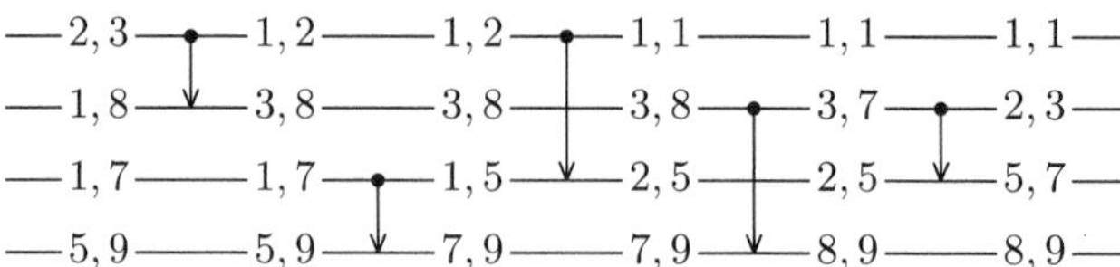

FIGURE 3. Another interpretation of the network in Fig. 1.

Theorem 1. *Let α be a comparator network for n elements, and let i and j be indices such that $(x\alpha)_i \not\leq (x\alpha)_j$ for some x and for some $m \geq 1$; in other words, the m elements of the multiset $(x\alpha)_i$ are not all less than or equal to the m elements of $(x\alpha)_j$. Then there is a sequence $y = \langle y_1, \ldots, y_n \rangle$ of zeros and ones such that $(y\alpha)_i = 1$ and $(y\alpha)_j = 0$.*

Proof. Let $\alpha = f_1 \ldots f_t$, where each f_s is an exchange or a comparator. Let u be the smallest element of $(x\alpha)_j$; we shall use the name A to stand for any number $\leq u$, and B for any number $> u$. By hypothesis, at least one element of $(x\alpha)_i$ is a B. We shall define $t+1$ sequences $y^{(s)}$ of zeros and ones, for $0 \leq s \leq t$, such that, for $1 \leq p \leq n$,

$$y_p^{(s)} = 0 \text{ implies that } (xf_1 \ldots f_s)_p \text{ contains an } A, \tag{6}$$

$$y_p^{(s)} = 1 \text{ implies that } (xf_1 \ldots f_s)_p \text{ contains a } B. \tag{7}$$

First for $s = t$ we define $y_i^{(t)} = 1$ and $y_j^{(t)} = 0$; the other elements $y_k^{(t)}$ are defined in any manner consistent with conventions (6) and (7) above.

Assuming that $y^{(s)}$ has been defined for some $s \geq 1$, we define $y^{(s-1)}$ as follows:

Case 1, $f_s = (pq)$. Then $y^{(s-1)} = y^{(s)}(pq)$.

Case 2, $f_s = [pq]$ and $y_p^{(s)} = y_q^{(s)}$. Then $y^{(s-1)} = y^{(s)}$. This definition fulfills the conditions above, since $y_q^{(s)} = 0$ implies that $(xf_1 \ldots f_s)_q$ contains at least one A, hence $(xf_1 \ldots f_s)_p$ contains *all* A's, hence there are more than m A's in all; some A's must be present in both $(xf_1 \ldots f_{s-1})_p$ and $(xf_1 \ldots f_{s-1})_q$. Similarly, $y_p^{(s)} = 1$ implies that $(xf_1 \ldots f_{s-1})_p$ and $(xf_1 \ldots f_{s-1})_q$ both contain at least one B.

Case 3, $f_s = [pq]$ and $y_p^{(s)} \neq y_q^{(s)}$. Then we define the coordinates $(y_p^{(s-1)}, y_q^{(s-1)})$ to be either $(0, 1)$ or $(1, 0)$, in any manner consistent with the conventions above; and we let $y_r^{(s-1)} = y_r^{(s)}$ for $p \neq r \neq q$. This definition of $y^{(s-1)}$ is justified because $(xf_1 \ldots f_{s-1})_p$ and $(xf_1 \ldots f_{s-1})_q$ are not both all A's or both all B's. Note that the condition $y_q^{(s)} = 0$ is impossible, since it implies as in case 2 that $(xf_1 \ldots f_s)_p$ is all A's, contradicting our convention; thus $y_p^{(s)} = 0$ and $y_q^{(s)} = 1$.

According to this definition, $y^{(s-1)} f_s = y^{(s)}$; hence $y^{(0)}\alpha = y^{(t)}$. Therefore $y = y^{(0)}$ satisfies the conditions of the theorem. □

When $m = 1$, Theorem 1 implies that a network will necessarily sort all possible inputs if we can prove that it sorts the 2^n sequences of zeros and ones:

Corollary 1. *A comparator network is a sorting network if and only if it sorts all sequences of zeros and ones.* □

Corollary 2. *Let $\hat{M}(m)$ be the minimum number of comparators needed to merge two sets of m elements, i.e., to sort all sequences $\langle x_1, \ldots, x_{2m} \rangle$ such that $x_1 \leq \cdots \leq x_m$ and $x_{m+1} \leq \cdots \leq x_{2m}$. Then*

$$\hat{S}(mn) \leq n\hat{S}(m) + \hat{M}(m)\hat{S}(n); \tag{8}$$

$$\hat{M}(mn) \leq \hat{M}(m)\hat{M}(n). \tag{9}$$

Proof. Replace each line in an n-element sorting network by m parallel lines, and replace each comparator by $\hat{M}(m)$ comparators that merge the $2m$ lines corresponding to the original 2 lines. Append n sorting networks for m elements at the left, in order to sort each of the groups; this yields a sorting network for mn elements having $\hat{M}(m)\hat{S}(n)+n\hat{S}(m)$ comparators.

If we start with a sorting network that was constructed in this way for $n = 2$, the right-hand part of the network has $\hat{M}(m)$ comparators; expanding each line to m' lines makes the $\hat{M}(m)\hat{M}(m')$ comparators of the right-hand part capable of merging two ordered groups of mm' elements. □

An example of the construction in Corollary 2 appears in Fig. 4. Bose and Nelson proved Corollary 2 in the special case of binary merging, $n = 2$; consequently we have $\hat{M}(2^n) \leq 3^n$ and $\hat{S}(2^n) \leq 3^n - 2^n$.

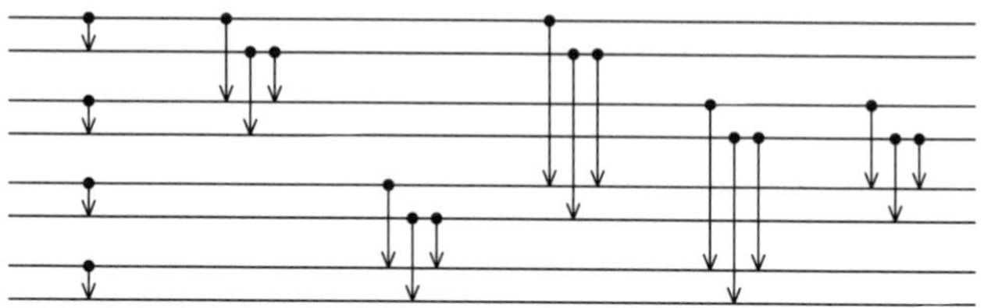

FIGURE 4. An 8-element sorting network, constructed from Fig. 1 and Corollary 2 ($m = 2$).

When $\hat{S}(n) \approx n^\beta - n$ and $\hat{M}(n) \approx n^\beta$, the inequalities in Corollary 2 do not allow us to lower the exponent β; and in fact those inequalities do not lead to an especially efficient way to construct sorting networks, compared to other known methods. Yet the special case $m = 7$, $n = 3$ shows that 21 elements can be sorted with one comparator less than predicted by the Bose–Nelson conjecture; this observation first showed the authors that the conjecture could not hold in general (see Floyd [4]). We went on to find that the conjecture is false for all $n > 8$. (This result was, perhaps, poetic justice, since Bose made the conjecture shortly after he had helped to disprove Euler's famous Latin-squares conjecture for all $n > 6$ — after having first disproved it for $n = 50$. And our own sorting networks have by now been shown to be nonoptimal for all $n > 9$.)

The algebraic conventions above lead to several useful simplifications. In the first place, we can use identities (3) and (4) to transform any comparator network so that all comparators precede all exchanges, and so that all comparators $[ij]$ have $i < j$. (Working from left to right, we replace $[ji]$ by $[ij](ij)$ when $i < j$, and we permute exchanges with comparators. This process clearly converges in a finite number of steps.) In this way, a network α is transformed into $\alpha'\alpha''$, where α' has only "downward" comparators and α'' has only exchanges. If α is a sorting network, we can see by considering the effect of α on $\langle 1, 2, \ldots, n\rangle$ that α'' must be the identity transformation; hence $\alpha \equiv \alpha'$.

Let us say that a sorting network is in *standard form* if it consists entirely of comparators $[ij]$ with $i < j$, and no exchanges. We have proved

Lemma 1. *Every sorting network is equivalent to a network in standard form, having the same number of comparators.* □

When the network is in standard form and when x_n exceeds $x_1, \ldots, x_{n-1}$, all comparators $[in]$ are essentially inoperative; hence we can construct a sorting network for $n-1$ elements by simply removing all such comparators $[in]$ from an n-element sorting network in standard form. Hibbard [5] observed that sorting networks having the same number of comparators as those originally constructed by Bose and Nelson can be obtained in this way by starting with a Bose–Nelson network for $2^k > n$ elements and deleting all comparators that involve $x_{n+1}, \ldots, x_{2^k}$.

We can now obtain a lower bound for the merging problem:

Theorem 2. $\hat{M}(2n) \geq 2\hat{M}(n) + n$.

Proof. Consider a network with $\hat{M}(2n)$ comparators in standard form, which sorts $\langle x_1, \ldots, x_{4n} \rangle$ whenever $x_1 \leq x_3 \leq \cdots \leq x_{4n-1}$ and $x_2 \leq x_4 \leq \cdots \leq x_{4n}$. We separate the comparators $[ij]$ into three types,

$$\begin{aligned} &\text{A: } i \leq 2n \text{ and } j \leq 2n;\\ &\text{B: } i \leq 2n \text{ and } j > 2n;\\ &\text{C: } i > 2n \text{ and } j > 2n. \end{aligned}$$

Since $x_{2n+1}, \ldots, x_{4n}$ may be very large, there must be at least $\hat{M}(n)$ comparators of type A; similarly there must be at least $\hat{M}(n)$ of type C. And since we might have $x_i = 1$ when i is odd, $x_i = 0$ when i is even, there must be at least n comparators of type B in order to let n zeros rise to the top half of the diagram. □

A similar proof shows that $\hat{M}(2n+1) \geq \hat{M}(n) + \hat{M}(n+1) + n$; and the same relations also hold with S in place of M. It follows that $\hat{M}(n) \geq \frac{1}{2} n \log_2 n + O(n)$ for all n. Sorting networks that are based recursively on binary merging must therefore involve the order of $n(\log n)^2$ comparators, at least.

The best sorting networks known for $n > 8$ do not use binary merging, so Theorem 2 does not give us useful information about lower bounds for $\hat{S}(n)$. When n is comparatively small, exact lower bounds can however be found, as we shall now see. First we shall examine a general commutativity condition:

Lemma 2. *If* $\alpha = [i_1 j_1][i_2 j_2] \ldots [i_t j_t]$, *where the multisets* $\{i_1, i_2, \ldots, i_t\}$ *and* $\{j_1, j_2, \ldots, j_t\}$ *are disjoint, and if* β *is any rearrangement of the comparators of* α, *then* $\alpha \equiv \beta$.

Proof. We shall show that $U\alpha$ is the set $\hat{S}(\alpha)$ of all vectors $\langle x_1, \ldots, x_n \rangle$ such that $x_{i_s} \leq x_{j_s}$ for $1 \leq s \leq t$. All vectors $x \in \hat{S}(\alpha)$ satisfy $x\alpha = x$, hence $\hat{S}(\alpha) \subseteq U\alpha$.

Conversely, suppose that $x\alpha \notin \hat{S}(\alpha)$, so that $(x\alpha)_{i_s} > (x\alpha)_{j_s}$ for some s. Clearly s must be less than t; let $\alpha' = [i_1 j_1] \ldots [i_{t-1} j_{t-1}]$.

By induction on t, we have $(x\alpha')_{i_s} \leq (x\alpha')_{j_s}$; hence $[i_t j_t]$ has either increased the i_s component or decreased the j_s component of $x\alpha'$. But then either $i_s = j_t$ or $j_s = i_t$, contradicting the hypothesis. □

Lemma 3. *Let α be a sorting network for three or more elements. Then there is a sorting network, with no more comparators than α, in which the first three operations belong to one of the following cases:*

$$\mathrm{A} = [12][13][23];\quad \mathrm{B} = [12][13][45];\quad \mathrm{C} = [12][34][13];\quad \mathrm{D} = [12][34][14].$$

Proof. Clearly α includes at least three comparators, or it could not sort. Since $(ij)\alpha \equiv \alpha$, we may use (3) and (4) to transform α into a sorting network α' in which the first operation is [12]. For example, if $\alpha = [47]\beta$, we may take $\alpha' = (14)(27)\alpha = [12](14)(27)\beta$; and if $\alpha = [21]\beta$ we may take $\alpha' = \alpha = [12](12)\beta$. Similarly we may assume that the second operation is either [12] or [13] or [23] or [34]; and since $[12][12] = [12]$, we may rule out the case [12][12]. If the second operation is [23], we may observe that whenever $\alpha = [i_1 j_1] \ldots [i_t j_t]$ is a sorting network, so is the "dual" network $\alpha' = [j_1 i_1] \ldots [j_t i_t]\tau$, where τ is any sequence of exchanges that transforms $\langle x_1, x_2, \ldots, x_n\rangle$ into $\langle x_n, \ldots, x_2, x_1\rangle$. Hence, when $\alpha = [12][23]\beta$, we may consider the sorting network $\alpha' = [21][32]\beta'\tau = [12][13](13)(12)\beta'\tau$. Therefore we may assume that the first two operations are [12][13] or [12][34].

Proceeding in this way, we can analyze the possibilities for the third comparator, as follows:

$$[12][13][12] = [12][13].$$
$$[12][13][23] = \mathrm{A}.$$
$$[12][13][14] \supseteq [12][34][13].$$
$$[12][13][24] \rightarrow [13][12][34](23) \equiv [12][13][34](23).$$
$$[12][13][34] \rightarrow [12][14][34] \equiv [12][34][14].$$
$$[12][13][45] = \mathrm{B}.$$
$$[12][34][13] = \mathrm{C}.$$
$$[12][34][14] = \mathrm{D}.$$
$$[12][34][23] \rightarrow [34][12][41](13)(24) = [12][34][14](14)(13)(24).$$
$$[12][34][24] \sim [21][43][42] = [12][34][13](13)(34)(12).$$
$$[12][34][15] = [12][15][34] \rightarrow [12][13][45](45)(35).$$
$$[12][34][25] = [12][25][34] \sim [21][52][43] \rightarrow [21][32][45](35) = [12][13][45](13)(12)(35).$$
$$[12][34][35] = [34][35][12] \rightarrow [12][13][45](45)(35)(24)(13).$$
$$[12][34][45] \rightarrow [34][12][25](24)(13) = [12][34][25](24)(13).$$

Here "$\rightarrow$" denotes an appropriate left-multiplication by one or more exchanges; "$\equiv$" denotes an application of Lemma 2; and "$\sim$" denotes dualization as above.

Finally, if the first three operations are [12][34][56], we may consider the first comparator that has an index in common with a previous one; this will reduce to a case already considered. □

The exhaustive method in this proof can be extended to show that there are essentially eleven ways to choose the first four comparators, when $n \geq 4$, namely

$$\text{A1} = [12][13][23][24], \quad \text{A2} = [12][13][23][45];$$
$$\text{B1} = [12][13][45][14], \quad \text{B2} = [12][13][45][46], \quad \text{B3} = [12][13][45][56];$$
$$\text{C1} = [12][34][13][24], \quad \text{C2} = [12][34][13][35],$$
$$\text{C3} = [12][34][13][45], \quad \text{C4} = [12][34][13][56];$$
$$\text{D1} = [12][34][14][35], \quad \text{D2} = [12][34][14][56].$$

Details are omitted here, since we shall not need that fact.

Theorem 3. $\hat{S}(n) \geq \hat{S}(n-1) + 3$, *for all* $n \geq 5$.

Proof. There is a sorting network with $\hat{S}(n)$ comparators, in standard form, beginning with one of the four sequences in Lemma 3. If we suppress all comparators $[ij]$ with $i = 1$ we have a sorting network for $x_2, \ldots, x_n$, so we must show that at least three comparators have $i = 1$. This is obvious, since in each case we already know two of the comparators; and at least one more is required to bring the smallest element to the required position. □

Theorem 3 probably possesses the unique property that it has exactly two applications, no more and no less! Once $\hat{S}(5)$ has been shown to equal 9, we can use Theorem 3 to show that $\hat{S}(6) = 12$; and $\hat{S}(7) = 16$ will imply that $\hat{S}(8) = 19$. Besides these results, the theorem appears to be quite useless.

We always have $\hat{S}(n) \geq \log_2 n!$, by an elementary information-theoretic argument. Hence the values of $\hat{S}(1)$, $\hat{S}(2)$, $\hat{S}(3)$, and $\hat{S}(4)$ are immediately established. But information theory tells us only that $\hat{S}(5) \geq 7$, and Theorem 3 shows that $\hat{S}(5) \geq 8$; the following theorem shows how to strengthen Theorem 3 when $n = 5$:

Theorem 4. $\hat{S}(5) = 9$.

Proof. We need only show that $\hat{S}(5) \geq 9$, in view of Bose and Nelson's construction. Proceeding as in Theorem 3, if the sorting network begins as in case D, we may permute the lines so that the first three comparators are [14][25][15]. Then we must have at least $\hat{S}(3)$ more comparators $[ij]$ with $i < j \le 3$, and at least $\hat{S}(3)$ more with $3 \le i < j$, to complete the sort. This makes 9 comparators.

For cases A, B, and C we may permute the lines to obtain a sorting network in standard form in which the first three comparators are respectively

$$\begin{aligned}&[12][15][25] \text{ in case A,}\\&[13][14][25] \text{ in case B,}\\&[14][25][12] \text{ in case C.}\end{aligned}$$

Applying these to all 32 combinations $\langle x_1, x_2, x_3, x_4, x_5\rangle$ of zeros and ones (see Corollary 1), then replacing all zeros at the left and all ones at the right by asterisks, discarding all duplicates and all sequences that are nothing but asterisks, we obtain the 5-tuples

$$\begin{matrix}* * * 1 0\\ * * 1 0 0\\ * * 1 1 0\\ * * 1 0 *\\ * 1 0 0 *\\ * 1 1 0 *\\ * 1 0 * *\end{matrix} \tag{10}$$

plus the "special" 5-tuples

$$\begin{matrix}1 1 0 0 *, & 1 1 0 * *, & 1 1 1 0 * & \text{in case A,}\\ 1 0 1 1 0, & 1 0 * * * & & \text{in case B,}\\ 1 1 0 * *, & * 1 0 1 0, & * 1 1 1 0 & \text{in case C.}\end{matrix} \tag{11}$$

In order to sort (10), we need at least $\hat{S}(3)$ comparators with $2 \le i < j \le 4$ and $\hat{S}(3)$ with $3 \le i < j \le 5$; and there must also be another with $i = 1$. The only way to do this with five more comparators is to use the sequence [34][23][45][34] or [34][45][23][34], with an additional $[1j]$ inserted somewhere. But then it is not difficult to verify that the special 5-tuples in (11) cannot all be sorted. □

Theorem 5. $\hat{S}(7) = 16$.

Proof. This theorem was proved by exhaustive enumeration on a CDC G-21 computer at Carnegie Institute of Technology in 1966. The program was written by Mr. Richard Grove, and its running time was approximately 20 hours. The algorithm consisted of constructing a set

S_t of sequences such that, for all α of the form $[i_1 j_1] \ldots [i_t j_t]$, there exist permutations π and ρ with $\pi\alpha\rho \supseteq \beta$ for some $\beta \in S_t$. The sets S_t were generated successively for $t = 1, 2, \ldots, 16$, taking care to keep each set rather small; for this purpose, a 128-bit vector was maintained for each element of S_t, characterizing those 7-tuples of zeros and ones that are output by the network. Most of the computation (about 13 hours) was spent in the cases $t = 8$ and 9, since S_9 had 729 elements. None of the six elements in S_{15} was a sorting network. □

The methods of proof used to establish these lower bounds on $\hat{S}(n)$ are of course quite unsatisfactory for larger values of n. We have no idea how to prove that $\hat{S}(n)$ grows as $cn(\log n)^2$, although the best upper bounds known to date have this asymptotic behavior.

David Van Voorhis [9] has recently generalized Theorem 3, proving that $\hat{S}(n) \geq \hat{S}(n-1) + \lceil \log_2 n \rceil$.

The preparation of this report has been supported in part by the National Science Foundation, and in part by the Office of Naval Research.

References

[1] K. E. Batcher, "A new internal sorting method," Rep. No. GER-11759 (Akron, Ohio: Goodyear Aerospace Corporation, 29 September 1964), ii + 22 pages.

[2] K. E. Batcher, "Sorting networks and their applications," *Proceedings of the AFIPS Spring Joint Computer Conference* **32** (1968), 307–314.

[3] R. C. Bose and R. J. Nelson, "A sorting problem," *Journal of the Association for Computing Machinery* **9** (1962), 282–296.

[4] R. W. Floyd, "A minute improvement in the Bose–Nelson sorting procedure," Memorandum (Wakefield, Massachusetts: Computer Associates, Inc., February 1964).

[5] Thomas N. Hibbard, "A simple sorting algorithm," *Journal of the Association for Computing Machinery* **10** (1963), 142–150.

[6] D. E. Knuth and R. W. Floyd, "Improved constructions for the Bose–Nelson sorting problem," *Notices of the American Mathematical Society* **14** (February 1967), 283, Abstract 67T-228.

[7] Donald E. Knuth, *Sorting and Searching*, Volume 3 of *The Art of Computer Programming* (Reading, Massachusetts: Addison–Wesley, 1973).

[8] Daniel G. O'Connor and Raymond J. Nelson, "Sorting system with n-line sorting switch," *U.S. Patent 3,029,413* (10 April 1962), 22 columns plus 6 figures.

[9] David C. Van Voorhis, "An improved lower bound for sorting networks," *IEEE Transactions on Computers* **C-21** (1972), 612–613.

Addendum

The best lower bounds for $\hat{S}(n)$ that were known in 1971, as listed in Table 1, have still never been improved. And the upper bounds in that table have been decreased only in two cases: Hugues Juillé ["Incremental co-evolution of organisms: A new approach for optimization and discovery of strategies," *Lecture Notes in Computer Science* **929** (1995), 246–260] used a genetic algorithm to prove that $\hat{S}(13) \leq 45$. And the asymptotic value of $\hat{S}(n)$ as $n \to \infty$ was (astonishingly) shown to be only $O(n \log n)$, by M. Ajtai, J. Komlós, and E. Szemerédi, "Sorting in $c \log n$ parallel steps," *Combinatorica* **3** (1983), 1–19.

The constant hidden by 'O' in the networks constructed by Ajtai, Komlós, and Szemerédi was, however, enormous. M. S. Paterson ["Improved sorting networks with $O(\log N)$ depth," *Algorithmica* **5** (1990), 75–92] showed how to reduce it greatly, proving that $\hat{S}(n) < 3050\, n \lg n$. However, even Paterson's networks don't begin to win over those of Batcher until n exceeds 2^{6099}, because Batcher proved that $\hat{S}(n) < n((\lg n)^2 - \lg n + 4)/4$ when n is a power of 2. A construction that needs only about $5\, n \lg n$ comparators will be needed to improve upon Batcher's method when n is approximately 1,000,000.

Thus the gap between the upper and lower bounds for the Bose–Nelson sorting problem remains tantalizingly large, for all $n > 8$. We still aren't even close to knowing, as yet, the true nature of optimum sorting networks except when n is extremely small.

Chapter 3

A One-Way, Stackless Quicksort Algorithm

[Written with Huang Bing-Chao. Originally published in BIT **26** *(1986), 127–130.]*

This note describes a sorting technique that is similar to the well-known "Quicksort" algorithm, but it is unidirectional and avoids recursion. The new approach, which assumes that the keys to be sorted are positive numbers, leads to a much shorter program.

Hoare's "Quicksort" [1] has for many years been the method of choice for general purpose sorting, because its fast inner loop leads to efficient average running time on most machines [3]. Given an array of N records $R_1 \ldots R_N$, where the records include keys $K_1 \ldots K_N$, we wish to rearrange things so that the keys are in increasing order. The main idea of Quicksort is to pick a "pivot" record R and to rearrange records so that R moves to its correct destination R_i; then the same process is applied recursively to the subarrays $R_1 \ldots R_{i-1}$ and $R_{i+1} \ldots R_N$.

If all the keys K_j are positive, we can negate the key in a record to indicate that it appears in its final position. Furthermore there's a way to do the basic pivoting operation with pointers that travel strictly from left to right in the array, instead of "burning the candle at both ends" as in the classical Quicksort method. The combination of these two ideas leads to a surprisingly simple stackless algorithm that runs only about 60 percent slower than the fastest implementations of Quicksort.

The method described below uses three indices i, j, l for which the following relations will be invariant: Records $R_1 \ldots R_{l-1}$ will be fully sorted; keys $K_l \ldots K_{i-1}$ will all be less than the current pivot key K; keys $K_{i+1} \ldots K_j$ will be greater than or equal to K. Furthermore $K_1 \ldots K_{i-1}\, K_{i+1} \ldots K_N\, K$ will be a permutation of the N original keys; hence we can regard position R_i of the array to be vacant. If $j < N$, it is easy to increase j without destroying these invariant conditions: For if $K \leq K_{j+1}$, we can simply set $j \leftarrow j + 1$; otherwise we set $j \leftarrow j + 1$,

$R_i \leftarrow R_j$, $i \leftarrow i+1$, and $R_j \leftarrow R_i$. (Notice that two records are moved, not one; this is why the one-way scheme loses some of the effectiveness of classical Quicksort.)

In order to keep the inner loop reasonably fast, we shall assume that the array contains an additional record R_{N+1} whose key K_{N+1} is negative. The algorithm can be described as follows in the style of [2]:

Algorithm O *(One-way quicksort).* Records $R_1 \ldots R_N$ are rearranged in place; after sorting is complete their keys will be in order, $K_1 \leq \cdots \leq K_N$. We assume that all keys are greater than zero, and that there is an additional sentinel key $K_{N+1} < 0$.

O1. [Initialize.] Set $l \leftarrow 1$. (Variable l represents our progress through the array; records $R_1 \ldots R_{l-1}$ will be in their final form.)

O2. [Begin pivot operation.] If $K_l < 0$, go to step O6. Otherwise set $i \leftarrow l$, $j \leftarrow l$, $R \leftarrow R_l$, $K \leftarrow K_l$.

O3. [Advance and compare.] Increase j by one; then repeat this step if $K \leq K_j$. (Since $K_{N+1} < 0$ and $K > 0$, we must eventually find $K > K_j$.)

O4. [Rearrange.] If $K_j > 0$, set $R_i \leftarrow R_j$, $i \leftarrow i + 1$, $R_j \leftarrow R_i$, and go back to step O3. (The invariant relations stated above are being maintained.)

O5. [Finish pivot operation.] Set $R_i \leftarrow R$, $K_i \leftarrow -K$, and go back to step O2. (The minus sign indicates that record R_i is in its final position.)

O6. [Restore the sign.] Set $K_l \leftarrow -K_l$, $l \leftarrow l + 1$. Then if $l \leq N$, return to step O2; otherwise the algorithm terminates. □

Here is a Pascal implementation, assuming that R is an **array** $[1\,..\,M]$ **of** *entry*, where *entry* is a record type containing an integer *key* field and where $M > N$:

```
procedure OSORT(N: integer);
  label 3;
  var L: 1..M; {progress indicator}
    J: 1..M; {end of current pivoting region}
    I: 1..M; {gap in current pivoting region}
    P: entry; {the pivot record}
  begin R[N + 1].key := -1;
  for L := 1 to N do
    begin while R[L].key > 0 do
      begin J := L; I := L; P := R[L];
```

```
    3: repeat J := J + 1 until P.key > R[J].key;
       if R[J].key > 0 then
          begin R[I] := R[J]; I := I + 1;
          R[J] := R[I]; goto 3;
          end;
       P.key := -P.key; R[I] := P;
       end;
     R[L].key := -R[L].key;
     end;
end;
```

We can compare the new algorithm to the old by studying its implementation on the hypothetical `MIX` computer and analyzing the running time as in [2]:

Program O *(One-way quicksort).* Records to be sorted consist of keys only. They appear in locations `INPUT+1` through `INPUT+N`; and location `INPUT+N+1` contains a negative value. Register assignments: $\mathrm{rI1} \equiv l-N$, $\mathrm{rI2} \equiv j$, $\mathrm{rI3} \equiv i$, and $\mathrm{rA} \equiv K = R$.

```
START ENT1  1-N                 1      O1. Initialize. l ← 1.
      JMP   2F                  1      To O2.
5H    STA   INPUT,3             N      O5. Finish pivot operation. R_i ← R.
      STX   INPUT,3(0:0)        N      K_i ← −K.
2H    LDA   INPUT+N,1          2N      O2. Begin pivot operation. K ← K_l.
      JAN   6F                 2N      To O6 if K < 0.
      ENT2  N,1                 N      j ← l.
      ENT3  0,2                 N      i ← j.
3H    INC2  1                   C      O3. Advance and compare. j ← j + 1.
      CMPA  INPUT,2             C
      JLE   3B                  C      Repeat if K ≤ K_j.
4H    LDX   INPUT,2           X + N    O4. Rearrange.
      JXN   5B                X + N    To O5 if K_j < 0.
      STX   INPUT,3             X      R_i ← R_j.
      INC3  1                   X      i ← i + 1.
      LDX   INPUT,3             X
      STX   INPUT,2             X      R_j ← R_i.
      JMP   3B                  X      Return to O3.
6H    STZ   INPUT+N,1(0:0)      N      O6. Restore the sign. K_l ← −K_l.
      INC1  1                   N      l ← l + 1.
      J1NP  2B                  N      To O2 if l ≤ N. □
```

Using Kirchhoff's law and the fact that every key is negated exactly once, it is easy to deduce that the running time of this program is $4C + 11X + 19N + 2$ (in `MIX` units), where C is the number of comparisons and X is the number of rearrangement steps. The techniques of [2] show that the average values C_N and X_N of C and X satisfy

$$C_N = \frac{1}{N} \sum_{1 \le i \le N} (N + C_{i-1} + C_{N-i}) ,$$

$$X_N = \frac{1}{N} \sum_{1 \le i \le N} (i - 1 + X_{i-1} + X_{N-i}) ,$$

when $N > 0$; $C_0 = X_0 = 0$. The solution to these recurrences is

$$C_N = 2(N+1)H_N - 3N ; \qquad X_N = (N+1)H_N - 2N .$$

Therefore the average running time of Program O is

$$19(N+1)H_N - 15N + 2 \approx 19N \ln N - 4.033N + O(\log N).$$

This compares with approximately $11.667N \ln N - 1.74N + O(\log N)$ units of time for the tuned-up Quicksort Program 5.2.2Q in [2]. The latter program is faster, but it requires 63 `MIX` instructions and an auxiliary stack; the new program is remarkably fast, considering that it is only 21 instructions long.

As in ordinary Quicksort, the worst-case running time is of order N^2. Indeed, the algorithm runs slowly when the keys are already in order.

The preparation of this paper was supported in part by National Science Foundation grant DCR-83-00984. The authors thank the referees for several important suggestions.

References

[1] C. A. R. Hoare, "Quicksort," *The Computer Journal* **5** (1962), 10–15.

[2] Donald E. Knuth, *Sorting and Searching*, Volume 3 of *The Art of Computer Programming*, second printing (Reading, Massachusetts: Addison–Wesley, 1975).

[3] Robert Sedgewick, *Quicksort* (New York: Garland, 1980).

Addendum

From a 21st-century perspective, we can say that the algorithm above is pleasantly "cache friendly." Its worst case, however, disqualifies it from general use. The worst case can be avoided by first permuting the data randomly, although Floyd's Theorem 5.4.9F in [2] proves that such randomization cannot be done in "cache friendly linear time."

Chapter 4

Optimum Binary Search Trees

[Originally published in Acta Informatica **1** *(1971), 14–25, 270.]*

One of the popular methods for retrieving information by its "name" is to store the names in a binary tree. To find if a given name is in the tree, we compare it to the name at the root, and four cases arise:

1. There is *no* root (the binary tree is empty): The given name is not in the tree, and the search terminates *unsuccessfully.*

2. The given name *matches* the name at the root: The search terminates *successfully.*

3. The given name is *less* than the name at the root: The search continues by examining the *left subtree* of the root in the same way.

4. The given name is *greater* than the name at the root: The search continues by examining the *right subtree* of the root in the same way.

Special cases of this method are the binary search and its variants (uncentered binary search; Fibonacci search) and the search-sort scheme of Wheeler, Berners-Lee, Booth, Hibbard, Windley, et al. (see [1, 3, 7, 10]).

When all names in the tree are equally probable, it is not difficult to see that a best possible binary tree from the standpoint of average search time is one with minimum path length, namely the *complete binary tree* (see [9, pages 400–401]). This is the tree that is implicitly present in one of the variants of the binary search method.

But when some names are known to be much more likely to occur than others, the best possible binary tree will not necessarily be balanced. For example, consider the following words and frequencies:

a	32	for	15	on	22
an	7	from	10	the	79
and	69	high	8	to	18
by	13	in	64	with	9
effects	6	of	142		

(These are the words to be ignored in a certain KWIC indexing application [6, page 124].) The best possible tree for those words turns out to be

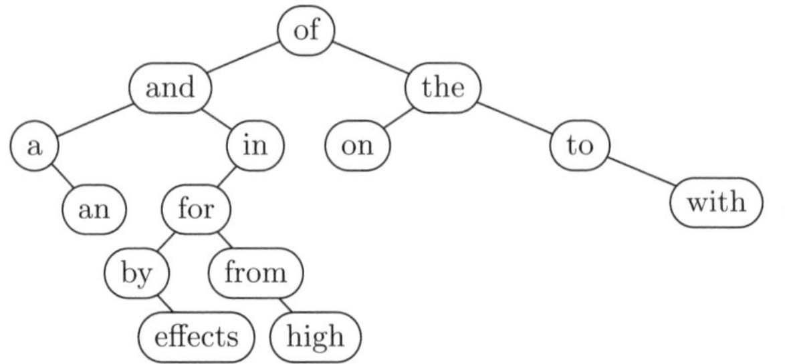

In this paper we discuss the question of finding such "optimum binary trees," when frequencies are given. The ordering property of the tree makes this problem more difficult than the standard "Huffman coding problem" (see [9, Section 2.3.4.5]).

For example, suppose that our words are A, B, C and the frequencies are α, β, γ. There are five binary trees with three nodes:

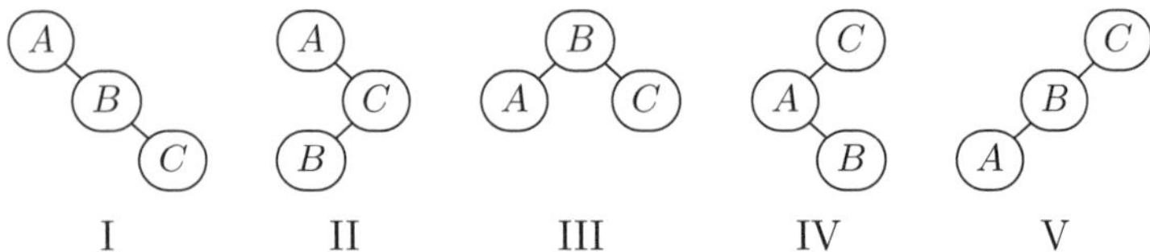

The following diagram shows the ranges of (α, β, γ) in which each of those trees is optimum, assuming that $\alpha + \beta + \gamma = 1$:

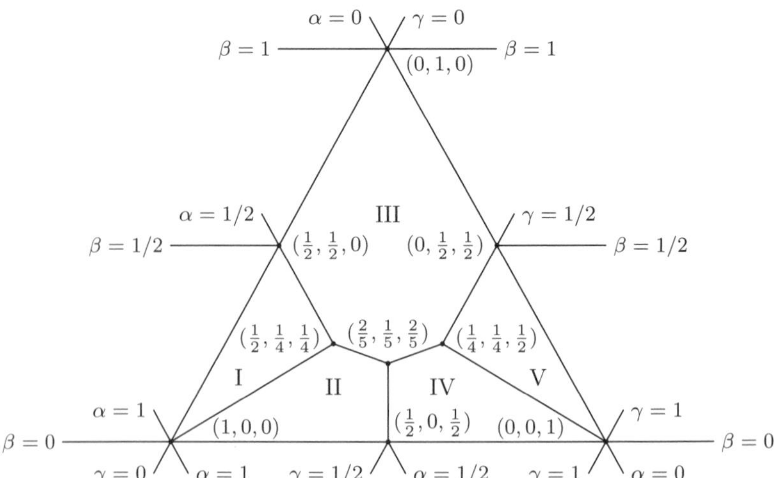

Note that it is sometimes best to put B at the root even when both A and C occur more frequently. And on the other hand, it is not sufficient simply to choose the root so as to equalize the left and right search probabilities as much as possible, contrary to a remark of Iverson (see [8, page 144; 2, page 318]).

In general, there are $\binom{2n}{n}/(n+1) \sim 4^n/(n\sqrt{\pi n})$ binary trees with n nodes, so an exhaustive search for the optimum is out of the question. However, we shall show below that an elementary application of "dynamic programming," which is essentially the same as the idea used as the basis of the Cocke–Kasami–Younger–Earley parsing algorithm for context-free grammars [4], can be used to find an optimum binary search tree in order n^3 steps. By refining the method we will in fact cut the running time to order n^2.

In practice we want to generalize the problem, considering not only the frequencies with which a successful search is completed, but also the frequencies where unsuccessful searches occur. Thus we are given n names A_1, A_2, ..., A_n, together with $2n+1$ frequencies α_0, α_1, ..., α_n and β_1, β_2, ..., β_n. Here β_i is the frequency of encountering name A_i, and α_i is the frequency of encountering a name that lies between A_i and A_{i+1}; the borderline frequencies α_0 and α_n have obvious interpretations.

The key fact that makes this problem amenable to dynamic programming is that all subtrees of an optimum tree are optimum. If A_i appears at the root, then its left subtree is an optimum solution for frequencies α_0, ..., α_{i-1} and β_1, ..., β_{i-1}; its right subtree, similarly, is optimum for α_i, ..., α_n and β_{i+1}, ..., β_n. Therefore we can build up optimum trees for all "frequency intervals" α_i, ..., α_j and β_{i+1}, ..., β_j when $i \le j$, starting from the smallest intervals and working toward the largest. Since there are only $(n+2)(n+1)/2$ choices of i and j with $0 \le i \le j \le n$, the total amount of computation is not excessive.

Consider the following binary tree:

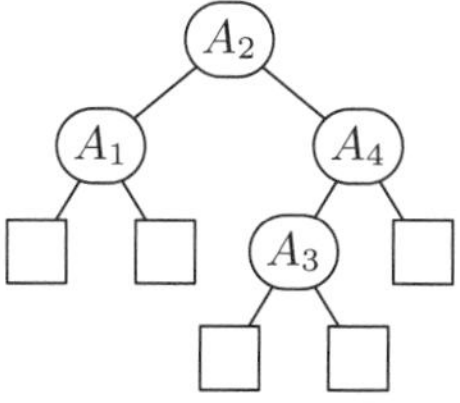

(Square nodes denote empty or terminal positions where no names are stored.) The "weighted path length" P of a binary tree is the sum of frequencies times the level of the corresponding nodes, with the root at level 1; in the example above the weighted path length is

$$3\alpha_0 + 2\beta_1 + 3\alpha_1 + \beta_2 + 4\alpha_2 + 3\beta_3 + 4\alpha_3 + 2\beta_4 + 3\alpha_4.$$

In general, we can see that the weighted path length satisfies the equation

$$P = P_L + P_R + W,$$

where P_L and P_R are the weighted path lengths of the left and right subtrees, and $W = \alpha_0 + \alpha_1 + \cdots + \alpha_n + \beta_1 + \cdots + \beta_n$ is the "weight" of the tree, the sum of all frequencies. The weighted path length measures the relative amount of work needed to search the tree, when the α's and β's are chosen appropriately; therefore the problem of finding an optimum search tree is the problem of finding a binary tree of minimum weighted path length, with the weights applied from left to right in the tree.

These remarks lead immediately to a straightforward way to determine an optimum search tree. Let $P_{i,j}$ and $W_{i,j}$ denote the weighted path length and the total weight of an optimum search tree for all words that lie strictly between A_i and A_{j+1} when $i \le j$; and let $R_{i,j}$ denote the index of the root of this tree, when $i < j$. The following formulas now yield the desired algorithm:

$$\begin{aligned} P_{i,i} &= W_{i,i} = \alpha_i, \quad \text{for } 0 \le i \le n; \\ W_{i,j} &= W_{i,j-1} + \beta_j + \alpha_j, \quad \text{for } 0 \le i < j \le n; \qquad (**) \\ P_{i,R_{i,j}-1} + P_{R_{i,j},j} &= \min_{i<k\le j} (P_{i,k-1} + P_{k,j}) = P_{i,j} - W_{i,j}, \quad \text{for } 0 \le i < j \le n. \end{aligned}$$

The problem of finding "best alphabetical encodings," considered by Gilbert and Moore in their classic paper [5], is easily seen to be a special case of the problem considered here, with $\beta_1 = \beta_2 = \cdots = \beta_n = 0$. Another closely related (but not identical) problem has been discussed by Wong [12]. In both cases the authors have suggested an algorithm for finding an optimum tree that is essentially identical to (**). Gilbert and Moore observed that the algorithm takes about $n^3/6$ iterations of the inner loop (choosing $R_{i,j}$ from among $j - i$ possibilities).

By studying the combinatorial properties of optimum binary trees more carefully, we can refine the algorithm somewhat.

Lemma. *If $n > 0$ and $\alpha_n = \beta_n = 0$, an optimum binary tree can be obtained by replacing the rightmost terminal node*

$$\diagdown\!\square$$

of the optimum tree for $\alpha_0, \ldots, \alpha_{n-1}$ *and* $\beta_1, \ldots, \beta_{n-1}$ *by the subtree*

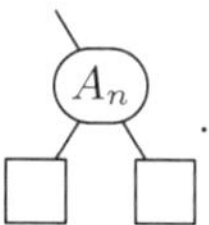

Proof. The lemma is obviously true when $n = 1$. Otherwise we may assume by induction that formulas (**) hold with $R_{k,n} = R_{k,n-1}$ for $0 \le k < n-1$ and $R_{n-1,n} = n$, except possibly when $i = 0$ and $j = n$.

In the latter case we must show that the minimum of the terms $P_{0,k-1} + P_{k,n}$ for $1 \le k \le n$ occurs when $k = R_{0,n-1}$. But we have $P_{k,n} = P_{k,n-1} + \alpha_{n-1}$ for $1 \le k < n$, by induction. Consequently

$$\begin{aligned}\min_{1\le k<n}(P_{0,k-1} + P_{k,n}) &= \min_{1\le k<n}(P_{0,k-1} + P_{k,n-1}) + \alpha_{n-1}\\ &= P_{0,n-1} - W_{0,n-1} + \alpha_{n-1} \le P_{0,n-1}.\end{aligned}$$

And $P_{0,n-1}$ is the term for $k = n$, because $P_{n,n} = 0$. □

Theorem. *Adding a new name to an optimum tree, which is greater than all other names, never forces the root of the optimum tree to move to the left. In other words, there is always a solution to the equations above such that*

$$R_{0,n} \ge R_{0,n-1},$$

when $n \ge 2$.

*Proof.** We use induction on n, the result being vacuous when $n = 1$. The weighted path length $P(\mathcal{T})$ of a binary search tree $\mathcal{T}$ is

$$l(\alpha_0)\alpha_0 + l(\beta_1)\beta_1 + l(\alpha_1)\alpha_1 + \cdots + l(\beta_n)\beta_n + l(\alpha_n)\alpha_n,$$

where $l(x)$ denotes the level associated with x. Thus if $\mathcal{T}'$ is another tree, with levels $l'(x)$, we have

$$\begin{aligned}P(\mathcal{T}) - P(\mathcal{T}') = (l(\alpha_0) - l'(\alpha_0))\alpha_0 + (l(\beta_1) - l'(\beta_1))\beta_1 + \cdots\\ + (l(\beta_n) - l'(\beta_n))\beta_n + (l(\alpha_n) - l'(\alpha_n))\alpha_n.\end{aligned}$$

This difference is a function of $\alpha_n + \beta_n$, because $l(\alpha_n) = l(\beta_n) + 1$. Consequently the optimum tree depends only on the sum $\alpha_n + \beta_n$, and we may assume that $\beta_n = 0$, $\alpha_n = \alpha$. In the special case $\alpha = 0$, the lemma assures us of a solution matrix $(R_{i,j})$ that satisfies $R_{0,n} = R_{0,n-1}$.

* *A complete proof of this theorem has been substituted here for the author's original proof, which contained a serious gap.*

We will show that, as α increases from 0 to ∞, there will be solution matrices such that $R_{0,n}$ increases from $R_{0,n-1}$ to n.

Consider all binary search trees with n names for which the external node associated with α_n appears on level l, and let $\mathcal{T}_l$ be the subset of those trees for which the weighted path length is as small as possible, given the weights $\alpha_0, \ldots, \alpha_{n-1}, \beta_1, \ldots, \beta_{n-1}$. Then every tree of $\mathcal{T}_l$ has weighted path length $C_l + l\alpha$, for some constant C_l. The weighted path length of an optimum tree for all $2n+1$ of the weights is therefore $\min_{1\le l-1\le n}(C_l + l\alpha)$, a piecewise linear (and concave) function of α.

Let $\mathcal{T}$ be a tree that is optimum when $\alpha = t$ but not when $\alpha > t$, and let $\mathcal{T}'$ be a tree that is optimum when $\alpha = t+\epsilon$ for all sufficiently small $\epsilon > 0$. Then $\mathcal{T}'$ is optimum also when $\alpha = t$. Thus we have $\mathcal{T} \in \mathcal{T}_l$ and $\mathcal{T}' \in \mathcal{T}_{l'}$ for some levels l and l', where $C_l + lt = C_{l'} + l't$. Consequently $P(\mathcal{T}) - P(\mathcal{T}') = (l-l')\epsilon$ when $\alpha = t+\epsilon$, and we must have $l > l'$.

If r is the root of $\mathcal{T}$, the theorem will therefore follow if we can prove that some tree in $\mathcal{T}_{l'}$ has a root $r' \ge r$. Consider the following diagrams:

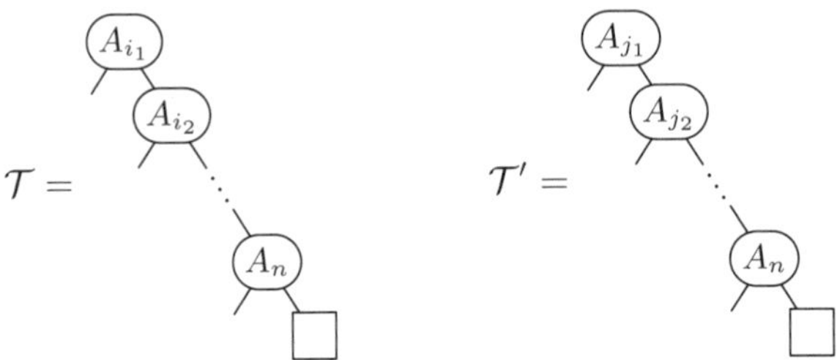

We have $r = i_1 < i_2 < \cdots < i_{l-1} = n$ and $r' = j_1 < j_2 < \cdots < j_{l'-1} = n$. Suppose r' is as large as possible, over all trees $\mathcal{T}' \in \mathcal{T}_{l'}$. If $r' \ge r$, we're done. Otherwise choose $\mathcal{T}'$ with root r' so that j_2 is as *small* as possible. Then, by induction and by the left-right symmetry of the theorem, we must have $j_2 \le i_2$. If $j_2 < i_2$, a similar argument shows that we can choose $\mathcal{T}'$ with $j_3 \le i_3$, etc. Since $n = j_{l'-1} > i_{l'-1}$, we must eventually find a $\mathcal{T}'$ with $j_k = i_k$ for some k. But then we can replace the right subtree of A_{i_k} in $\mathcal{T}$ by the corresponding subtree of $\mathcal{T}'$, obtaining a tree of $\mathcal{T}_{l'}$ whose root is r, contrary to our assumptions. □

Corollary. *Conditions (**) above can always be satisfied with*

$$R_{i,j-1} \le R_{i,j} \quad \text{and} \quad R_{i,j} \le R_{i+1,j}, \qquad \text{for } 0 \le i < j-1 < n.$$

Proof. This is simply the result of applying the theorem to all subtrees and using left-right symmetry. □

The corollary suggests an algorithm that is much faster than the previous one, since we usually will not have to search the entire range $i < r \leq j$ when determining $R_{i,j}$. In fact, only $R_{i+1,j} - R_{i,j-1} + 1$ cases need to be examined when $R_{i,j}$ is being calculated. Summing for fixed $j - i$ gives a telescoping series, which shows that the total amount of work is at worst proportional to n^2.

Summary, and Open Problems

The formulas above amount to a systematic method for finding optimum search trees, given the frequency of occurrence of each name in the tree as well as the frequencies of occurrence of names not in the tree. The number of steps is essentially proportional to the square of the number of names. An ALGOL program for the algorithm appears in the appendix, together with a detailed example from a compiler application.

Several open problems remain to be solved. Perhaps the most interesting is to obtain the best possible bound on the weighted path length in the optimum tree as a function of n, given arbitrary frequencies such that

$$\alpha_0 + \alpha_1 + \cdots + \alpha_n + \beta_1 + \cdots + \beta_n = 1.$$

For example, when $n = 2$ the weighted path length is ≤ 3, and the worst case occurs when $\alpha_1 = 1$, $\alpha_0 = \alpha_2 = \beta_1 = \beta_2 = 0$. The same bound applies when $n = 3$ since the tree

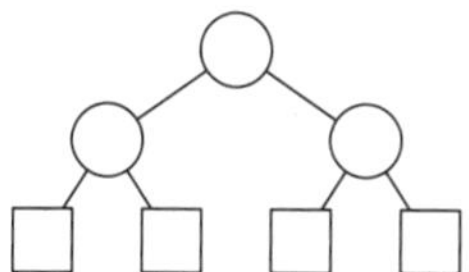

obviously has weighted path length ≤ 3. It is not obvious what the best possible bounds are when $n > 3$, although it is easy to see that the optimum weighted path length never exceeds $\lfloor \log_2(n+1) \rfloor + 1$.

Another problem concerns the efficiency of the algorithm. Our $O(n^2)$ algorithm essentially finds all of the optimum trees for $0 \leq i \leq j \leq n$. But if we discover by some means that $R_{0,n-1} \geq 5$, it is unnecessary to determine $R_{i,n}$, for $1 \leq i \leq 4$ when we compute $R_{0,n}$. There may be some way to arrange the calculation so that the running time is less than order n^2 on the average.*

* *Note Added in Proof.* T. C. Hu and A. C. Tucker have recently discovered a completely different way to find optimum binary search trees, in the special

A harder problem, but perhaps solvable, is to devise an algorithm that keeps its own frequency counts empirically, maintaining the tree in optimum form depending on the past history of the searches. Names that occur most frequently gradually move towards the root, etc. Perhaps some such updating method could be devised that would save more time than it consumes.

Another interesting problem is related to our first example. The optimum in the KWIC-index application discussed in the introduction case turned out to be obtainable by the following "top-down" rule: Place the most frequently occurring name at the root of the tree, then proceed similarly on the subtrees. Another plausible rule is to choose the root so as to equalize the total weight of the left and right subtrees as much as possible. Our discussion of the complete solution when $n = 3$ shows that neither of these rules will produce an optimum tree in all cases, but it might be possible to give some quantitative estimate of how far from the optimum these methods can be.

The solution to any of these problems should provide further insight into the nature of optimum search trees.

The research reported here was supported by IBM Corporation. I wish to thank Ronald L. Rivest for formulating a conjecture that led to the theorem in this paper, and John Bruno for correcting an error in my original proof of the lemma.

Appendix: A Detailed Implementation

The program below is written in ALGOL W, a refinement of ALGOL 60 due to Wirth and Hoare [11]. More than half of the code (the procedure *display*) is actually devoted to printing out the optimum tree in a reasonable pictorial fashion, once it has been found.

```
begin comment Finding an "optimum" binary search tree;
  string(10) array wd(1 :: 100); integer array a, b(0 :: 100);
  integer n; comment The number of names (at most 100);
  record node(string(10) info; integer col;
        reference(node) left, right);

  procedure display(integer value n; reference(node) value root);
    begin comment Draw a picture of binary tree referenced by 'root';
      reference(node) array active, waiting(1 :: n);
```

case that the β's are all zero. Their algorithm requires only $O(n)$ units of memory and $O(n \log n)$ units of time, when suitable data structures are employed. [See T. C. Hu and A. C. Tucker, "Optimal computer search trees and variable-length alphabetical codes," *SIAM Journal on Applied Mathematics* **21** (1971), 514–532.]

```
string(132) line;  comment One line to be printed;
integer k, newk;  comment The number of nodes waiting;
reference(node) p;  comment The current node of interest;
integer j; comment Counter used in the colno procedure;

procedure colno(reference(node) value r);
   begin comment Assign a column number to each
         node of the binary tree referenced by r;
      if r ≠ null then
         begin colno(left(r));
            col(r) := round(123 * j/(n − 1)) + 4;  j := j + 1;
            colno(right(r));
         end;
   end colno;

j := 0;  colno(root);
waiting(1) := root;  k := 1;
while k > 0 do
begin line := "";  comment Set line to 132 blanks;
   for j := 1 until k do
   begin comment Move waiting node to active area,
         and draw '|' lines down to it;
      active(j) := p := waiting(j);
      line(col(p) | 1) := "|";
   end;
   write(line, line);  comment Write two identical lines;
   newk := 0;
   for j := 1 until k do
   begin comment Put nodes descended from active nodes
         onto the waiting list, and prepare an appropriate
         line containing the info of active nodes;
      integer cl, cr;  comment Left and right extents;
      p := active(j);  cl := cr := col(p);
      if left(p) ≠ null then
         begin cl := col(left(p)); newk := newk + 1;
            waiting(newk) := left(p);
         end;
      if right(p) ≠ null then
         begin cr := col(right(p)); newk := newk + 1;
            waiting(newk) := right(p);
         end;
      for i := cl until cr do line(i | 1) := "-";
      begin comment Center info(p) on line, about col(p);
         integer s;  s := 0;
         while info(p)(s + 1 | 1) ≠ " " do s := s + 1;
         cl := col(p) − s div 2;
```

```
                for i := 0 until s do line(cl + i | 1) := info(p)(i | 1);
              end;
            end;
            write(line); comment Publish all the active nodes;
            k := newk;
          end;
        end display;

      n := 0;
      intfieldsize := 5; comment Set the output field size for integers;
      write("THE GIVEN FREQUENCIES ARE:");
    rloop: read(a(n), wd(n + 1), b(n + 1));
      write("                                   ", a(n));
      if wd(n + 1)(0 | 1) ≠ "." then
        begin n := n + 1;
          write("                                        ", wd(n), b(n));
          go to rloop;
        end;

      begin comment Find an n-node optimum tree, given relative
          frequency b(i) of encountering wd(i) and relative frequency a(i)
          of being between wd(i) and wd(i + 1);
        integer array p, w, r(0 :: n, 0 :: n);
        comment p(i, j), w(i, j), and r(i, j) denote respectively the weighted
          path length, the total weight, and the root of an optimum tree
          for the words lying between wd(i) and wd(j + 1), when i < j + 1.
          The average search length in this tree is p(i, j)/w(i, j);

        reference(node) procedure createtree(integer value i, j);
          if i = j then null
          else node(wd(r(i, j)), 0, createtree(i, r(i, j)−1), createtree(r(i, j), j));

        for i := 0 until n do p(i, i) := w(i, i) := a(i);
        for i := 0 until n do for j := i + 1 until n do
          w(i, j) := w(i, j − 1) + b(j) + a(j);
        for k := 1 until n do for i := 0 until n − k do
        begin integer ik, mn, mx; ik := i + k;
          mx := if k = 1 then ik else r(i, ik − 1);
          mn := p(i, mx − 1) + p(mx, ik);
          if k > 1 then for j := mx + 1 until r(i + 1, ik) do
            if p(i, j − 1) + p(j, ik) < mn then
              begin mn := p(i, j − 1) + p(j, ik); mx := j end;
          p(i, ik) := mn + w(i, ik); r(i, ik) := mx;
        end;
        write("AVERAGE PATH LENGTH IS", p(0, n)/w(0, n));
        iocontrol(3); comment Begin new page of output;
        display(n, createtree(0, n));
```

```
      iocontrol(3); intfieldsize := 2;
      for i := 0 until n do
      begin iocontrol(2); comment Begin new line of output;
         for j := 0 until n do writeon(if i < j then r(i, j) else 0);
      end;
   end;
end.
```

In order to try the algorithm on a nontrivial test case, a count was made of all identifiers in about 25 example ALGOL W programs prepared by the author for an introductory programming course. The frequency of occurrence of each reserved word was counted, as well as the frequency of occurrence of identifiers lying between adjacent reserved words. This led to the following data ($n = 36$):

33		
	`abs`	1
5		
	`and`	6
0		
	`array`	9
26		
	`begin`	77
37		
	`case`	5
12		
	`comment`	95
54		
	`div`	12
0		
	`do`	50
23		
	`else`	16
0		
	`end`	77
15		
	`false`	2
0		
	`for`	35
36		
	`go`	1
0		
	`goto`	1
57		
	`if`	34
7		
	`integer`	37
142		
	`logical`	2
0		
	`long`	5
113		

113		
	`null`	8
2		
	`of`	5
0		
	`or`	5
30		
	`procedure`	16
38		
	`real`	29
0		
	`record`	2
0		
	`reference`	13
0		
	`rem`	9
0		
	`result`	0
23		
	`short`	0
11		
	`step`	5
0		
	`string`	5
99		
	`then`	34
2		
	`to`	1
4		
	`true`	8
5		
	`until`	34
4		
	`value`	8
0		
	`while`	16
111		

For example, any identifier starting with the letter `J`, `K`, or `L` would be among the 142 identifiers that fell between `integer` and `logical` in the examples. The program inputs such data as a sequence of lines beginning with '`33 abs 1`', '`5 and 6`', etc., ending with '`111 . 0`'.

The R matrix computed by the program is shown on the next page, opposite the optimum tree as printed out by the program's *display* procedure. The average search length for this fairly large tree came to only $7329/1552 \approx 4.72$.

A quite different optimum tree is obtained when the frequencies $\alpha_0, \alpha_1, \ldots, \alpha_{36}$ are set to zero. Thus the "betweenness" frequencies can

0	1	1	3	4	4	4	6	6	6	6	6	6	6	6	6	10	10	10	10	10	10	10	10	10	10	10	10	10	10	10	16	16	16	16	16	16
0	0	2	3	4	4	6	6	6	6	6	6	6	6	6	6	10	10	10	10	10	10	10	10	10	10	10	10	10	10	10	16	16	16	16	16	16
0	0	0	3	4	4	6	6	6	6	6	6	6	6	6	10	10	10	10	10	10	10	10	10	10	10	10	10	10	10	10	16	16	16	16	16	16
0	0	0	0	4	4	6	6	6	6	6	6	6	6	6	10	10	10	10	10	10	10	10	10	10	10	10	10	10	10	16	16	16	16	16	16	16
0	0	0	0	0	5	6	6	6	6	8	8	8	8	10	10	10	10	10	10	10	10	16	16	16	16	16	16	16	16	16	16	16	16	16	16	16
0	0	0	0	0	0	6	6	6	6	8	8	8	8	10	10	10	10	10	10	10	16	16	16	16	16	16	16	16	16	16	16	16	16	16	16	16
0	0	0	0	0	0	0	7	8	8	8	8	10	10	10	10	10	10	16	16	16	16	16	16	16	16	16	16	16	16	16	16	16	16	16	16	16
0	0	0	0	0	0	0	0	8	8	10	10	10	10	10	10	14	14	16	16	16	16	16	16	16	16	16	16	16	16	16	16	16	16	16	16	19
0	0	0	0	0	0	0	0	0	9	10	10	10	10	12	12	16	16	16	16	16	16	16	16	16	16	16	16	16	16	16	17	17	17	19	19	19
0	0	0	0	0	0	0	0	0	0	10	10	10	10	12	12	16	16	16	16	16	16	16	16	16	16	16	16	16	16	17	17	17	17	19	19	19
0	0	0	0	0	0	0	0	0	0	0	11	12	12	14	14	16	16	16	16	16	16	16	17	17	17	17	17	17	17	17	17	17	17	19	19	19
0	0	0	0	0	0	0	0	0	0	0	0	12	12	14	14	16	16	16	16	16	16	17	17	17	17	17	17	17	17	17	17	17	17	19	19	23
0	0	0	0	0	0	0	0	0	0	0	0	0	13	14	15	16	16	16	16	16	16	17	17	17	17	17	17	17	17	17	17	17	17	19	19	23
0	0	0	0	0	0	0	0	0	0	0	0	0	0	14	15	16	16	16	16	16	17	17	17	17	17	17	17	17	17	17	17	17	17	19	23	23
0	0	0	0	0	0	0	0	0	0	0	0	0	0	0	15	16	16	16	16	16	17	17	17	17	17	17	17	17	17	17	17	17	17	19	23	23
0	0	0	0	0	0	0	0	0	0	0	0	0	0	0	0	16	16	17	17	17	17	17	17	17	17	17	17	19	19	19	19	19	19	23	23	23
0	0	0	0	0	0	0	0	0	0	0	0	0	0	0	0	0	17	17	17	17	17	17	17	17	19	19	19	19	19	19	19	19	19	23	23	23
0	0	0	0	0	0	0	0	0	0	0	0	0	0	0	0	0	0	18	19	19	19	19	19	19	19	19	19	19	19	23	23	23	23	23	23	23
0	0	0	0	0	0	0	0	0	0	0	0	0	0	0	0	0	0	0	19	19	19	19	19	19	19	19	19	19	19	23	23	23	23	23	23	23
0	0	0	0	0	0	0	0	0	0	0	0	0	0	0	0	0	0	0	0	20	21	22	22	22	22	22	23	23	23	23	30	30	30	30	30	31
0	0	0	0	0	0	0	0	0	0	0	0	0	0	0	0	0	0	0	0	0	21	22	22	22	22	23	23	23	23	23	30	30	30	30	30	31
0	0	0	0	0	0	0	0	0	0	0	0	0	0	0	0	0	0	0	0	0	0	22	22	22	22	23	23	23	23	23	30	30	30	30	30	31
0	0	0	0	0	0	0	0	0	0	0	0	0	0	0	0	0	0	0	0	0	0	0	23	23	23	23	23	23	23	30	30	30	30	30	30	31
0	0	0	0	0	0	0	0	0	0	0	0	0	0	0	0	0	0	0	0	0	0	0	0	24	25	25	27	27	27	30	30	30	30	31	31	31
0	0	0	0	0	0	0	0	0	0	0	0	0	0	0	0	0	0	0	0	0	0	0	0	0	25	25	27	27	27	30	30	30	30	31	31	31
0	0	0	0	0	0	0	0	0	0	0	0	0	0	0	0	0	0	0	0	0	0	0	0	0	0	26	27	28	28	30	30	30	31	31	31	31
0	0	0	0	0	0	0	0	0	0	0	0	0	0	0	0	0	0	0	0	0	0	0	0	0	0	0	27	28	28	30	30	30	31	31	31	31
0	0	0	0	0	0	0	0	0	0	0	0	0	0	0	0	0	0	0	0	0	0	0	0	0	0	0	0	28	28	30	30	30	31	31	31	31
0	0	0	0	0	0	0	0	0	0	0	0	0	0	0	0	0	0	0	0	0	0	0	0	0	0	0	0	0	29	30	31	31	31	31	31	31
0	0	0	0	0	0	0	0	0	0	0	0	0	0	0	0	0	0	0	0	0	0	0	0	0	0	0	0	0	0	30	31	31	31	31	31	31
0	0	0	0	0	0	0	0	0	0	0	0	0	0	0	0	0	0	0	0	0	0	0	0	0	0	0	0	0	0	0	31	31	31	31	31	31
0	0	0	0	0	0	0	0	0	0	0	0	0	0	0	0	0	0	0	0	0	0	0	0	0	0	0	0	0	0	0	0	32	33	34	34	36
0	0	0	0	0	0	0	0	0	0	0	0	0	0	0	0	0	0	0	0	0	0	0	0	0	0	0	0	0	0	0	0	0	33	34	34	36
0	0	0	0	0	0	0	0	0	0	0	0	0	0	0	0	0	0	0	0	0	0	0	0	0	0	0	0	0	0	0	0	0	0	34	34	36
0	0	0	0	0	0	0	0	0	0	0	0	0	0	0	0	0	0	0	0	0	0	0	0	0	0	0	0	0	0	0	0	0	0	0	35	36
0	0	0	0	0	0	0	0	0	0	0	0	0	0	0	0	0	0	0	0	0	0	0	0	0	0	0	0	0	0	0	0	0	0	0	0	36
0	0	0	0	0	0	0	0	0	0	0	0	0	0	0	0	0	0	0	0	0	0	0	0	0	0	0	0	0	0	0	0	0	0	0	0	0

FIGURE 1. R matrix for the ALGOL-reserved-word application.

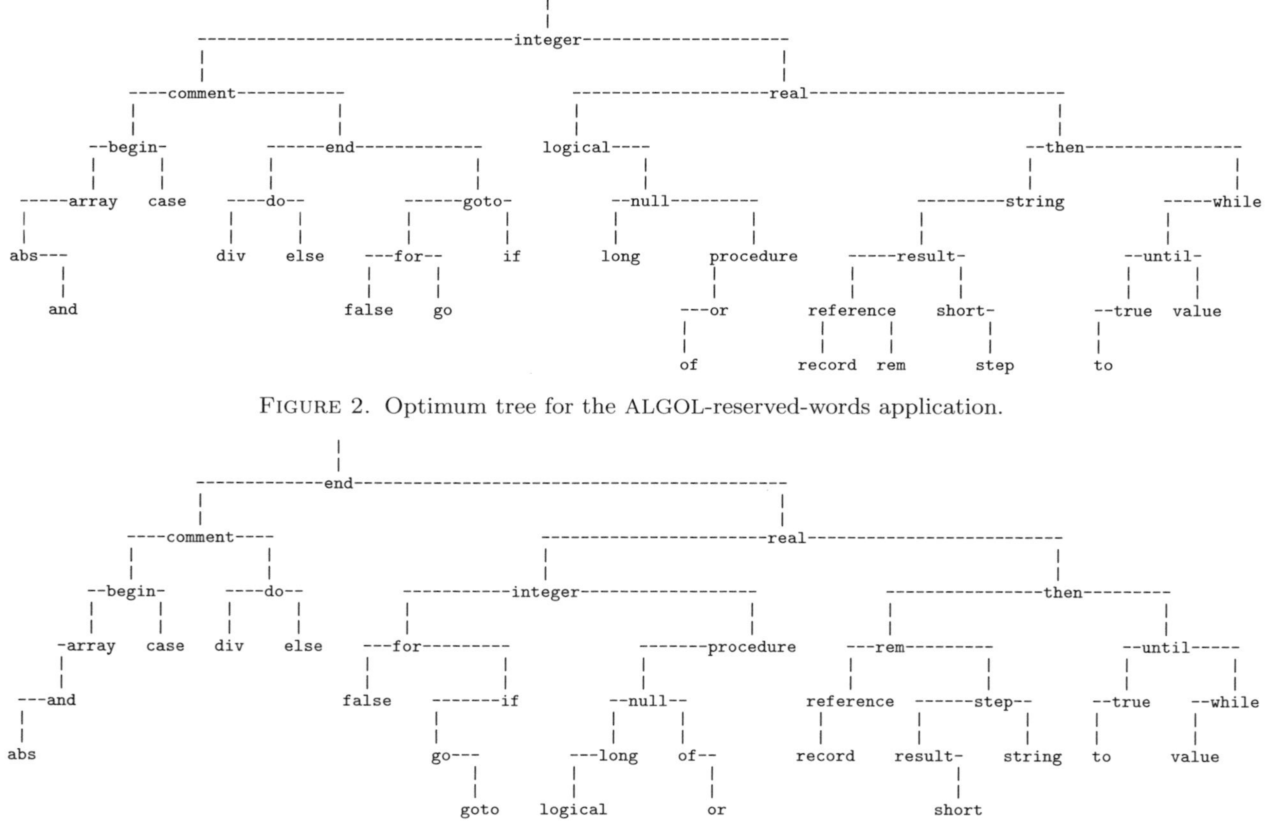

FIGURE 2. Optimum tree for the ALGOL-reserved-words application.

FIGURE 3. Optimum tree when the α frequencies are ignored.

profoundly influence the nature of the optimum tree, and it is important to consider them.

References

[1] A. D. Booth and A. J. T. Colin, "On the efficiency of a new method of dictionary construction," *Information and Control* **3** (1960), 327–334.

[2] Frederick P. Brooks, Jr., and Kenneth E. Iverson, *Automatic Data Processing*, System/360 edition (New York: Wiley, 1969).

[3] A. S. Douglas, "Techniques for the recording of, and reference to data in a computer," *The Computer Journal* **2** (1959), 1–9.

[4] Jay Earley, "An efficient context-free parsing algorithm," *Communications of the ACM* **13** (1970), 94–102.

[5] E. N. Gilbert and E. F. Moore, "Variable-length binary encodings," *The Bell System Technical Journal* **38** (1959), 933–968.

[6] Jan Helbich, "Direct selection of keywords for the KWIC index," *Information Storage and Retrieval* **5** (1969), 123–128.

[7] Thomas N. Hibbard, "Some combinatorial properties of certain trees with applications to searching and sorting," *Journal of the Association for Computing Machinery* **9** (1962), 13–28.

[8] Kenneth E. Iverson, *A Programming Language* (New York: Wiley, 1962).

[9] Donald E. Knuth, *Fundamental Algorithms*, Volume 1 of *The Art of Computer Programming* (Reading, Massachusetts: Addison–Wesley, 1968).

[10] P. F. Windley, "Trees, forests and rearranging," *The Computer Journal* **3** (1960), 84–88, 174, 184.

[11] Niklaus Wirth and C. A. R. Hoare, "A contribution to the development of ALGOL," *Communications of the ACM* **9** (1966), 413–432.

[12] Eugene Wong, "A linear search problem," *SIAM Review* **6** (1964), 168–174.

Addendum

When the $2n+1$ frequencies α_j and β_j sum to 1, the largest possible optimum weighted path length occurs when $\alpha_j = [j \text{ is odd}]/\lceil n/2 \rceil$ for $0 \le j \le n$ and $\beta_j = 0$ for $1 \le j \le n$. See T. C. Hu and K. C. Tan, "Least upper bound on the cost of optimum binary search trees," *Acta Informatica* **1** (1972), 307–310.

No simple solution has apparently been found to any of the other open problems discussed above, but many other facts about optimum alphabetic trees have been discovered. See, for example, Russell L. Wessner, "Optimal alphabetic search trees with restricted maximal height," *Information Processing Letters* **4** (1976), 90–94; F. Frances Yao, "Efficient dynamic programming using quadrangle inequalities," *ACM Symposium on Theory of Computing* **12** (1980), 429–435. Later developments are summarized in D. E. Knuth, *Sorting and Searching*, Volume 3 of *The Art of Computer Programming*, second edition (Reading, Massachusetts: Addison–Wesley, 1998), §6.2.2.

Chapter 5

Dynamic Huffman Coding

[Originally published in Journal of Algorithms **6** *(1985), 163–180.]*

This note shows how to maintain a prefix code that remains optimum as the weights change. A Huffman tree with nonnegative integer weights can be represented in such a way that any weight w at level l can be increased or decreased by unity in $O(l)$ steps, preserving minimality of the weighted path length. One-pass algorithms for file compression can be based on such a representation.

1. Introduction

Given nonnegative weights $(w_1, \ldots, w_n)$, the well-known algorithm of David Huffman [2] can be used to construct a binary tree with n external nodes and $n-1$ internal nodes, where the external nodes are labeled with the weights $(w_1, \ldots, w_n)$ in some order. Huffman's tree has the minimum value of $w_1 l_1 + \cdots + w_n l_n$ over all such binary trees, where l_j is the level at which w_j occurs in the tree.

Binary trees with n external nodes are in one-to-one correspondence with sets of n strings on $\{0, 1\}$ that form a "minimal prefix code." A prefix code is a set of strings in which no string is a proper prefix of another; and a minimal prefix code is a prefix code such that, if α is a proper prefix of some string in the set, then $\alpha 0$ is either in the set or a proper prefix of some string in the set, and so is $\alpha 1$. The correspondence between trees and codes is simply to represent the path from the root to each external node as a string of 0s and 1s, where 0 corresponds to a left branch and 1 to a right branch. An external node at level l corresponds in this way to a string of length l. For example, the binary tree

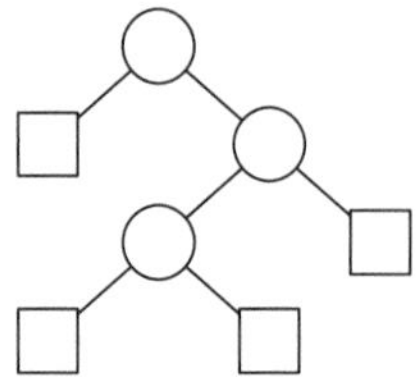

corresponds to the minimal prefix code $\{0, 100, 101, 11\}$.

If the strings $\{\alpha_1, \ldots, \alpha_n\}$ form a prefix code, it is easy to factor any string $\alpha_{i_1}\alpha_{i_2}\ldots\alpha_{i_m}$ into its component αs, by reading the 0s and 1s from left to right, traversing the binary tree until an external node is reached. Therefore if we have a string $a_{i_1}a_{i_2}\ldots a_{i_m}$ in some n-letter alphabet $\{a_1, \ldots, a_n\}$, we can encode that string as $\alpha_{i_1}\alpha_{i_2}\ldots\alpha_{i_m}$, a sequence of 0s and 1s.

Suppose the letter a_j occurs exactly w_j times in the original string $a_{i_1}a_{i_2}\ldots a_{i_m}$; then the total length of $\alpha_{i_1}\alpha_{i_2}\ldots\alpha_{i_m}$ after encoding with a prefix code is $w_1l_1+\cdots+w_nl_n$, where l_j is the length of α_j. Hence Huffman's algorithm produces a prefix code for which the resulting sequence of 0s and 1s has minimum length.

All of this is, of course, well known. But in practice there are times when the weights w_j are not given, since the string $a_{i_1}a_{i_2}\ldots a_{i_m}$ is to be encoded "on line." In this case we can imagine a dynamically changing prefix code in which the encoding of a_{i_k} is based on a Huffman tree for the frequencies of occurrence in the previously encoded portion $a_{i_1}\ldots a_{i_{k-1}}$; the encoding process thereby "learns" the frequencies of the encoded message as it proceeds.

When the code is changing, we might naturally wonder how the output could be unraveled later. But the decoding process can also be learning how the code is evolving, in exactly the same way as the encoding process does, since the decoder also knows the frequencies of occurrence in $a_{i_1}\ldots a_{i_{k-1}}$ when it is decoding a_{i_k}. Thus, by continually updating a Huffman code, both sender and receiver can keep synchronized with each other. A nearly optimum encoding will be obtained, assuming that the transmission is error-free.

In particular, this scheme avoids the need to transmit any information about the weights w_j. Such additional transmission of weights would be necessary if a two-pass strategy were adopted under which the sender first counted the frequencies of each letter and then used a globally optimum Huffman code.

Section 2 of this paper develops the basic ideas by which continuously varying Huffman trees can be maintained, where the updating is done in real time (that is, in time proportional to the length of transmission). Section 3 presents an example, and detailed algorithms for the encoding and decoding appear in Sections 4, 5, and 6. The concluding section presents empirical results and discusses various ways to modify the procedure.

The basic principles of adaptive Huffman coding were discovered by Robert G. Gallager [1]. This paper extends Gallager's work in three ways: (1) Pointer manipulations necessary for real-time operation are described explicitly. (2) Zero-weight letters are taken into account. (3) Positive weights may be decreased as well as increased.

2. A Strategy

Recall that Huffman's algorithm combines the two smallest weights w_i and w_j, replacing them by their sum $w_i + w_j$, and repeats this process until only one weight is left. For example, given the weights $(2, 3, 4, 5)$, the first step combines $2 + 3 = 5$; the remaining weights are $(4, 5, 5)$. The next step combines $4 + 5 = 9$, and then comes $5 + 9 = 14$. There is some ambiguity about which '5' is which, so this procedure might form two different trees

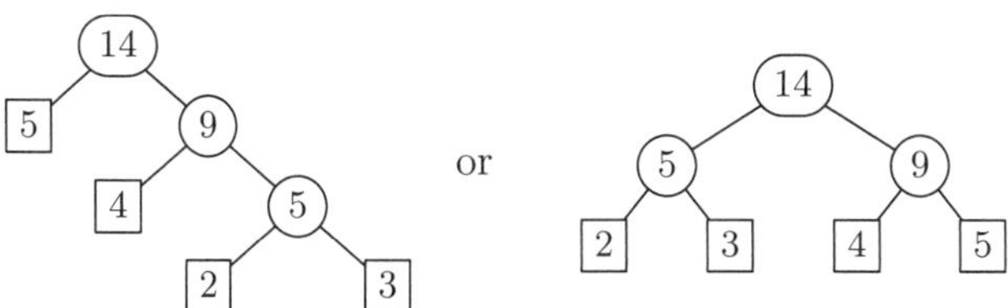

depending on where the 5 from '$2+3=5$' is placed. Both of these trees are optimum for the given weights, since $5 \cdot 1 + 4 \cdot 2 + 2 \cdot 3 + 3 \cdot 3 = 2 \cdot 2 + 3 \cdot 2 + 4 \cdot 2 + 5 \cdot 2$. However, if we now increase the given weight 5 to 6, the left-hand tree is better, while if we increase the given weight 2 to 3 the right-hand tree is better. A procedure for updating Huffman trees dynamically must therefore be able to convert from each of these possibilities to the other.

The weight combination process of Huffman's algorithm leads to a nondecreasing sequence

$$(x_1, x_2, \ldots, x_{2n-1})$$

of node weights for the internal and external nodes, and this sequence is the same for all Huffman trees on the given weights $(w_1, \ldots, w_n)$.

For example, both of the trees above correspond to the x-sequence $(2, 3, 4, 5, 5, 9, 14)$. The weighted path length $w_1 l_1 + \cdots + w_n l_n$ is easily seen to be equal to the sum $x_1 + \cdots + x_{2n-2}$.

The $n - 1$ internal nodes of each Huffman tree that correspond to a particular sequence $(x_1, x_2, \ldots, x_{2n-1})$ have the weights $(x_1 + x_2, x_3 + x_4, \ldots, x_{2n-3} + x_{2n-2})$. Conversely, a binary tree is a Huffman tree on the weights $(w_1, \ldots, w_n)$ if it satisfies the following two conditions, which Gallager [1] called the *sibling property*:

i) The n external nodes have been assigned the weights $(w_1, \ldots, w_n)$ in some order, and each internal node has been assigned a weight equal to the sum of the weights of its two children;
ii) The $2n - 1$ nodes (external and internal) can be arranged in a sequence $y_1, \ldots, y_{2n-1}$ such that if x_j is the weight of node y_j we have $x_1 \le \cdots \le x_{2n-1}$, and such that nodes y_{2j-1} and y_{2j} are siblings (children of the same parent), for $1 \le j < n$, where this common parent node does not precede y_{2j-1} and y_{2j} in the sequence.

For we can show inductively that such a tree is one that Huffman's algorithm might construct: The values of x_1 and x_2 must equal the values of the smallest two weights of $(w_1, \ldots, w_n)$; and the remainder of the tree satisfies the same conditions on $n - 1$ weights, if the weights x_1 and x_2 are replaced by $x_1 + x_2$ and if the parent of y_1 and y_2 is regarded as an external node.

Given a tree satisfying (i) and (ii), let $y_{i_0}, y_{i_1}, \ldots, y_{i_l}$ be a sequence of nodes leading from some external node of weight w to the root. If w is replaced by $w + 1$, we must increase each of $x_{i_0}, x_{i_1}, \ldots, x_{i_l}$ by unity; the resulting tree will still satisfy (i) and (ii), provided that we had

$$x_{i_j} < x_{i_j+1} \qquad \text{for } 0 \le j < l \qquad (*)$$

in the original tree. Thus, the same tree will be optimum both for w and $w + 1$, whenever condition $(*)$ holds.

Suppose that all the weights w_i are positive. Then we can always transform a given Huffman tree into another one that satisfies condition $(*)$, by interchanging subtrees of equal weight: First let i'_0 be maximum such that $x_{i'_0} = x_{i_0}$, and when i'_k has been defined let i'_{k+1} be maximum such that $x_{i'_{k+1}}$ has the weight of the parent of $y_{i'_k}$. (In the special case that $i'_k = 2n - 1$, however, we let $l' = k$ and terminate the construction.) Now the tree can be permuted by interchanging the subtree rooted at y_{i_0} with the subtree rooted at $y_{i'_0}$, then interchanging the subtree rooted at the parent of $y_{i'_0}$ with the subtree rooted at $y_{i'_1}$, and so on. We obtain a tree satisfying (i) and (ii), where the path from w to the root is $y_{i'_0}, y_{i'_1}, \ldots, y_{i'_{l'}}$, and where $x_{i'_j} < x_{i'_j+1}$ for $0 \le j < l'$.

It is not difficult to verify by induction that $i_j \le i'_j$ for $0 \le j \le l'$ in this construction, hence $l' \le l$; in other words, at most l interchanges are necessary to obtain a Huffman tree satisfying (∗).

The construction just described is the key to an efficient algorithm for maintaining optimal Huffman trees, so it will be helpful to illustrate it with an example. Let $(w_1, \ldots, w_6) = (2, 3, 5, 5, 6, 11)$, and consider the Huffman tree

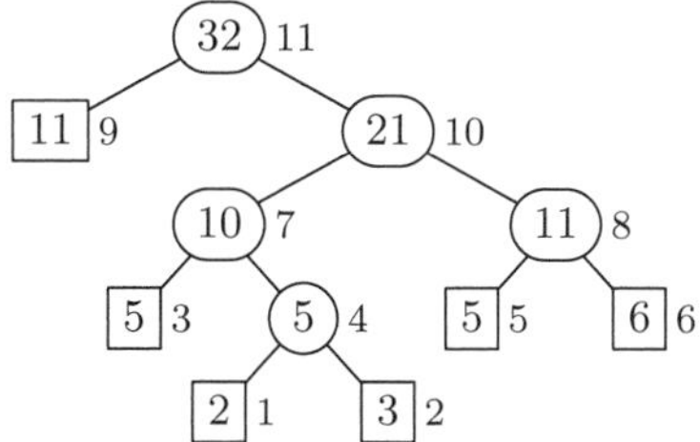

in which the nodes have been labeled with their weights and with indices from 1 to 11. The path from the weight-3 node to the root is defined by $(i_0, i_1, i_2, i_3, i_4) = (2, 4, 7, 10, 11)$; the construction above yields $(i'_0, i'_1, i'_2, i'_3) = (2, 5, 9, 11)$ and $l' = 3$. After carrying out appropriate interchanges of subtrees, we obtain two Huffman trees

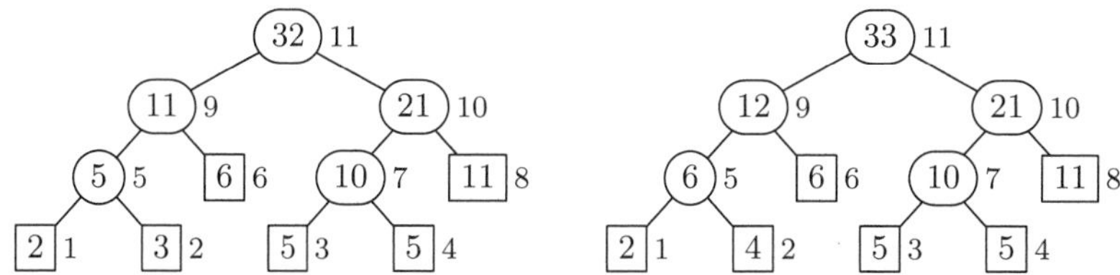

that are identical in structure, but the weight of 3 has been increased to 4 in the right-hand tree.

A similar procedure can be used when weights of 0 are present, but the maneuvering is more delicate because of the chance that a node and its parent both have the same weight. We can assume that there is at most one weight equal to 0, because Huffman's algorithm will begin by putting all of the 0-weight nodes into a subtree of total weight 0.

When the weights are all 0, every binary tree is "optimum"; but in our application it is more appropriate to make the external node levels be as equal as possible so that the encoded first appearance of each letter will not be too long. Therefore we shall use the following minimal prefix code, when there are m letters $(a_1, \ldots, a_m)$ of weight zero, assuming that $m = 2^e + r$ and that $0 \le r < 2^e$: Letter a_k is encoded as the $(e+1)$-bit binary representation of $k-1$, if $1 \le k \le 2r$, otherwise as the e-bit binary representation of $k - r - 1$. For example, let $m = 5$; then $e = 2$, $r = 1$, and the encoding is

$$a_1 \mapsto 000\,, \quad a_2 \mapsto 001\,, \quad a_3 \mapsto 01\,, \quad a_4 \mapsto 10\,, \quad a_5 \mapsto 11\,.$$

This encoding is optimum when all letters have the same weight ϵ, for any $\epsilon > 0$.

3. An Example

The methods sketched above lead to a real-time algorithm for maintaining Huffman trees as the weights change, as we shall see in Sections 4 and 5. But first it will be useful to study a worked example of dynamic Huffman coding, so that the detailed constructions are more readily understood.

Before transmission begins, both sender and receiver know only the size n of the alphabet being encoded. Let us assume for now that there are just 27 characters, namely ‘a’, ‘b’, …, ‘z’, and ‘!’; the last symbol will be used only as the final character of the message. Since this adaptive encoding scheme seems to have a somewhat magic flavor, we shall attempt to encode the message ‘abracadabra!’.

Initially all weights are zero, so the first letter is encoded by the 0-weight scheme described at the end of Section 2, where $m = 27 = 2^4 + 11$:

$$\texttt{a} \mapsto 00000\,.$$

Now ‘a’ has weight 1 and the other letters $(\texttt{!}, \texttt{b}, \ldots, \texttt{z})$ have weight 0. In the list of remaining letters, ‘!’ has swapped places with ‘a’, so that minimal changes need to be made to the data structure. The coding scheme at this point is represented by the tree

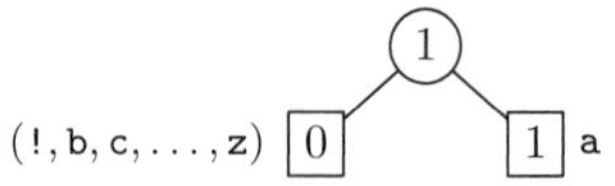

(!, b, c, …, z) [0] —(1)— [1] a ,

hence the second letter of our message would be encoded simply as ‘1’ if it were another ‘a’. However, it is a ‘b’, which comes out as an encoded 0-weight letter, prefixed by ‘0’:

$$\texttt{b} \mapsto 000001\,.$$

At this point the Huffman tree is

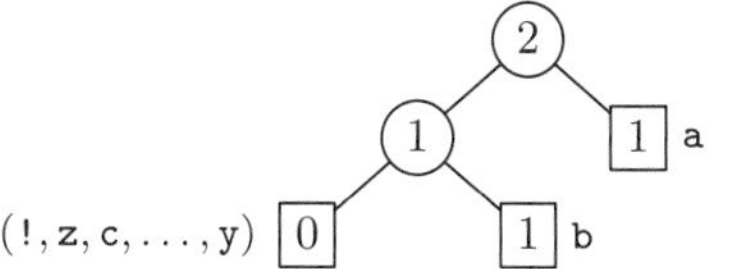

The third letter is therefore encoded as

$$\texttt{r} \mapsto 0010001\,,$$

after which the Huffman tree has changed to

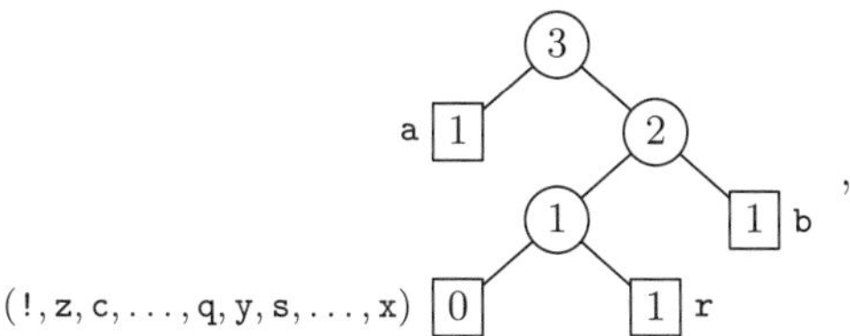

assuming that the algorithms of Section 5 have been used. The encodings continue as follows:

$$\begin{aligned}
\texttt{a} &\mapsto 0\\
\texttt{c} &\mapsto 10000010\\
\texttt{a} &\mapsto 0\\
\texttt{d} &\mapsto 110000011\\
\texttt{a} &\mapsto 0\\
\texttt{b} &\mapsto 110\\
\texttt{r} &\mapsto 110\\
\texttt{a} &\mapsto 0
\end{aligned}$$

and by this time the tree has grown to

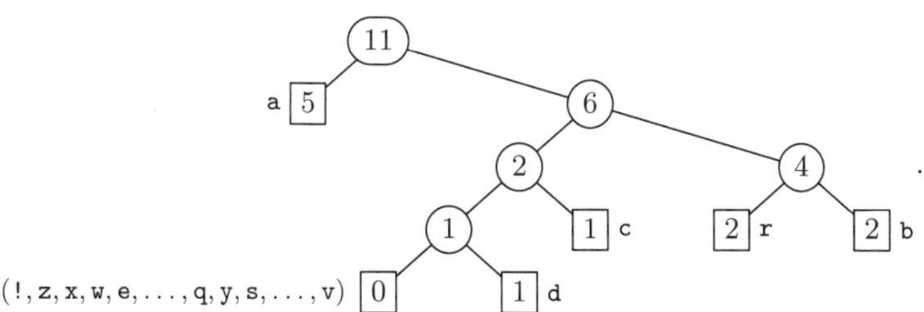

The final "shriek" is now transmitted:

$$! \mapsto 100000000\,.$$

4. Data Structures

In order to update the Huffman tree and to encode and decode messages, we need to do the following things:

a) Represent a binary tree with weights in each node;
b) Maintain a linear list of nodes, in nondecreasing order by weight;
c) Given a node in this linear list, belonging to a block of nodes that have the same weight, find the last node in its block;
d) Interchange two subtrees of the same weight;
e) Increase the weight of the last node in some block by unity;
f) Represent the correspondence between letters and external nodes.

The zero-weight letters must also be dealt with. Each of the necessary operations can be done in real time with the following combination of data structures:

boolean array $S[1:n-1]$; a stack that holds bits before they are transmitted.

integer M, E, R; zero-weight counters. The number of zero-weight elements is $M = 2^E + R$ where $0 \le R < 2^E$, except that $M = 0$ implies $R = -1$.

integer array $P[1:n]$; pointers to the parents of nodes. The parent of nodes $2j-1$ and $2j$ is node $P[j]$, except that $P[n] = 0$.

integer array $C[1:2n-1]$; pointers to the children of nodes. If node k is internal, and if its children are nodes $2j-1$ and $2j$, then $C[k] = j$, while if node k is external it represents letter $a_{C[k]}$.

integer array $A[0:n]$; representation of the alphabet. If letter a_k is currently the jth letter of weight zero, then $A[k] = j \le M$; otherwise $A[k]$ is the number of the external node corresponding to a_k, except that $A[0]$ has a special meaning.

integer array $B[1:2n-1]$; pointers to the blocks of nodes. All nodes j of a given weight have the same value of $B[j]$.

integer array $W[1:2n-1]$; the weights. Block k has weight $W[k]$.

integer array $L[1:2n-1]$; left pointer for blocks. The nearest block whose weight is less than that of block k is block $L[k]$, unless block k has the smallest weight, in which case $L[k]$ is the block of largest weight.

integer array $G[1:2n-1]$; right pointer for blocks. The nearest block whose weight is greater than that of block k is block $G[k]$, unless block k has the largest weight, in which case $G[k]$ is the block of smallest weight.

integer H; pointer to the block of smallest weight.

integer array $D[1:2n-1]$; pointer to the largest node number in a given block.

integer V; pointer to the head of the list of array positions not currently used as block numbers. The available positions are V, $G[V]$, $G[G[V]]$, etc.

integer Z; constant equal to $2n-1$. This value also points to the root of the tree.

These data structures combine a variety of standard ideas. For example, the L, W, and G arrays define a doubly linked circular list ordered on the W field; the list head is H, and V is the available-space pointer. The C and P arrays define the tree structure, taking advantage of the sequential nature of the nodes. The B and D arrays connect the tree to the list of weights.

If there are $M > 1$ letters of weight zero, the nodes of the trees will be in positions $2M-1$ through $Z = 2n-1$, inclusive. The node in position $k \geq 2M-1$ is external if and only if $A[C[k]] = k$; we have $A[0] = 2M-1$ and $C[2M-1] = 0$, to represent the zero-weight nodes. Nodes are in order by weight (that is, $W[B[2M-1]] \leq W[B[2M]] \leq \cdots \leq W[B[Z]]$), and the root of the tree is node Z. Blocks are also in order by weight; that is,

$$W[H] < W[G[H]] < W[G[G[H]]] < \cdots < W[B[Z]].$$

If letter a_k is currently the jth letter of weight zero, we have $A[k] = j$ and $C[j] = k$. When there is at most one letter of weight zero, the nodes are in positions 1 through $2n-1$ and essentially the same conventions hold; but $A[0]$ will be 1 and $C[1]$ will not be zero. There is a convenient ambiguity when $M = 1$, which allows a simple transition between $M = 0$ and $M = 1$.

Notice that most of the integer values in the data structures are nonnegative and less than $2n$, so we need just $\lceil \lg n \rceil + 1$ bits to represent them. The only exceptions are the integer weights in the W table, which may be as large as the length of the message being transmitted. Thus, the total memory requirements are about $12n \lg n + 2n \lg w$ bits, where w is a bound on the total weight of all n letters.

If a node and its parent are in the same block, the parent node should immediately follow its children, thereby comprising the first three nodes of the node sequence.

5. Algorithms

Given the data structures specified in Section 4, it is not difficult to design algorithms that maintain the necessary invariants. The algorithms will be presented here in an ALGOL-like language that uses '**fi**' and '**od**' to close '**if**' and '**do**', so that other delimiters like '**begin**' and '**end**' are not needed very often.

At the beginning we can get all the data structures off to a good start as follows:

```
procedure initialize;
begin integer i;
M ← 0; E ← 0; R ← −1; Z ← 2 × n − 1;
for i ← 1 to n do
   M ← M + 1; R ← R + 1;
   if 2 × R = M then E ← E + 1; R ← 0 fi;
   A[i] ← C[i] ← i od;
H ← 1; L[H] ← G[H] ← H; W[H] ← 0; D[H] ← A[0] ← Z;
V ← 2; for i ← V to Z − 1 do G[i] ← i + 1 od; G[Z] ← 0;
P[n] ← 0; C[Z] ← 0; B[Z] ← H;
end;
```

Let us consider next a procedure that encodes the letter a_k and sends out the corresponding bits, based on the present state of the tree.

```
procedure encode(integer k);
begin integer i, j, q, t;
i ← 0; q ← A[k];
if q ≤ M then comment encode zero weight;
   q ← q − 1;
   if q < 2 × R then t ← E + 1 else q ← q − R; t ← E fi;
   for j ← 1 to t do i ← i + 1; S[i] ← q mod 2; q ← q div 2 od;
   q ← A[0] fi;
while q < Z do
   i ← i + 1; S[i] ← (q + 1) mod 2;
   q ← P[(q + 1) div 2] od;
while i > 0 do transmit(S[i]); i ← i − 1 od;
end;
```

Here '*transmit*' is a system output procedure that sends one bit.

The decoding process does essentially the same thing as the encoder, but backwards. It uses a system procedure '*receive*' that reads one bit of input and returns that value:

```
integer procedure decode;
begin integer j, q;
q ← Z;
while A[C[q]] ≠ q do q ← 2 × C[q] − 1 + receive od;
if C[q] = 0 then comment decode zero weight;
   q ← 0;
   for j ← 1 to E do q ← 2 × q + receive od;
   if q < R then q ← 2 × q + receive else q ← q + R fi;
   q ← q + 1 fi;
return C[q];
end;
```

The following subroutine interchanges two subtrees of the same weight, assuming that neither is a child of the other:

```
procedure exchange(integer q, t);
begin integer ct, cq, acq;
ct ← C[t];  cq ← C[q];  acq ← A[cq];
if A[ct] ≠ t then P[ct] ← q else A[ct] ← q fi;
if acq ≠ q then P[cq] ← t else A[cq] ← t fi;
C[t] ← cq;  C[q] ← ct;
end;
```

The heart of dynamic Huffman tree processing is the '*update*' procedure, which is used by both sender and receiver after a_k has been encoded or decoded. It has the following overall form:

```
procedure update(integer k);
begin integer q;
⟨Set q to the external node whose weight should increase⟩;
while q > 0 do
   ⟨Move q to the right of its block⟩;
   ⟨Transfer q to the next block, with weight one higher⟩;
   q ← P[(q + 1) div 2] od;
end;
```

Thus, '*update*' performs three main functions. The first of these is quite simple unless the letter a_k is appearing for the first time. In the latter case a new external node of zero weight is appended to the tree, together with a new internal node:

```
⟨Set q to the external node whose weight should increase⟩ =
   begin q ← A[k];
   if q ≤ M then comment a zero weight will become positive;
```

```
    A[C[M]] ← q;  C[q] ← C[M];
    if R = 0 then R ← M div 2;  if R > 0 then E ← E − 1 fi fi;
    M ← M − 1;  R ← R − 1;
    if M > 0 then
      q ← A[0] − 1;  A[0] ← q − 1;  A[k] ← q;  C[q] ← k;
      if M > 1 then C[q − 1] ← 0 fi;
      P[M] ← q + 1;  C[q + 1] ← M;
      B[q] ← B[q − 1] ← H fi;
    fi;
  end
```

The next portion of '*update*' moves node q to the right of its block, unless both q and its parent are already at the right of the block.

```
  ⟨Move q to the right of its block⟩ =
    if q < D[B[q]] and D[B[P[(q + 1) div 2]]] > q + 1 then
      exchange(q, D[B[q]]);  q ← D[B[q]] fi
```

Finally, the remaining job is to augment the weight of q (and to augment the weight of q's parent as well, if they currently have the same weight):

```
  ⟨Transfer q to the next block, with weight one higher⟩ =
    begin integer u, gu, lu, x, t, qq;
    u ← B[q];  gu ← G[u];  lu ← L[u];  x ← W[u];  qq ← D[u];
    if W[gu] = x + 1 then
      B[q] ← B[qq] ← gu;
      if D[lu] = q − 1 or (u = H and q = A[0]) then
        comment block u disappears;
        G[lu] ← gu;  L[gu] ← lu;  if H = u then H ← gu fi;
        G[u] ← V;  V ← u;
      else D[u] ← q − 1 fi;
    else if D[lu] = q − 1 or (u = H and q = A[0]) then W[u] ← x + 1;
      else comment a new block appears;
        t ← V;  V ← G[V];
        L[t] ← u;  G[t] ← gu;  L[gu] ← G[u] ← t;
        W[t] ← x + 1;  D[t] ← D[u];  D[u] ← q − 1;
        B[q] ← B[qq] ← t fi;
    fi;
    q ← qq;
    end
```

The total time is $O(l)$ when updating an element at level l of the tree. It is important to realize that the algorithm does not assume that the parent of nodes $2j-1$ and $2j$ is less than the parent of nodes $2j+1$ and

$2j + 2$. Such monotonicity cannot be guaranteed without violating the $O(l)$ time bound; fortunately it is not necessary.

6. Decreasing a Weight

The data structures developed in Section 4 for adding unity to a nonnegative weight can also be used to subtract unity from a positive weight. Since the ideas are similar to those discussed above, it suffices to present the algorithm here without further comment.

```
procedure downdate(integer k);
begin integer q;
q ← A[k]; comment q > M;
while q > 0 do
   ⟨Move q to the left of its block⟩;
   ⟨Transfer q to the previous block, with weight one lower⟩;
   q ← P[(q + 1) div 2] od;
if W[B[A[k]]] = 0 then ⟨Absorb zero-weight letter a_k⟩ fi;
end;

⟨Move q to the left of its block⟩ =
   begin integer u, t;
   u ← B[q]; if u = H then t ← 1 else t ← D[L[u]] + 1 fi;
   comment q is not the parent of t;
   if q ≠ t then exchange(q, t); q ← t fi;
   end

⟨Transfer q to the previous block, with weight one lower⟩ =
   begin integer u, lu, x, t;
   u ← B[q]; lu ← L[u]; x ← W[u];
   if W[lu] = x − 1 then
      B[q] ← lu; D[lu] ← q;
      if D[u] = q then comment block u disappears;
         G[lu] ← G[u]; L[G[u]] ← lu;
         G[u] ← V; V ← u fi;
   else if D[u] = q then W[u] ← x − 1
      else comment a new block appears;
         t ← V; V ← G[V]; if u = H then H ← t fi;
         L[t] ← lu; G[t] ← u; G[lu] ← L[u] ← t;
         W[t] ← x − 1; D[t] ← q; B[q] ← t fi;
   fi;
   if A[C[q]] ≠ q and B[q] = B[2 × C[q]] and q > 2 × C[q] + 1 then
      exchange(2 × C[q] + 1, q) fi;
   end
```

```
⟨Absorb zero-weight letter a_k⟩ =
   begin integer q;
   M ← M + 1;  R ← R + 1;
   if M > 1 then comment there are three nodes of weight zero;
      if 2 × R = M then E ← E + 1;  R ← 0 fi;
      q ← A[0] + 2;  A[0] ← q;  C[q] ← 0;
      A[k] ← M;  C[M] ← k fi;
   end
```

It is well known (and easy to prove directly) that the levels of two equal-weight nodes in a Huffman tree cannot differ by more than one, if the nodes have positive weight, except that a difference of two is possible when one node is the sibling of the zero-weight node. Hence the number of iterations of '*downdate*' cannot be more than three times the original level of the external node that corresponds to a_k, plus one. For example, if a_k is at level 4, four steps of exchanging with the leftmost member of its block and moving to the parent node must lead to a node on level ≤ 3. Three more steps lead to level ≤ 2, and so on.

7. Empirical Tests

The algorithms above were applied to three kinds of data, and in each case the online approach to encoding produced results comparable to the "optimum" results that would be obtainable from two passes over the data.

The '`abracadabra!`' example of Section 3 can be used to illustrate the criteria by which larger examples may be judged: Altogether 54 bits were produced by the online encoding of '`abracadabra!`', while the optimum weighted path length for the frequencies of letters $(1,1,1,2,2,5)$ is 28. It is not fair simply to compare 54 to 28, since the 28-bit encoding would be achieved only if the receiver knew the optimum code; further bits must be added to the 28 in order to specify what code is being used. Since it takes $2k - O(\log k)$ bits to specify a binary tree on k external nodes, and since each of the 27 possible letters of this example requires at least 4 bits for identification, we might expect to transmit at least $6k - O(\log k)$ bits to identify an optimum coding scheme that uses k of the letters. In the example, $k = 6$ and $28 + 6k = 64$; so the 54-bit result of the online method looks quite good.

The following statistics appear to be the most relevant numbers governing the performance of dynamic Huffman encoding, based on these considerations:

Σb (the total number of bits sent to encode a file);

$\Sigma b_{\rm opt}$ (the minimum weighted path length for the file); and

$\rho = (\Sigma b - \Sigma b_{\rm opt} - 2k)/k$ (the "overhead ratio"), where k is the number of letters with nonzero weight.

For example, the '`abracadabra!`' statistics are $\Sigma b = 54$, $\Sigma b_{\rm opt} = 28$, and $\rho \approx 2.333$. If ρ is less than the binary logarithm of the alphabet size n, or if the ratio $\Sigma b/\Sigma b_{\rm opt}$ is near 1, we can say that the online method is doing well.

The first file subjected to experiments was textual data from Grimm's first ten *Fairy Tales*, since the Grimm file has been available for many years at Stanford. In this case the relevant alphabet size is $n = 128$, because the data appears on the disk in 7-bit ASCII form. The carriage-returns and line-feeds at the end of each line, and the form-feeds at the end of each Tale, were encoded together with the text itself; the text, of course, consisted primarily of lowercase and uppercase letters together with blank spaces and a few punctuation marks. Only 58 of the 128 possible codes actually appeared in the file.

After 1000 characters had been transmitted, the statistics were

$$\Sigma b = 4558\,, \quad \Sigma b_{\rm opt} = 4297\,, \quad \rho = 5.68\,;$$

after the first 10000 characters, the statistics were

$$\Sigma b = 44733\,, \quad \Sigma b_{\rm opt} = 44272\,, \quad \rho = 6.70\,;$$

and after the first 100000 characters, they were

$$\Sigma b = 440164\,, \quad \Sigma b_{\rm opt} = 439613\,, \quad \rho = 7.50\,.$$

As the file gets larger, the overhead ratio grows to the point where a two-pass scheme would transmit fewer bits, yet the online method is not far from optimum.

Note that the Grimm data have about 4.4 bits of information per character, when we consider each character separately. Another experiment was made on the same file, but with 14-bit *character pairs* replacing the individual 7-bit characters. The Grimm tales used 680 of the $n = 16384$ possible character pairs in this 'alphabet'; and the statistics corresponding to the three sets of numbers above were

$$\begin{aligned}\Sigma b &= 5981\,, & \Sigma b_{\rm opt} &= 3447\,, & \rho &= 12.40\,;\\ \Sigma b &= 44704\,, & \Sigma b_{\rm opt} &= 38124\,, & \rho &= 13.06\,;\\ \Sigma b &= 393969\,, & \Sigma b_{\rm opt} &= 383264\,, & \rho &= 13.95\,.\end{aligned}$$

Thus the larger alphabet size meant that the one-pass scheme had $\rho < \lg n$ for the entire file; but this improvement was probably not enough to justify the 128-fold increase in memory requirements.

A second set of experiments was based on the computer representation of a technical book. This file differed from the Grimm text primarily because it contained mathematical symbols and numerical data, as well as special formatting characters to control the typesetting; 106 of the 128 possible ASCII characters were present. The respective statistics corresponding to the first 1000, 10000, and 100000 characters of this file were

$$\begin{aligned}
&\Sigma b = 5913\,, && \Sigma b_{\rm opt} = 5321\,, && \rho = 6.00\,;\\
&\Sigma b = 50440\,, && \Sigma b_{\rm opt} = 49619\,, && \rho = 7.33\,;\\
&\Sigma b = 519561\,, && \Sigma b_{\rm opt} = 518361\,, && \rho = 9.32\,.
\end{aligned}$$

When 14-bit character pairs were considered, the numbers were

$$\begin{aligned}
&\Sigma b = 7899\,, && \Sigma b_{\rm opt} = 3874\,, && \rho = 12.32\,;\\
&\Sigma b = 53204\,, && \Sigma b_{\rm opt} = 41970\,, && \rho = 12.80\,;\\
&\Sigma b = 472534\,, && \Sigma b_{\rm opt} = 442564\,, && \rho = 13.23\,;
\end{aligned}$$

and 1968 different character pairs were present. A technical file like this is clearly more difficult to compress than a file of Fairy Tales.

The third set of experiments was based on graphical data for the boundaries of alphabetic characters that were about 500 pixels tall. Each boundary was encoded as a sequence of "king moves," where each move represents a choice between seven possibilities: the boundary either continues in the previous direction, or it turns $\pm 45°$, $\pm 90°$, or $\pm 135°$. The king moves were grouped into triples, so the effective alphabet size was $7^3 = 343$. Seven different boundaries were studied, having respective lengths of 2508, 2736, 2172, 1236, 1008, 1632, and 612 moves; each boundary was treated as a separate file, with the tree initialized to zero at the beginning. The results were

	Σb	$\Sigma b_{\rm opt}$	ρ	$\Sigma b/(\text{total moves})$
Boundary 1	3048	2877	10.21	1.22
Boundary 2	3284	3110	8.88	1.20
Boundary 3	2448	2278	9.30	1.13
Boundary 4	1495	1350	10.02	1.21
Boundary 5	1155	1020	8.38	1.15
Boundary 6	1258	1093	7.71	0.77
Boundary 7	490	394	7.60	0.80

The last two boundaries consisted mostly of straight diagonal lines; the number of different king-move patterns of length t that appear on a straight line is at most $t + 1$ (see [3]), hence the encoding was (predictably) more efficient in this case.

The same experiment was carried out also with quadruples instead of triples, using an alphabet of size $7^4 = 2401$, with the following results:

	Σb	Σb_{opt}	ρ	Σb/(total moves)
Boundary 1	2935	2598	9.62	1.17
Boundary 2	3179	2805	10.90	1.16
Boundary 3	2369	2044	10.50	1.09
Boundary 4	1501	1205	10.87	1.21
Boundary 5	1191	916	10.50	1.18
Boundary 6	1205	993	9.83	0.74
Boundary 7	449	315	9.17	0.73

However, the overhead ratio in these experiments was higher than expected, and the reason seems to be that comparatively few of the "letters" actually occurred in the character boundaries. Only 25 of 343 triples and 39 of 2401 quadruples were present, so it is clear that bits were being wasted to identify the triples that occurred. There is no necessity for an online Huffman algorithm to start out with a *tabula rasa* in which all weights are zero, so another experiment was attempted in which the 25 triples and 39 quadruples were initially given weight 1. (In other words, each boundary had effectively been preceded by a message containing the "important" letters, but this preamble message did not need to be sent because the decoder already knew what it was.) The resulting numbers of bits per king move were thereby reduced as follows:

	Triples	Quadruples
Boundary 1	1.19	1.10
Boundary 2	1.16	1.08
Boundary 3	1.10	1.00
Boundary 4	1.16	1.06
Boundary 5	1.10	1.01
Boundary 6	0.73	0.65
Boundary 7	0.79	0.72

The algorithm of Section 6, which allows weights to decrease as well as to increase, suggests another dynamic coding scheme based on "local" properties of the source text: Given a message $a_{i_1} a_{i_2} \ldots$ and a buffer size d, we can increase the weight of a_{i_k} by 1 and decrease the weight of $a_{i_{k-d}}$ by 1 just after encoding a_{i_k}, so that the coding scheme for $a_{i_{k+1}}$ depends on the letter frequencies in the previous d characters $a_{i_{k-d+1}} \ldots a_{i_k}$. The encoder and decoder now need to maintain a buffer of length d in order to stay synchronized, and the computing time is approximately doubled. But this "d-window method" has two advantages that might compensate for the extra effort: (a) The weights will be bounded by d, so there will be no chance of overflow in the W array. (b) The windowing method gets rid of the "garbage" that occasionally appears, minimizing the effect of anomalous weights that might otherwise penalize an entire file.

A d-window method seems especially suited to the boundary data, since the path around a character shape typically runs through distinct phases. For example, boundaries 1 and 2 in the author's experiments were curves that sometimes went left for awhile then changed to rightward-turning curves; boundary 3 was a combination of curves and straight lines. However, closer examination revealed that windowing helps only if the phases are substantially longer than d, since it is necessary for the tree to go through a somewhat traumatic transitional period between phases. Furthermore d needs to be large enough that the tree weights have a significant effect on the encoding. Therefore the windowing method actually had a negative effect on the boundary data, except in the case of boundary 3 which was very slightly improved.

Windowing was of no use with the Grimm text either, since that data was quite uniform in character.

A small improvement was, however, noted when windowing was applied to the technical file, because its beginning portions involved material of somewhat special nature; unfortunately the early advantage of windowing disappeared later when a more uniform part of the file was encountered. Here are the Σb figures for the first 10000, 25000, and 100000 characters, with various window sizes:

	$d = 500$	$d = 750$	$d = 1000$	$d = 1250$	$d = 1500$	$d = \infty$
10000	50823	50456	50257	50248	50256	50440
25000	129083	128202	127900	128045	128375	129120
100000	526736	523541	521102	520248	520706	519561

The methods of this paper work in "real time," but the constant of proportionality is rather large, so they would be most advantageous if embedded in hardware. Their one-pass character means that the number of disk accesses is cut in half by comparison with two-pass schemes; in many applications, this fact outweighs the time for internal processing.

Acknowledgment

I wish to thank Joel Bion, whose term paper suggesting the use of dynamically changing codes for computer-to-computer file transfers led me to think of the problem treated here.

This research was supported in part by National Science Foundation grant MCS 83-00984 and by Office of Naval Research contract N00014-81-K-0269.

References

[1] Robert G. Gallager, "Variations on a theme by Huffman," *IEEE Transactions on Information Theory* **IT-24** (1978), 668–674.

[2] David A. Huffman, "A method for the construction of minimum-redundancy codes," *Proceedings of the IRE* **40** (1952), 1098–1101.

[3] Donald E. Knuth, "Solution to problem E2307," *American Mathematical Monthly* **79** (1972), 773–774.

Addendum

A similar procedure had been presented several years before Gallager's paper by Newton Faller, "An adaptive system for data compression," in the conference record of the *Asilomar Conference on Circuits, Systems, and Computers* **7** (1973), 593–597. Therefore the method of this chapter has become known as "Algorithm FGK." Independently, Gordon V. Cormack and R. Nigel Horspool published an algorithm that will change the weight of any external node by any desired amount, although $\Omega(n^2)$ steps might be needed in the worst case ["Algorithms for adaptive Huffman codes," *Information Processing Letters* **18** (1984), 159–165].

By introducing a somewhat more complicated algorithm, in which equal-weight leaves and branches of the tree are maintained in separate blocks, Jeffrey Scott Vitter ["Design and analysis of dynamic Huffman codes," *Journal of the Association for Computing Machinery* **34** (1987), 825–845] showed that it is possible to maintain dynamic trees that not only minimize $\sum w_j l_j$, they also minimize $\sum l_j$ and $\max l_j$ among all Huffman trees. He defined $\sum b'$ to be the net number of bits transmitted when we do not count the encodings of new characters when they split off from the 0-node, and proved that his algorithm always gives $\sum b' \leq \sum b_{\text{opt}} + t$, where t is the length of the message (the total number of letters encoded).

Vitter showed that his algorithm actually minimizes $\sum b'$ over *all* dynamic Huffman schemes. He also found empirically that the values of $\sum b'$ produced by the simpler Algorithm FGK were not much larger. Theoretical justification for this good behavior was eventually found by Ruy Luiz Milidiú, Eduardo Sany Laber, and Artur Alves Pessoa ["Bounding the compression loss of the FGK algorithm," *Journal of Algorithms* **32** (1999), 195–211], who proved that $\sum b'$ is always less than $\sum b_{\text{opt}} + 2t$ with the FGK method. They gave contrived examples to show that this upper bound is asymptotically best possible, although much better results are usually obtained in practice.

Chapter 6

Inhomogeneous Sorting

[Written with A. V. Anisimov. Originally published in International Journal of Computer and Information Sciences **8** *(1979), 255–260. Russian translation by A. B. Анисимов, «Неоднородная Сортировка», in Программирование* **5***, 1 (1979), 11–14. English retranslation in Programming and Computer Software* **5** *(1979), 7–10.]*

The purpose of this paper is to consider a special kind of sorting problem in which a sequence $a_1 \ldots a_n$ is to be rearranged into an optimal order by interchanging pairs of adjacent elements, where certain pairs of elements are not allowed to commute with each other. Section 1 presents an informal example intended to clarify the nature of the problem, while Sections 2 and 3 give some formal definitions and characterizations. Section 4 shows that the problem we are considering is equivalent to finding the lexicographically smallest topological sorting in a partial order.

1. An Informal Example

Consider a hypothetical firm in which there are n different job positions occupied by n different employees. These job positions are listed in order of importance:

$$J_1 J_2 \ldots J_n .$$

For example, J_1 is president, J_2 is first vice-president, …, J_n is garbage collector.

Some of the employees have equivalent skills, but some are considered more competent than others. Each employee is represented by a letter from some ordered alphabet $X = \langle x_1, x_2, \ldots, x_m \rangle$, where two equivalent employees are represented by the same letter, but an employee represented by x_i is more competent than an employee represented by x_j when $i < j$.

Suppose two people from classes x_i and x_j can fulfill the same duties; then we write

$$x_i x_j \equiv x_j x_i$$

meaning that they can interchange jobs. Of course if $i < j$, an x_i person will do the job better than an x_j, so we prefer to have x_i in the more important job.

In this firm it is possible for employees to change jobs only by going from J_k to an adjacent job $J_{k\pm1}$. Thus if people move to more important jobs, their left neighbors must take over their previous duties.

Suppose the current employee list is

$$a_1 a_2 \ldots a_n ,$$

where each $a_i \in X$. The problem of optimal reorganization in this firm is to find a way to rearrange the employees by permissible interchanges so that, after the rearrangement, each job J_k is being performed by the most competent possible person, subject to the condition that jobs $J_1 \ldots J_{k-1}$ have already been filled as well as possible.

For example, let $m = n = 3$ and suppose we have

$$x_1 x_2 \equiv x_2 x_1 \quad \text{and} \quad x_2 x_3 \equiv x_3 x_2 .$$

If the three jobs are originally filled in the order $x_3 x_2 x_1$, it is possible to improve the arrangement either to $x_2 x_3 x_1$ or to $x_3 x_1 x_2$. But once the first change has been made, no further permutations are possible without going back to $x_3 x_2 x_1$. The optimum arrangement in this case is $x_2 x_3 x_1$, since the other possibilities $x_3 x_1 x_2$ and $x_3 x_2 x_1$ have a less competent president.

2. Formal Definitions

Let $X = \langle x_1, \ldots, x_m \rangle$ be a finite, ordered alphabet and let Ω be an irreflexive, symmetric relation on X. In other words, Ω is a subset of $X \times X$, containing no pairs (a, a) for any $a \in X$, and containing (b, a) whenever $(a, b) \in \Omega$. We may regard Ω as an undirected graph on the vertices X, with an edge from vertex a to vertex b if and only if $(a, b) \in \Omega$. For example, the graph

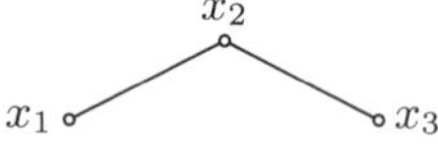

corresponds to the example considered in the preceding section.

Each graph Ω yields a congruence relation on the free monoid X^*. Let $\Pi = X^*/\Omega$ be the semigroup defined by Ω, namely the set of all equivalence classes of strings in X^* that can be proved equivalent to a given string by successive use of the relations

$$ab \equiv ba \qquad \Longleftrightarrow \qquad (a,b) \in \Omega$$

within strings.

Lemma 1. *The semigroup $\Pi = X^*/\Omega$ satisfies the left cancellation law. In other words, if $a\alpha \equiv a\beta$, then $\alpha \equiv \beta$.*

Proof. Suppose

$$a\alpha = \theta_1 \equiv \cdots \equiv \theta_t = a\beta$$

where each θ_{j+1} is obtained from θ_j by an interchange in Ω. We define a sequence of equivalences

$$\alpha = \theta'_1 \equiv \cdots \equiv \theta'_t = \beta$$

where each θ'_{j+1} is either identically equal to θ'_j or is obtained from θ'_j by an interchange in Ω.

The idea is simply to obtain θ'_j by eliminating the leftmost a in θ_j. If θ_{j+1} is obtained from θ_j by an interchange that involves this leftmost a, then $\theta'_{j+1} = \theta'_j$; and if it doesn't, then θ'_{j+1} is obtained from θ'_j by interchanging the same letters that changed places when θ_j became θ_{j+1}. □

If $\alpha = a_1 \ldots a_n$ and $\beta = b_1 \ldots b_{n'}$ and $\alpha \equiv \beta$, it is clear that $n = n'$ and $\beta = a_{p(1)} \ldots a_{p(n)}$ for some permutation $p(1) \ldots p(n)$ of $\{1, \ldots, n\}$. In this case we write

$$\beta = \alpha_p.$$

When α has repeated letters, there will be several permutations p such that $\beta = \alpha_p$, but only one of these permutations is really of interest to us. We say that a permutation $p(1) \ldots p(n)$ is *stable* with respect to α when the following condition holds for $1 \le i, j \le n$:

$$\text{If } p(i) < p(j) \text{ and } a_{p(i)} = a_{p(j)}, \text{ then } i < j.$$

Each word obtained from α by permutation of letters is equal to α_p for a *unique* stable permutation p.

Lemma 2. *Let p be a stable permutation with respect to α, such that $\alpha \equiv \alpha_p$. If $i < j$ and $(a_i, a_j) \notin \Omega$, then $p^{-1}(i) < p^{-1}(j)$.*

Proof. This property is true when p is the identity permutation, and it is preserved by adjacent interchanges of commuting elements. □

3. Normal Forms

So far we have made no use of the fact that X is ordered. Lexicographic order on words of equal length in X^* is defined by the rule

$$\alpha a\beta < \alpha b\beta' \quad \text{when } a < b,$$

for all words $\alpha, \beta, \beta' \in X^*$ and all letters $a, b \in X$. We also say that the word $\alpha = a_1 \ldots a_n$ is in *normal form* if it satisfies the condition

$$a_i \le a_{i+1} \quad \text{whenever } (a_i, a_{i+1}) \in \Omega.$$

It is clear that the lexicographically smallest word α_0 equivalent to a given word α must be in normal form, but a stronger condition is necessary to characterize α_0 completely. Let us say that the word $\alpha = a_1 \ldots a_n$ is in *strong normal form* when the following condition holds for $1 \le i < j \le n$:

$$\text{If } (a_i, a_j) \in \Omega,\ \ (a_{i+1}, a_j) \in \Omega,\ \ \ldots,\ \ (a_{j-1}, a_j) \in \Omega,\ \ \text{then } a_i \le a_j.$$

Putting this definition another way, let us say that the pair (a_i, a_j) is an *inversion* in α if $i < j$ and $a_i > a_j$. Then the word α is in strong normal form if and only if, for each inversion (a_i, a_j), at least one of the elements a_i, a_{i+1}, $\ldots$, a_{j-1} fails to commute with a_j.

Theorem 1. *If words α and β are both in strong normal form and $\alpha \equiv \beta$, then $\alpha = \beta$.*

Proof. Let $\alpha = a_1 \ldots a_n$ and β be in strong normal form, where $\alpha \equiv \beta$. Then $\beta = \alpha_p$ for some stable permutation $p(1) \ldots p(n)$ of $\{1, \ldots, n\}$.

The proof is by induction on n, the result being obvious when $n = 0$. If $a_1 = a_{p(1)}$, then the shorter words $a_2 \ldots a_n$ and $b_2 \ldots b_n$ are both in strong normal form, and they are equivalent by Lemma 1, hence they are equal by induction.

If $a_1 \ne a_{p(1)}$, we may assume without loss of generality that $a_1 > a_{p(1)}$. Then $(a_1, a_{p(1)})$ is an inversion of α, so $(a_k, a_{p(1)}) \notin \Omega$ for some $k < p(1)$. But this is impossible, because $p^{-1}(k) < p^{-1}(p(1)) = 1$ by Lemma 2. □

Theorem 2. *The lexicographically smallest word α_0 equivalent to a given word α is in strong normal form.*

Proof. It suffices to show that any word not in strong normal form is equivalent to a lexicographically smaller word, since equivalence classes are finite. Suppose $\alpha = a_1 \ldots a_n$ is not in strong normal form; then there is some inversion (a_i, a_j) such that a_i, a_{i+1}, $\ldots$, a_{j-1} all commute with a_j. But the equivalent word $a_1 \ldots a_{i-1} a_j a_i a_{i+1} \ldots a_{j-1} a_{j+1} \ldots a_n$ is lexicographically smaller. □

Theorems 1 and 2 show that Π can be represented as the set of all words in strong normal form, since there is precisely one such word in each equivalence class (namely the lexicographically smallest).

4. Connection with Topological Sorting

Let us now construct a directed graph $\Gamma(\alpha)$ on n vertices from each word $\alpha = a_1 \dots a_n$ in the following way: The vertices of $\Gamma(\alpha)$ are the integers $\{1, \dots, n\}$, and there is an arc $i \to j$ if and only if

$$i < j \quad \text{and} \quad (a_i, a_j) \notin \Omega.$$

It is clear that $\Gamma(\alpha)$ is acyclic.

A *topological sorting* $p(1) \dots p(n)$ of $\Gamma(\alpha)$ is any embedding of its vertices into a linear order, i.e., any permutation of $\{1, \dots, n\}$ such that

$$p(i) \to p(j) \quad \text{implies} \quad i < j.$$

It follows that every topological sorting is a stable permutation.

Theorem 3. *Let $\alpha = a_1 \dots a_n$. If $p(1) \dots p(n)$ is a topological sorting of $\Gamma(\alpha)$, then the word $\alpha_p = a_{p(1)} a_{p(2)} \dots a_{p(n)}$ is equivalent to α.*

Proof. The result is clear if $n = 0$. For $n > 0$, suppose $p(1) \dots p(n)$ is a topological sorting and let $p(j) = n$. Then $p(i) \to p(j)$ if and only if $a_{p(i)}$ does not commute with a_n, hence a_n commutes with each of $a_{p(j+1)}, \dots, a_{p(n)}$. Let $\alpha' = a_1 \dots a_{n-1}$ and let $\alpha'_q = a_{q(1)} \dots a_{q(n-1)}$, where $q(1) \dots q(n-1) = p(1) \dots p(j-1) p(j+1) \dots p(n)$. It is easy to verify that $q(1) \dots q(n-1)$ is a topological sorting of $\Gamma(\alpha')$; hence $\alpha' \equiv \alpha'_q$ by induction, and we have $\alpha = \alpha' a_n \equiv \alpha'_q a_n \equiv \alpha_p$. □

The converse of Theorem 3 is also true:

Theorem 4. *Let $\alpha = a_1 \dots a_n$, and let $p(1) \dots p(n)$ be a stable permutation of $\{1, \dots, n\}$ with respect to α. If $\alpha_p = a_{p(1)} \dots a_{p(n)}$ is equivalent to α, then $p(1) \dots p(n)$ is a topological sorting of $\Gamma(\alpha)$.*

Proof. If p is a stable permutation such that $\alpha \equiv \alpha_p$, and if $p(i) \to p(j)$ in $\Gamma(\alpha)$, we have

$$p(i) < p(j) \quad \text{and} \quad (a_{p(i)}, a_{p(j)}) \notin \Omega.$$

Lemma 2 now says that $p^{-1}(p(i)) < p^{-1}(p(j))$, namely that $i < j$. Hence p is a topological sorting of $\Gamma(\alpha)$. □

5. Conclusions

By combining the results of Sections 3 and 4, we see that the problem of optimum reorganization subject to commutativity relations is equivalent to the problem of finding strong normal forms, which in turn is equivalent to the problem of finding the lexicographically smallest topological sorting of a partial order. When the letters of α are distinct, one can construct an irreflexive, symmetric relation Ω such that the transitive closure of the directed graph $\Gamma(\alpha)$ is isomorphic to any given partial order.

One way to solve the sorting problem stated in the introduction would therefore be to use a topological sorting algorithm such as that described in [1, §2.2.3], where the "source" vertex j with *smallest* a_j is chosen at each step. This algorithm is somewhat analogous to selection sorting ([2, §5.2.3]).

It is interesting also to develop an algorithm that is analogous to insertion sorting ([2, §5.3.1]), where elements $a_1 \ldots a_{n-1}$ have been rearranged before a_n is examined: Let $j \geq 1$ be minimal such that $(a_k, a_n) \in \Omega$ for $j \leq k < n$, and let $k \geq j$ be minimal such that $a_k \geq a_n$. Then a_n should be inserted between a_{k-1} and a_k (or at the very left if $k = 1$, at the very right if $k = n$).

The work of A. V. Anisimov was performed while visiting Stanford University, with the support of the International Research and Exchanges Board (IREX).

References

[1] Donald E. Knuth, *Fundamental Algorithms*, Volume 1 of *The Art of Computer Programming* (Reading, Massachusetts: Addison–Wesley, 1968).

[2] Donald E. Knuth, *Sorting and Searching*, Volume 3 of *The Art of Computer Programming* (Reading, Massachusetts: Addison–Wesley, 1973).

Addendum

The semigroup X^*/Ω was introduced by P. Cartier and D. Foata in "Problèmes combinatoires de commutation et réarrangements," *Lecture Notes in Mathematics* **85** (1969), iv+88 pages. Their work has spawned a large literature in recent years, because many commutativity graphs Ω have turned out to have important applications, and X^*/Ω has become known among mathematicians as a *Cartier–Foata monoid.*

Meanwhile the same semigroup was independently introduced by computer scientists who studied sequences of computational steps that are equivalent under parallelism, and it has become known to them as a *trace monoid.* (See Antoni Mazurkiewicz, "Concurrent program schemes and their interpretations," DAIMI Report PB 78 (Århus, Denmark: Århus University, July 1977), 51 pages.)

The Book of Traces, edited by V. Diekert and G. Rozenberg (Singapore: World Scientific, 1995), x + 576 pages, is a comprehensive survey of the early work of both "schools," with 278 entries in its bibliography.

An intuitively appealing way to understand the elements of X^*/Ω, due to G. X. Viennot, is to regard each equivalence class as a "heap of pieces": If $\alpha = a_1 \ldots a_n$, we place the pieces $a_1, a_2, \ldots, a_n$ successively on the floor in such a way that each a_j drops down to rest on the topmost already-placed piece a_i for which $(a_i, a_j) \notin \Omega$, or on the floor if no such a_i exists. (See Gérard Xavier Viennot, "Heaps of pieces, I: Basic definitions and combinatorial lemmas," *Lecture Notes in Mathematics* **1234** (1986), 321–350.)

Let c_j be the number of cliques of size j in the graph Ω, and let $C(\Omega)$ be the polynomial $c_0 - c_1 z + c_2 z^2 - \cdots + (-1)^n c_n z^n$. For example, the simple graph used as an example in Sections 1 and 2 has $C(\Omega) = 1 - 3z + 2z^2$. Cartier and Foata proved that the number of equivalence classes in X^*/Ω for words of length n is the coefficient of z^n in the generating function $1/C(\Omega)$. Viennot proved, more generally, that the number of heaps for which all elements on the floor belong to $\{x_1, \ldots, x_k\}$ is $C(\Omega_k)/C(\Omega)$, where Ω_k is Ω restricted to the vertices $X \backslash \{x_1, \ldots, x_k\}$. (In our example, $C(\Omega_1) = 1 - 2z + z^2$ and $C(\Omega_2) = 1 - z$.)

Volker Diekert, Yuri Matiyasevich, and Anca Muscholl ["Solving word equations modulo partial commutations," *Theoretical Computer Science* **224** (1999), 215–235] have given an algorithm to decide whether or not an equation of the form $\alpha \equiv \beta$, where α and β may include the names of variables as well as letters of X, has a solution in a given trace monoid.

By restricting the graph $\Gamma(\alpha)$ defined in Section 4 above so that the arc $i \to j$ is included only when $a_i \neq a_k$ for all k between i and j, we can construct $\Gamma(\alpha)$ in $O(mn)$ steps, and complete the lexicographic topological sort in another $O(mn)$ steps. Is there a faster algorithm for the inhomogeneous sorting problem that is considered in this chapter?

Chapter 7

Lexicographic Permutations with Restrictions

[Originally published in Discrete Applied Mathematics **1** *(1979), 117–125.]*

An efficient algorithm is developed to generate all partitions of a set of elements into subsets of specified sizes, in lexicographic order, and more generally to generate all permutations whose elements are arranged consistently with a partial ordering that has a tree structure.

Let $\prec$ denote a partial ordering relation on the elements $\{1, 2, \ldots, n\}$ such that $i \prec j$ implies $i < j$. We shall consider the problem of generating all permutations $a_1 a_2 \ldots a_n$ of $\{1, 2, \ldots, n\}$ such that $i \prec j$ implies $a_i < a_j$; permutations satisfying this condition will be called *admissible.* Furthermore we shall insist that the permutations be generated in lexicographic order. Thus, the first admissible permutation will be $1\,2 \ldots n$, and we seek an algorithm to find the lexicographically smallest admissible permutation $b_1 b_2 \ldots b_n$ greater than a given one $a_1 a_2 \ldots a_n$.

In the special case that the partial ordering is a tree or forest, we shall see that an efficient and rather pleasant algorithm is possible. This algorithm has "optimum" running time, in the sense that it takes $O(m)$ steps, where m is the number of positions that change when we go from $a_1 a_2 \ldots a_n$ to $b_1 b_2 \ldots b_n$. More precisely, m is defined as follows: The leading elements $b_1 \ldots b_{n-m}$ are equal to $a_1 \ldots a_{n-m}$, but b_{n-m+1} is greater than a_{n-m+1}.

The case of trees and forests is not as esoteric as it may seem at first, because it includes the following important special case: Find all ways to partition a set of n elements into s subsets of given sizes n_1, n_2, …, n_s. It is easy to see that this problem is equivalent to finding all permutations that are partially ordered by the "ancestor" relation in a certain oriented forest. For example, if we want to find all ways to partition the set $\{1, 2, \ldots, 12\}$ into five subsets of sizes $\{2, 2, 2, 3, 3\}$, we can convert this task to the problem of finding all permutations $a_1 a_2 \ldots a_{12}$ of $\{1, 2, \ldots, 12\}$ such that

$$a_1 < a_2,\ a_1 < a_3 < a_4,\ a_3 < a_5 < a_6,\ a_7 < a_8 < a_9,\ a_7 < a_{10} < a_{11} < a_{12}. \quad (1)$$

The forest is

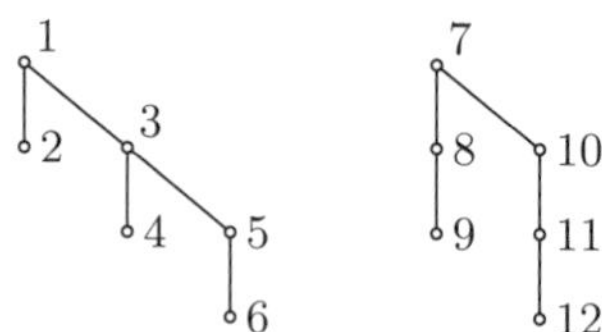

and the respective subsets are $\{a_1, a_2\}$, $\{a_3, a_4\}$, $\{a_5, a_6\}$, $\{a_7, a_8, a_9\}$, $\{a_{10}, a_{11}, a_{12}\}$. Since there are 138,600 ways to do the partitioning, an efficient algorithm is desirable.

It may be of interest to point out a connection between the problem of generating admissible permutations and the "topological sorting" problem. The latter problem asks for all permutations $z_1 z_2 \ldots z_n$ of $\{1, 2, \ldots, n\}$ such that i appears to the left of j whenever $i \prec j$; so it is equivalent to saying that $z_1 z_2 \ldots z_n$ is the *inverse* of an admissible permutation. A fairly efficient way to generate all topological sortings of *any* given partially ordered set, without restriction to lexicographic order, appears in [4].

Determining the Number of Changes

The first step in going from $a_1 \ldots a_n$ to its lexicographic successor $b_1 \ldots b_n$ is to determine the leftmost position that must change. Let this position be m steps from the right, that is, element a_{n-m+1}. We shall say that m is the number of changes (even though it may turn out by coincidence that $b_j = a_j$ for some $j > n - m$). If $a_1 \ldots a_n$ is the lexicographically largest permutation, we may define $m = n + 1$; our algorithm will report "all perms generated" when this condition occurs. It turns out that m can be determined very simply when we have a suitably numbered partial ordering of dimension two:

Lemma 1. *Let $a_1 \ldots a_n$ be admissible with respect to the partial ordering $\prec$, and let $a_0 = a_{n+1} = 0$. The number of changes to get to the lexicographically next admissible permutation can be determined by a program of the form*

$$i \leftarrow n;\ \textbf{while}\ a_i > a_{g(i)}\ \textbf{do}\ i \leftarrow i - 1;\ \ m \leftarrow n + 1 - i \quad (2)$$

for some function $g : \{0, 1, \ldots, n\} \to \{1, \ldots, n, n + 1\}$ if and only if the partial ordering can be characterized by a permutation $p_1 p_2 \ldots p_n$ of

$\{1, 2, \ldots, n\}$ *in the following way:*

$$i \prec j \quad \text{if and only if} \quad i < j \text{ and } p_i > p_j. \tag{3}$$

For example, in the forest ordering (1) we can let

$$(g(0), g(1), \ldots, g(12)) = (13, 9, 4, 9, 6, 9, 9, 13, 12, 12, 13, 13, 13),$$

and this gives a simple and fast way to determine the number of changes. In practice we would of course start (2) with $i \leftarrow 9$ instead of $i \leftarrow 12$, since $a_{g(12)} = a_{g(11)} = a_{g(10)} = 0$. The permutation $p_1 p_2 \ldots p_{12}$ that yields order (1) is

$$6\ 1\ 5\ 2\ 4\ 3\ 12\ 8\ 7\ 11\ 10\ 9.$$

Proof. If the ordering is defined by (3), let $q_1 \ldots q_n$ be the inverse permutation (e.g., 2 4 6 5 3 1 9 8 12 11 10 7 in case (1)). We let $g(0) = n+1$, and for $i > 0$ we let $g(i)$ be the first element larger than i appearing to the right of i in the q array. If no such element exists, $g(i) = n + 1$.

We shall prove that if $a_j > a_{g(j)}$ for $n - k < j \leq n$, then $m > k$, and the elements $a_{n-k+1} \ldots a_n$ must satisfy

$$a_{r(1)} > a_{r(2)} > \cdots > a_{r(k)} \tag{4}$$

where $r(1) \ldots r(k)$ is the permutation of $\{n - k + 1, \ldots, n\}$ obtained by striking out all values $\leq n - k$ from $q_1 q_2 \ldots q_n$. This permutation r depends on k; but we will write $r(j)$ instead of $r(k, j)$ or $r_k(j)$, in order to avoid making the notation even more complicated. For example, if $k = 9$ we have $r(1) \ldots r(9) = 4\ 6\ 5\ 9\ 8\ 12\ 11\ 10\ 7$. The stated condition is vacuously true when $k = 0$, and we will prove it by induction on k.

Assume that $0 \leq k < n$ and $a_j > a_{g(j)}$ for $n - k < j \leq n$, so that (4) holds; and let $i = n - k$. It is convenient to define $r(k + 1) = n + 1$, so that we have $g(i) = r(l)$ for some l depending on k. We know from the definitions that $p_i > p_j$ for $j = r(1)$, $r(2)$, ..., $r(l - 1)$, because these values of j appear to the left of i in q; hence $a_i < a_j$ for all such j. Now if $a_i > a_{g(i)}$, we'll have $a_{r(l-1)} > a_i > a_{r(l)}$; hence we can increase k by 1 and (4) will still hold as desired.

Furthermore we need to verify that $m \neq k + 1$ when $a_i > a_{g(i)}$. But it is clear from the discussion that a_i must be less than $l - 1$ elements of the set $\{a_{i+1}, \ldots, a_n\}$; so it cannot be increased unless some element to the left of a_i changes.

On the other hand if $a_i < a_{g(i)}$ we do have $m = k + 1$, because we can find an admissible permutation with a_i increased. For example, if

$a_{r(t-1)} > a_i > a_{r(t)}$ where $t > l$, we can interchange the values of $a_{r(t-1)}$ and a_i. This interchange does not introduce any inadmissibilities, since $r(t-1)$ appears to the right of i in the q array. (More precisely, let $u = r(t-1)$; then $p_i < p_u$, so we do not need to have $a_i < a_u$. And if we need $a_v < a_u$ for some $v < i$, then $p_v > p_u > p_i$ implies that $a_v < a_i$, so a_v will be less than a_u after the interchange $a_i \leftrightarrow a_u$.) This completes the proof that program (2) correctly determines m.

Conversely, suppose g is a function for which (2) correctly determines m whenever $a_1a_2 \ldots a_n$ is admissible; we want to show that the partial ordering satisfies (3) for some p. In the first place we can't have $g(i) = i$. Furthermore if $g(i) < i$ for some i we can replace it by $g(i) = n+1$. For example, suppose $g(4) = 2$. Then if $a_1a_2 \ldots a_n$ is the lexicographically largest admissible permutation beginning 1 2 3 4..., we have $a_4 > a_2$ so there are no admissible permutations beginning 1 2 3 5.... The inadmissibility of 1 2 3 5 4..., 1 2 3 5 6 4..., 1 2 3 5 6 7 4..., etc., implies that $4 \prec j$ for all $j > 4$, so the function value $g(4) = n+1$ must work if anything does.

We may therefore assume that $g(i) > i$ for all i; hence the program in (2) makes decisions based entirely on the relative order of the elements $\{a_i, \ldots, a_n\}$.

We will show by induction on k that if $a_j > a_{g(j)}$ for $n-k < j \le n$, there must be a permutation $r(1) \ldots r(k)$ of $\{n-k+1, \ldots, n\}$, depending only on k, where the elements $a_{n-k+1} \cdots a_n$ satisfy (4). Furthermore if $i = n-k \ge 1$ and if $g(i) = r(l)$, where $r(k+1) = n+1$, we will show that $i \prec r(1), \ldots, i \prec r(l-1), i \not\prec r(l), \ldots, i \not\prec r(k)$; hence i appears immediately before $r(l)$ in the r permutation of order $k+1$.

It is clear that $i \not\prec r(t)$ for all $t \ge l$; for $i \prec r(t)$ would imply that $a_i < a_{g(i)}$, and repeated uses of program (2) would never terminate (the value of m would never exceed $k+1$). Therefore the following permutation is admissible by induction:

$$\begin{aligned}
a_j &= j, \quad \text{for} \quad j < i;\\
a_{r(t)} &= i+k-t, \quad \text{for} \quad l \le t \le k;\\
a_i &= i+k-l+1;\\
a_{r(t)} &= i+k-t+1, \quad \text{for} \quad 1 \le t < l.
\end{aligned}$$

This permutation causes (2) to bypass further increase of a_i, so it must be true that $i \prec r(t)$ for $1 \le t < l$ or we will miss admissible permutations. Namely, if there exists $1 \le s < l$ such that $i \not\prec r(s)$ but $i \prec r(t)$ for $s < t < l$, then we must have $r(t) \not\prec r(s)$ for $s < t < l$; so the following permutation is admissible:

$$\begin{aligned} a_j &= j, \quad \text{for} \quad 1 \le j < i; \\ a_{r(t)} &= i+k-t, \quad \text{for} \quad l \le t \le k; \\ a_{r(t)} &= i+k-t+2, \quad \text{for} \quad s < t < l; \\ a_{r(s)} &= i+k-l+1; \\ a_{r(t)} &= i+k-t+1, \quad \text{for} \quad 1 \le t < s; \\ a_i &= i+k-l+2. \end{aligned}$$

The inductive proof is complete, and when $k = n$ the permutation $r(1)\ldots r(n) = q_1 \ldots q_n$ is the inverse of a permutation $p_1 \ldots p_n$ that defines the partial order we have deduced. □

The proof of Lemma 1 seems unnecessarily complicated because the notation is messy, but there is no apparent "high-level" way to prove this result. Perhaps the fact that (3) holds if and only if the partial ordering satisfies "$i \not\prec j$ and $j \not\prec k$ and $i < j < k$ implies $i \not\prec k$," and/or the fact that the partial orderings satisfying (3) form a lattice, may lead to a simpler proof. (See [3, exercises 5.1.1–11 and 5.1.1–12].)

It should be pointed out that condition (3) is not equivalent to saying that the partial order has dimension 2; the numbering of elements must also be suitable. For example, suppose $n = 3$, and let $i \prec j$ if and only if $i = 1$ and $j = 3$. Then there is no permutation $p_1p_2p_3$, although this relation is certainly a partial order of dimension 2.

When the partial ordering is a forest and we number the nodes in "preorder," as in the example illustrated above for (1), the permutation $q_1 \ldots q_n$ (the inverse of $p_1 \ldots p_n$) is the sequence of node numbers in "postorder." (See [2, exercise 2.3.2–20].)

Making the Changes

Once m has been found, we want to rearrange $\{a_{n-m+1}, \ldots, a_n\}$ to achieve the lexicographically next-least admissible permutation $b_1 \ldots b_n$ of the form

$$a_1 \ldots a_{n-m} c_{n-m+1} \ldots c_n.$$

In the special case considered above, the algorithm not only determines m, it also determines the relative ordering of $\{a_{n-m+2}, \ldots, a_n\}$, and with a few more comparisons the entire set $\{a_{n-m+1}, \ldots, a_n\}$ will essentially be sorted; let us call it $\{d_1, \ldots, d_m\}$ where $d_1 < d_2 < \cdots < d_m$.

Let $d_0 = 0$. By looking at where the elements $\{a_1, \ldots, a_{n-m}\}$ fall among the d's, we can determine constraints on the c's. Suppose, for example, that $i \le n-m$ and $j > n-m$ and $i \prec j$. If $d_{k-1} < a_i < d_k$, then we need $c_j \ge d_k$. For each $j > n-m$, let $k(j)$ be the maximum constraint determined in this way; that is,

$$k(j) = \max\{k \mid \text{either } k=1, \text{ or } j=n-m+1 \text{ and } d_{k-1}=a_j, \\ \text{or } d_{k-1} < a_i \text{ for some } i \le n-m \text{ with } i \prec j\}.$$

It is interesting to observe that the numbers $k(j)$ are enough to determine the c's completely — we do not need to know how the partial order $\prec$ acts on pairs of elements that are both greater than $n-m$. This fact is demonstrated as follows:

Lemma 2. *Let the numbers* $d_1, \ldots, d_m$ *and* $k(n-m+1), \ldots, k(n)$ *be defined as above. If for* $j = n-m+1, n-m+2, \ldots, n$ *we set* c_j *to the smallest element of the set*

$$\{d_{k(j)}, \ldots, d_m\} \setminus \{c_{n-m+1}, \ldots, c_{j-1}\},$$

then the permutation

$$a_1 \ldots a_{n-m} c_{n-m+1} \ldots c_n = b_1 \ldots b_n$$

will be the lexicographically least admissible successor of $a_1 \ldots a_n$.

Proof. It is clear that no admissible successor of $a_1 \ldots a_n$ could be lexicographically smaller than $b_1 \ldots b_n$, so we must prove only that $b_1 \ldots b_n$ is admissible. If $i \prec j$ and $i \le n-m$, we have $b_i < b_j$ immediately by construction, so the only possible problem occurs if $i > n-m$. But then b_i is the smallest element of $S_i = \{d_{k(i)}, \ldots, d_m\} \setminus \{c_{n-m+1}, \ldots, c_{i-1}\}$ and b_j is the smallest element of $S_j = \{d_{k(j)}, \ldots, d_m\} \setminus \{c_{n-m+1}, \ldots, c_{j-1}\}$. We have $k(j) \ge k(i)$, since every k in the set being maximized for $k(i)$ appears in the corresponding set for $k(j)$. Therefore S_j is a subset of S_i; and S_j does not include c_i (the smallest element of S_i), so c_j must exceed c_i. □

The problem facing us in general, when lexicographic generation is required, boils down to the following task of interest in its own right: *Given integers* $1 \le u_j \le m$ *for* $1 \le j \le m$, *determine the lexicographically least permutation* $v_1 \ldots v_m$ *of* $\{1, \ldots, m\}$ *such that* $v_j \ge u_j$ *for* $1 \le j \le m$. This is essentially the "parking problem" discussed in [3, exercises 6.4–29, 6.4–30, 6.4–31], although the question of determining $v_1 \ldots v_m$ efficiently was not treated there. In our case, $u_j = k(n-m+j)$ and $c_{n-m+j} = d_{v_j}$ for $1 \le j \le m$. It appears that the fastest way to compute $v_1 \ldots v_m$ will take $O(m\alpha(m))$ steps in the worst case, where $\alpha(m)$ is the inverse of Ackermann's function [6]; we can sequentially choose $v_j = \min S_j$, where

$$S_j = \{u_j, u_j + 1, \ldots, m\} \setminus \{v_1, \ldots, v_{j-1}\},$$

by using a disjoint set union algorithm to keep track of the "holes" of consecutive elements in $\{v_1, \ldots, v_{j-1}\}$. Another way to approach the problem is to determine the inverse permutation $w_1 \ldots w_m$ of the v's: Let $T_i = \{j \mid u_j = i\}$ and for $i = 1, 2, \ldots, m$ let

$$w_i = \min((T_1 \cup \cdots \cup T_i) \setminus \{w_1, \ldots, w_{i-1}\}).$$

This is the "offline MIN" problem discussed in [1, page 140], where it is reduced to disjoint set union.

If we specialize the partial ordering to trees, however, the complications of the general case do not arise. We have the basic property

$$i_1 \prec j \text{ and } i_2 \prec j \text{ and } i_1 < i_2 \quad \text{implies} \quad i_1 \prec i_2, \tag{5}$$

since any two ancestors of a node must be related and we cannot have $i_2 \prec i_1$. This means we can compute the numbers $k(j)$ required by Lemma 2 in a very simple way:

Lemma 3. *If the partial ordering satisfies* (3) *and* (5), *let* $q_1 \ldots q_n$ *be defined as above. For each* $j > n - m$, *let* $i(j)$ *be the first element* $\leq n - m$ *appearing to the right of* j *in* $q_1 \ldots q_n$, *if such elements exist; then* $k(j)$ *is determined by the condition*

$$d_{k(j)-1} < a_{i(j)} < d_{k(j)}.$$

(If $i(j)$ *does not exist then* $k(j) = 1$.*) Furthermore*

$$k(n-m+1) \geq k(n-m+2) \geq \cdots \geq k(n).$$

Proof. In general when the partial ordering satisfies (3), the value of $k(j)$ is the maximum k such that $a_i > d_{k-1}$ maximized over all i appearing to the right of j in $q_1 \ldots q_n$. Under condition (5), such elements i must appear in decreasing order, and the first one will yield the largest k. Furthermore if we have $k(j) > k(j')$ for some j and j', the permutation $q_1 \ldots q_n$ must contain the substring $\ldots j \ldots i \ldots j' \ldots$ for some $i \leq n-m$, and (4) implies that $j < j'$. □

The fact that the k's are nonincreasing makes it very easy to compute the c's using an auxiliary stack. Let $k(n-m) = m+1$; we can proceed as follows:

```
⟨Set the stack empty⟩;
for j ← n − m + 1 to n do
   begin ⟨Push d_{k(j−1)−1}, ..., d_{k(j)} onto
                    the stack (in decreasing order)⟩;          (6)
   ⟨Pop the top of the stack to c_j⟩;
   end.
```

This simple algorithm also has another useful property: When c_j receives its new value, the former value a_j has already been moved to the stack, since the stack has received all elements $\geq d_{k(j)}$. Therefore the rearrangement of elements needs no further auxiliary memory.

The Final Algorithm

These rather elaborate preparations now yield the algorithm we have been seeking. All that is needed is to specify auxiliary tables, so that computations such as ⟨Push $d_{k(j-1)-1}, \ldots, d_{k(j)}$ onto the stack (in decreasing order)⟩ can be done without loss of time.

Suppose we are given a forest with nodes $1\,2\ldots n$ in preorder and $q_1 q_2 \ldots q_n$ in postorder, as defined in [2, §2.3.2]. The algorithm in Fig. 1 below deals with the following arrays and precomputed auxiliary tables:

$a[0]a[1]\ldots a[n]$, the permutation being manipulated. (Element $a[0]$ is always 0, and the algorithm finds the lexicographic successor of $a[1]\ldots a[n]$.)

$s[1]\ldots s[n]$, the auxiliary stack, used to hold elements in decreasing order as they are being permuted.

$g[0]g[1]\ldots g[n]$, used in the test (2). (Element $g[0]$ is always 0, and for $i > 0$ the value of $g[i]$ is the first element $> i$ that lies to the right of i in postorder. If no such element exists, we let $g[i] = 0$.)

$f[1]\ldots f[n]$, the "parent" pointers. (The value of $f[i]$ is the first element $< i$ that lies to the right of i in postorder. If no such element exists, we let $f[i] = 0$.)

$h[1]\ldots h[n-1]$, the "handle" pointers. (The value of $h[i]$ is the first element $> i$ to appear in postorder.)

$e[1]\ldots e[n]$, postorder pointers, either $f[i]$ or $g[i]$. (The value of $e[i]$ is the first element to the right of i in postorder.)

For example, the ordering (1) would be encoded as follows:

$i =$	0	1	2	3	4	5	6	7	8	9	10	11	12
$q[i] =$		2	4	6	5	3	1	9	8	12	11	10	7
$g[i] =$	0	9	4	9	6	9	9	0	12	12	0	0	0
$f[i] =$		0	1	1	3	3	5	0	7	8	7	10	11
$h[i] =$		2	4	4	6	6	9	9	9	12	12	12	
$e[i] =$		9	4	1	6	3	5	0	12	8	7	10	11

Note that the case $g[i] = 0$ in this representation corresponds to the case $g(i) = n + 1$ in (2).

The algorithm also uses the following variables:

i, the leftmost node that changes;
$i0$, the largest i such that $g[i] \neq 0$;
j, the node being filled with its new value;
k, the number of elements in the stack;
l, position of next node in postorder to put on the stack;
t, temporary storage of the old value of $a[i]$;
x, index for which $a[x]$ is less than everything in the stack.

```
i ← i0;                                                 1
while a[i] > a[g[i]] do i ← i − 1;                      m − (n − i0)
if i = 0 then go to all perms generated;                1
k ← 0;  t ← a[i];  l ← h[i];  x ← n;                    1
for j ← i step 1 until n do                             m + 1
   begin if f[j] < x then                               m
      begin if x ≥ i then                               A
         begin if x > i then x ← i                      2, 1
         else begin k ← k + 1;  s[k] ← t;  x ← f[j];    1
            end
         end
      else x ← f[j];                                    A − 2
      while a[l] > a[x] do                              m + A
         begin k ← k + 1;  s[k] ← a[l];                 m
         l ← e[l];  if l ≤ i then l ← g[l];             m, A
         end;
      end;
   a[j] ← s[k];  k ← k − 1;                             m
   end;
```

FIGURE 1. Algorithm for tree permutations in lexicographic order.

Lines *01–03* of Fig. 1 correspond to (2); the remaining lines correspond to (6). Lines *13–16* put elements onto the stack in decreasing order; lines *07–11* arrange to get the old value of $a[i]$ into its correct place with respect to the other elements, which appear in postorder according to (4).

The coding of the algorithm is slightly tricky in line *15*. If $l \le i$ after advancing l to its successor $e[l]$ in postorder, it follows that either $l = 0$ or $g[l]$ is greater than i (or $g[l] = 0$, which is treated as greater than i because it's like $g[l] = n + 1$ in Lemma 1). In the former case $g[0] = 0 = x$, so the **while** loop on line *13* will terminate immediately; in the latter case $g[l]$ is not only the first successor of l in postorder that is $> l$, it is also $> i$ because of (5). This trick ensures a running time of order m.

The right-hand margin of Fig. 1 shows how many times the statements on that line will be executed, assuming that $a[1] \ldots a[n]$ isn't lexicographically last. Here A is the number of times x changes, namely two more than the number of ancestors of $i = n - m + 1$ that are parents to nodes numbered $> i$.

Repeated Elements

The algorithm in Fig. 1 clearly does not require that $a[1] \ldots a[n]$ be a permutation of $\{1, \ldots, n\}$, since it operates entirely by comparisons. Whenever the elements $a[0]$, $a[1]$, $\ldots$, $a[n]$ are distinct and $a[0]$ is the smallest, all admissible permutations will be found.

In fact, the algorithm will work also when the elements being permuted are *not* distinct, if we make a few changes to account for the case of equalities. This is one of the advantages of lexicographic generation. Lines *02* and *03* should be replaced by

while $a[i] \ge a[g[i]]$ **do**
  begin $i \leftarrow i - 1$;
  if $i = 0$ **then go to** all perms generated;
  end;

we should also insert after line *04* the new statement

$a[i] \leftarrow t + 1$;

and then we should use the loop specification

while $a[l] \ge a[x]$ **and** $l \ne 0$ **do**

in place of line *13*. (The resulting program is a little kludgy; it would be cleaner to separate the case $x = i$ from the other cases.) It is not difficult to see that our arguments for the validity of Fig. 1 can now be extended to the general case with repeated elements; all permutations such that

$$i \prec j \quad \text{implies} \quad a_i \le a_j \tag{7}$$

will be generated, provided that $a[0] \le a[1] \le \cdots \le a[n]$ initially.

Unfortunately the problem of set partition with repeated elements ("multiset partition") no longer reduces to the generation of admissible permutations (7) under a partial ordering, if the subsets can be of equal size. For example, if we want to divide $\{1, 1, 2, 3\}$ into subsets of size two, both $\{1, 2\}\{1, 3\}$ and $\{1, 3\}\{1, 2\}$ will be generated. The algorithm above would properly partition the multisets $\{1, 2, 2, 3\}$ and $\{1, 2, 3, 3\}$, but it appears that no simple algorithm of the kind we have been considering will work in general.

This research was supported in part by National Science Foundation grant MCS-77-23738 and by Office of Naval Research contract N00014-76-C-0330.

References

[1] Alfred V. Aho, John E. Hopcroft, and Jeffrey D. Ullman, *The Design and Analysis of Computer Algorithms* (Reading, Massachusetts: Addison–Wesley, 1974).

[2] Donald E. Knuth, *Fundamental Algorithms*, Volume 1 of *The Art of Computer Programming* (Reading, Massachusetts: Addison–Wesley, 1968).

[3] Donald E. Knuth, *Sorting and Searching*, Volume 3 of *The Art of Computer Programming* (Reading, Massachusetts: Addison–Wesley, 1973).

[4] Donald E. Knuth and Jayme L. Szwarcfiter, "A structured program to generate all topological sorting arrangements," *Information Processing Letters* **2** (1974), 153–157; **3** (1974), 64. [Reprinted with revisions as Chapter 3 of *Literate Programming*, CSLI Lecture Notes 27 (Stanford, California: Center for the Study of Language and Information, 1992), 91–97.]

[5] Albert Nijenhuis and Herbert S. Wilf, *Combinatorial Algorithms* (New York: Academic Press, 1975).

[6] Robert Endre Tarjan, "Efficiency of a good but not linear set union algorithm," *Journal of the Association for Computing Machinery* **22** (1975), 215–225.

Addendum

Researchers who are trying to extend the capabilities of methods for mechanical verification of algorithms are invited to study this particular algorithm carefully, because it is brief yet subtle.

The algorithm actually works in more cases than claimed. For example, there are only 14 forests when $n = 4$, yet the program of Fig. 1 is correct for 20 of the 24 permutations $p_1p_2p_3p_4$ in (3). (It fails only when p is 2 3 1 4 or 2 3 4 1 or 2 4 1 3 or 3 4 1 2. For example, in the second case it gives 1 4 3 4 as the successor of 1 2 3 4; in the third case it gives 3 1 2 4 as the successor of 3 1 4 2.) Is there a nice way to characterize the set of all p for which this simple program generates the admissible permutations correctly?

Notice that when the elements $\{a_1, \ldots, a_n\}$ being permuted are all 0s and 1s, the "permutations" can be interpreted as *subsets* of $\{1, \ldots, n\}$. Thus, for example, the extended algorithm at the end of the paper will produce the thirty answers

```
0 0 0 0 0 0 1 1 1 1 1 1        0 1 0 0 0 1 0 1 1 0 1 1
0 0 0 0 0 1 0 1 1 1 1 1        0 1 0 0 1 1 0 0 0 1 1 1
0 0 0 0 1 1 0 0 1 1 1 1        0 1 0 0 1 1 0 0 1 0 1 1
0 0 0 0 1 1 0 1 1 0 1 1        0 1 0 0 1 1 0 1 1 0 0 1
0 0 0 1 0 0 0 1 1 1 1 1        0 1 0 1 0 0 0 0 1 1 1 1
0 0 0 1 0 1 0 0 1 1 1 1        0 1 0 1 0 0 0 1 1 0 1 1
0 0 0 1 0 1 0 1 1 0 1 1        0 1 0 1 0 1 0 0 0 1 1 1
0 0 0 1 1 1 0 0 0 1 1 1        0 1 0 1 0 1 0 0 1 0 1 1
0 0 0 1 1 1 0 0 1 0 1 1        0 1 0 1 0 1 0 1 1 0 0 1
0 0 0 1 1 1 0 1 1 0 0 1        0 1 0 1 1 1 0 0 0 0 1 1
0 0 1 1 1 1 0 0 0 0 1 1        0 1 0 1 1 1 0 0 1 0 0 1
0 0 1 1 1 1 0 0 1 0 0 1        0 1 0 1 1 1 0 1 1 0 0 0
0 0 1 1 1 1 0 1 1 0 0 0        0 1 1 1 1 1 0 0 0 0 0 1
0 1 0 0 0 0 0 1 1 1 1 1        0 1 1 1 1 1 0 0 1 0 0 0
0 1 0 0 0 1 0 0 1 1 1 1        1 1 1 1 1 1 0 0 0 0 0 0
```

which represent the order ideals that have cardinality 6 when '$<$' is replaced by '$\le$' in (1).

Chapter 8

Nested Satisfiability

[Originally published in Acta Informatica **28** *(1990), 1–6.]*

A special case of the satisfiability problem, in which the clauses have a hierarchical structure, is shown to be solvable in linear time, assuming that the clauses have been represented in a convenient way.

Let X be a finite alphabet linearly ordered by $<$; we will think of the elements of X as Boolean variables. As usual, we define the *literals* over X to be elements of the form x or $\overline{x}$, where $x \in X$. Literals that belong to X are called positive; the others are called negative.

The linear ordering of X can be extended to a linear preordering of all its literals in a natural way if we simply disregard the signs. For example, if $X = \{a, b, c\}$ has the usual alphabetic order, we have

$$a \equiv \overline{a} < b \equiv \overline{b} < c \equiv \overline{c}\,.$$

If σ and τ are literals, we write $\sigma \le \tau$ if $\sigma < \tau$ or $\sigma \equiv \tau$; this relation holds if and only if the relation $\sigma > \tau$ is false.

A *clause* over X is a set of literals on distinct variables. Thus, the literals of a clause can be written in increasing order,

$$\sigma_1 < \sigma_2 < \cdots < \sigma_k\,.$$

A set $\mathcal{C}$ of clauses over X is *satisfiable* if there exists a clause over X that has a nonempty intersection with every clause in $\mathcal{C}$. For example, the clauses

$$\{a, \overline{b}, c\} \qquad \{\overline{a}, \overline{c}\} \qquad \{\overline{a}, b, c\} \qquad \{\overline{b}, \overline{c}\} \qquad \{a, b\}$$

over $\{a, b, c\}$ are satisfiable uniquely by the clause $\{a, b, \overline{c}\}$.

We say that clause C *straddles* clause C' if there are literals σ, τ in C and ξ' in C' such that

$$\sigma < \xi' < \tau .$$

Two clauses *overlap* if they straddle each other. For example, $\{a, \bar{b}, c\}$ and $\{\bar{a}, b, c\}$ overlap; but the other nine pairs of clauses in the example above are nonoverlapping. Clauses on two elements each, like $\{a, c\}$ and $\{b, d\}$, can also be overlapping. A set of clauses in which no two overlap is called *nested.*

The general problem of deciding whether a given set of clauses is satisfiable is well known to be NP-complete. But we will see that the analogous question for nested clauses is efficiently decidable. The main reason for interest in nested clauses is David Lichtenstein's theorem of *planar satisfiability*, which can be restated in algebraic terms as follows: The joint satisfiability problem for *two* sets $\mathcal{C}, \mathcal{C}'$ of nested clauses is NP-complete. In fact, Lichtenstein proved [1, page 339] that this problem is NP-complete even if all clauses of $\mathcal{C}$ contain only positive literals and all clauses of $\mathcal{C}'$ contain only negative literals, with at most three literals per clause.

1. Structure of Nested Clauses

A clause over an ordered alphabet has a least literal σ and a greatest literal τ. Any variable that lies strictly between σ and τ is said to be *interior* to that clause. A variable can occur as an interior literal at most once in a set of nested clauses; for if it is an interior literal in two different clauses, those clauses overlap. Hence, the total number of elements among m nested clauses on n variables is at most $2m + n$.

Let us write $C \succ C'$ if C straddles C' but C' does not straddle C. This relation is transitive. For if $C \succ C'$ and C' straddles C'', we have literals

$$\sigma < \xi' < \tau , \qquad \sigma' < \xi'' < \tau'$$

in appropriate clauses; and we must have $\sigma \le \sigma'$ and $\tau' \le \tau$, or else C' would straddle C. Hence C straddles C''. Similarly if $C' \succ C''$ and C'' straddles C, then C' straddles C. Therefore $C \succ C' \succ C''$ implies that $C \succ C''$.

In a set of nested clauses, we have $C \succ C'$ if and only if C straddles C'. The transitivity of this relation implies that we can topologically sort any set of nested clauses into a linear arrangement in which each clause appears after every clause that it straddles. When such an arrangement is given, and when the elements of each clause are presented in order, we will show that satisfiability can be decided in $O(m+n)$ steps on a RAM (a random access machine), where m is the number of clauses and n is the number of variables.

(Incidentally, a set of nested clauses can be shown to have a treelike structure, although we do not need this characterization in the algorithm. Let us write $C \le C'$ if $\sigma \le \tau'$ for all $\sigma \in C$ and $\tau' \in C'$. If neither C nor C' straddles the other, it is easy to see that we must have either $C \le C'$ or $C' \le C$, unless C and C' are both clauses on the same two literals. Suppose we call such 2-element clauses equivalent. Then a nested set of clauses will satisfy the condition

$$(C \succ C'' \text{ and } C' \succ C'') \text{ implies } (C \succ C' \text{ or } C \equiv C' \text{ or } C' \succ C),$$

because we cannot have $C \succ C''$ and $C' \succ C''$ when $C \le C'$. This means that $\succ$ is the ancestor relation in a hierarchy.)

2. An Algorithm

Let us assume that the alphabet X is represented as the positive integers $\{1, 2, \ldots, n\}$, with $\overline{x} = -x$. The clauses will be specified in two arrays

$$\mathit{lit}[1 \,..\, 2m+n] \quad \text{and} \quad \mathit{start}[1 \,..\, m+1],$$

where the literals of clause i are

$$\mathit{lit}[j] \qquad \text{for } \mathit{start}[i] \le j < \mathit{start}[i+1],$$

in increasing order as j increases. The clauses are assumed to be arranged so that clause i does not straddle clause i' when $i < i'$. We can safely assume that all clauses contain at least two literals.

The key idea of the algorithm below is that the interior variables of a clause are not present in subsequent clauses. Therefore we only need to remember information about the dynamically changing set of all variables

$$1 = x_1 < x_2 < \cdots < x_k = n$$

that have not yet appeared as interior variables. Initially $k = n$ and $x_j = j$ for $1 \le j \le k$.

The set of all clauses seen so far, as the algorithm proceeds to consider the clauses in turn, can be conceptually partitioned into intervals

$$[x_1 \,..\, x_2],\ [x_2 \,..\, x_3],\ \ldots,\ [x_{k-1} \,..\, x_k],$$

such that all literals of each previously processed clause belong to one of these intervals. The current intervals are maintained in an array

$$next[1\,..\,n],$$

where $next[x_j] = x_{j+1}$ for $1 \le j < k$.

The only slightly complex data structure in the algorithm below is the array

$$sat[1\,..\,n, boolean, boolean]$$

which has the following interpretation: If $[x_j\,..\,x_{j+1}]$ is an interval of the current partition, then $sat[x_j, s, t]$ will be either 0 or 1 for each pair $s, t \in \{false, true\}$. It is 1 if and only if the clauses already processed, belonging to the interval $[x_j\,..\,x_{j+1}]$, are satisfiable by clauses in which the least and greatest literals are respectively $x_j|s$ and $x_{j+1}|t$, where

$$x|s = \begin{cases} -x\,, & s = false;\\ +x\,, & s = true. \end{cases}$$

For example, suppose we have seen only one clause, $\{1, -2\}$. Then we will have

$$sat[1, false, true] = 0;$$
$$sat[1, false, false] = sat[1, true, false] = sat[1, true, true] = 1\,.$$

It turns out that the *sat* array contains all the information necessary to continue processing, because literals that have appeared as interior variables will not be present in subsequent clauses.

The algorithm's main task is to maintain the *sat* array as it examines a new clause $C_i = \{\sigma_1, \ldots, \sigma_q\}$. The variables $|\sigma_1| < \cdots < |\sigma_q|$ will be a subset of the current partition variables $x_1, \ldots, x_k$. All of the current partition variables between $|\sigma_1|$ and $|\sigma_q|$, whether they appear in the new clause or not, are interior to the clause, so they will be removed.

Suppose $|\sigma_1| = x_p$. The algorithm proceeds by letting a variable x run through the values $x_p, x_{p+1}, \ldots, |\sigma_q|$, maintaining information needed to update the values of $sat[x_p, s, t]$ when the interior variables of C_i are eliminated from the partition. Let $C_i(x)$ be the literals of C_i that are strictly less than x, and let $\mathcal{C}(x)$ be the clauses preceding C_i whose literals are confined to the interval $[x_p\,..\,x]$. The updating process is carried out by computing auxiliary values $newsat_x[s, t]$ defined as follows:

$$newsat_x[s,t] = \begin{cases} 0, & \text{if } \mathcal{C}(x) \text{ is not satisfiable}(s,t);\\ 1, & \text{if } \mathcal{C}(x) \text{ is satisfiable}(s,t) \text{ but } \mathcal{C}(x) \cup \{C_i(x)\} \text{ isn't};\\ 2, & \text{if } \mathcal{C}(x) \cup \{C_i(x)\} \text{ is satisfiable}(s,t). \end{cases}$$

Here "satisfiable(s,t)" means that there's a clause containing $x_p|s$ and $x|t$ that has a nonempty intersection with each clause of the given set of clauses.

For example, suppose we have $C_i = \{-1,2,4\}$ and $\{x_1,x_2,x_3,x_4\} = \{1,2,3,4\}$, and suppose that the clauses $C_1, \ldots, C_{i-1}$ have led to the following values:

s	t	$sat[1,s,t]$	$sat[2,s,t]$	$sat[3,s,t]$
false	*false*	0	0	0
false	*true*	1	1	0
true	*false*	1	1	0
true	*true*	1	0	1

Then we have

s	t	$newsat_1[s,t]$	$newsat_2[s,t]$	$newsat_3[s,t]$	$newsat_4[s,t]$
false	*false*	1	0	2	0
false	*true*	0	2	0	0
true	*false*	0	1	2	0
true	*true*	1	1	1	1

and we will want to update the arrays by setting $next[1] \leftarrow 4$ and

$$sat[1, \mathit{false}, \mathit{false}] \leftarrow 0\,;$$
$$sat[1, \mathit{false}, \mathit{true}] \leftarrow 0\,;$$
$$sat[1, \mathit{true}, \mathit{false}] \leftarrow 0\,;$$
$$sat[1, \mathit{true}, \mathit{true}] \leftarrow 1\,.$$

If C_i were $\{-1,2,-4\}$ instead of $\{-1,2,4\}$, the computation of *newsat* would be the same, but the values of $sat[1,s,t]$ would all become 0; the clauses would be unsatisfiable, since $newsat_4[\mathit{true},\mathit{true}]$ is only 1, not 2. (The reader is encouraged to study this example carefully, because it reveals the key principles underlying the algorithm.)

3. Programming Details

It is convenient to assume that an artificial $(m+1)$st clause with the dummy variables $\{0, n+1\}$ has been added after C_m. Therefore we will declare arrays that are slightly larger than stated earlier:

$$lit[1\,..\,2m+n+2]\,;\; start[1\,..\,m+2]\,;$$
$$next[0\,..\,n]\,;\;\; sat[0\,..\,n, \mathit{boolean}, \mathit{boolean}]\,.$$

There also are two auxiliary arrays called $newsat[\mathit{boolean},\mathit{boolean}]$ and $tmp[\mathit{boolean},\mathit{boolean}]$. We can now decide the nested satisfiability problem as follows:

```
for x ← 0 to n do next[x] ← x + 1;
for x ← 0 to n do for s ← false to true do
        for t ← false to true do sat[x, s, t] ← 1;
for i ← 1 to m + 1 do
    begin l ← abs(lit[start[i]]);  r ← abs(lit[start[i + 1] − 1]);
    ⟨Compute the newsat table⟩;  next[l] ← r;
    for s ← false to true do for t ← false to true do
        sat[l, s, t] ← newsat[s, t] div 2;
    end;
if sat[0, true, true] = 1 then print("Satisfiable")
else print("Unsatisfiable").
```

The example in the previous section illustrates how the *newsat* table can be computed in general. We run the process slightly longer so that a good *newsat* value will be 2 (not 1) at the end. (The value of σ_q must be examined.)

```
⟨Compute the newsat table⟩ =
    j ← start[i]; sig ← lit[j]; x ← abs(sig);
    newsat[false, false] ← 1; newsat[true, true] ← 1;
    newsat[false, true] ← 0; newsat[true, false] ← 0;
    while true do
        begin if x = abs(sig) then
            begin ⟨Upgrade a newsat from 1 to 2, if possible⟩;
            j ← j + 1;  sig ← lit[j];
            if j = start[i + 1] then goto done;
            end;
        ⟨Modify newsat for the next x value⟩;
        x ← next[x];
        end;
    done:

⟨Upgrade a newsat from 1 to 2, if possible⟩ =
    t ← (x = sig);
    for s ← false to true do
        if newsat[s, t] = 1 then newsat[s, t] ← 2

⟨Modify newsat for the next x value⟩ =
    for s ← false to true do for t ← false to true do
        tmp[s, t] ← max(newsat[s, false] × sat[x, false, t],
                        newsat[s, true] × sat[x, true, t]);
    for s ← false to true do for t ← false to true do
            newsat[s, t] ← tmp[s, t]
```

The running time is $O(m+n)$, because each value of x is either first or last in the current clause (accounting for $2(m+1)$ cases) or it is being permanently removed from the partition (in exactly n cases, because of the dummy clause $\{0, n+1\}$ at the end).

We have not considered here the time that might be required to test if a given satisfiability problem is, in fact, nested under some ordering of its variables.

Concluding Remarks

This algorithm for nested satisfiability works by essentially replacing each clause by a clause containing only two literals, using a special form of "dynamic 2SAT" to justify the replacement. However, the instances of 2SAT that arise are not completely general. A somewhat larger special case of the satisfiability problem might therefore be solvable in linear time by similar techniques.

Acknowledgments

I wish to thank Andrew Goldberg for posing the problem of nested satisfiability during a conversation about Lichtenstein's theorem, and I wish to thank the referees for their helpful remarks.

This research was supported in part by National Science Foundation grant CCR-8610181.

References

[1] David Lichtenstein, "Planar formulæ and their uses," *SIAM Journal on Computing* **11** (1982), 329–343.

Addendum

Jan Kratochvíl and Mirko Křivánek, in "Satisfiability of co-nested formulas," *Acta Informatica* **30** (1993), 397–403, extended this work in several interesting directions. First, they showed that satisfiability can be tested efficiently also when variables are nested with respect to clauses: The clauses, not the variables, are linearly ordered, and the set of clauses in which variable x or $\bar{x}$ appears should not overlap the set of clauses in which another variable y or $\bar{y}$ appears. Second, they showed that it's possible to solve the *maximum satisfiability* problem for nested or co-nested clauses, namely to find the largest number of clauses that are simultaneously satisfiable, in time $O(m+n)$.

Kratochvíl and Křivánek also observed that a linear-time algorithm is available to test whether any given set of clauses is nested with respect to *some* linear ordering of the variables: Construct the bipartite graph with vertices X and $\mathcal{C}$, where edge x — C exists for $x \in X$ and $C \in \mathcal{C}$ if and only if $x \in C$ or $\bar{x} \in C$. Now add another vertex v, which is adjacent to all $x \in X$. This graph is planar if and only if the clauses of $\mathcal{C}$ are nested, using the order in which the vertices of X appear around v.

See also Pierre Hansen, Brigitte Jaumard, and Gérard Plateau, "An extension of nested satisfiability," RUTCOR Technical Report RRR-29-93 (New Brunswick, New Jersey: Rutgers University, 1993), ii + 13 pages.

Chapter 9

Fast Pattern Matching in Strings

[Written with James H. Morris, Jr. and Vaughan R. Pratt. Originally published in SIAM Journal on Computing **6** *(1977), 323–350.]*

An algorithm is presented that finds all occurrences of one given string within another, in running time proportional to the sum of the lengths of the strings. The constant of proportionality is low enough to make this algorithm of practical use, and the procedure can also be extended to deal with some more general pattern-matching problems. A theoretical application of the algorithm shows that the set of concatenations of even palindromes, namely the language $\{\alpha\alpha^R\}^*$*, can be recognized in linear time. Other algorithms that run even faster on the average are also considered.*

Text-editing programs are often required to search through a string of characters looking for instances of a given "pattern" string; we wish to find all positions, or perhaps only the leftmost position, in which the pattern occurs as a contiguous substring of the text. For example, the string $catenary$ contains the pattern ten, but we do not regard $canary$ as a substring.

The obvious way to search for a matching pattern is to try searching at every starting position of the text, abandoning the search as soon as an incorrect character is found. But this approach can be very inefficient, for example when we are looking for an occurrence of $aaaaaaab$ in $aaaaaaaaaaaaaab$. When the pattern is $a^n b$ and the text is $a^{2n}b$, we will find ourselves making $(n+1)^2$ comparisons of characters. Furthermore, the traditional approach involves "backing up" the input text as we go through it, and this can add annoying complications when we consider the buffering operations that are frequently involved.

In this paper we describe a pattern-matching algorithm that finds all occurrences of a pattern of length m within a text of length n in $O(m+n)$ units of time, without "backing up" the input text. The algorithm needs only $O(m)$ locations of internal memory if the text is read from an external file, and at most $O(\log m)$ units of time elapse between consecutive single-character inputs. All of the constants of proportionality implied by these "O" formulas are independent of the alphabet size.

We shall first consider the algorithm in a conceptually simple but somewhat inefficient form. Sections 3 and 4 of this paper discuss some ways to improve the efficiency and to adapt the algorithm to other problems. Section 5 develops the underlying theory, and Section 6 uses the algorithm to disprove the conjecture that a certain context-free language cannot be recognized in linear time. Section 7 discusses the origin of the algorithm and its relation to other recent work. Finally, Section 8 discusses still more recent work on pattern matching.

1. Informal Development

The idea behind this approach to pattern matching is perhaps easiest to grasp if we imagine placing the pattern over the text and sliding it to the right in a certain way. Consider for example a search for the pattern *abcabcacab* in the text *babcbabcabcaabcabcabcacabc*; initially we place the pattern at the extreme left and prepare to scan the leftmost character of the input text:

$$\begin{array}{l} a\,b\,c\,a\,b\,c\,a\,c\,a\,b \\ b\,a\,b\,c\,b\,a\,b\,c\,a\,b\,c\,a\,a\,b\,c\,a\,b\,c\,a\,b\,c\,a\,c\,a\,b\,c \\ \uparrow \end{array}$$

The arrow here indicates the current text character; since it points to b, which doesn't match the a above it, we shift the pattern one space right and move to the next input character:

$$\begin{array}{l} a\,b\,c\,a\,b\,c\,a\,c\,a\,b \\ b\,a\,b\,c\,b\,a\,b\,c\,a\,b\,c\,a\,a\,b\,c\,a\,b\,c\,a\,b\,c\,a\,c\,a\,b\,c \\ \uparrow \end{array}$$

Now we have a match, so the pattern stays put while the next several characters are scanned. Soon we come to another mismatch:

$$\begin{array}{l} a\,b\,c\,a\,b\,c\,a\,c\,a\,b \\ b\,a\,b\,c\,b\,a\,b\,c\,a\,b\,c\,a\,a\,b\,c\,a\,b\,c\,a\,b\,c\,a\,c\,a\,b\,c \\ \uparrow \end{array}$$

At this point we have matched the first three pattern characters but not the fourth, so we know that the last four characters of the input have been $abcx$ where $x \neq a$. We don't have to remember the previously scanned characters, since *our position in the pattern yields enough information to recreate them.* In this case, no matter what x is (as long as it's not a), we deduce that the pattern can immediately be shifted four more places to the right; one, two, or three shifts couldn't possibly lead to a match.

Before long we get to another partial match, this time with a failure on the eighth pattern character:

```
     abcabcacab
babcbabcabcaabcabcabcacabc
            ↑
```

Now we know that the last eight characters seen were *abcabcax*, where $x \neq c$. The pattern should therefore be shifted three places to the right:

```
        abcabcacab
babcbabcabcaabcabcabcacabc
            ↑
```

We try to match the new pattern character, but this fails too; so we shift the pattern four (not three or five) more places. That produces a match, and we continue scanning until reaching *another* mismatch on the eighth pattern character:

```
            abcabcacab
babcbabcabcaabcabcabcacabc
                   ↑
```

Again we shift the pattern three places to the right. This time a match is produced, and we eventually discover the full pattern:

```
               abcabcacab
babcbabcabcaabcabcabcacabc
                         ↑
```

The play-by-play description for this example indicates that the pattern-matching process will run efficiently if we have an auxiliary table that tells us exactly how far to slide the pattern, when we detect a mismatch at its jth character, $pattern[j]$. Let $next[j]$ be the character position in the pattern that should be checked next after such a mismatch, so that we are sliding the pattern $j - next[j]$ places relative to the text. The following table lists the appropriate values:

$j =$	1	2	3	4	5	6	7	8	9	10
$pattern[j] =$	a	b	c	a	b	c	a	c	a	b
$next[j] =$	0	1	1	0	1	1	0	5	0	1

(Note that when $next[j] = 0$, we are supposed to slide the pattern all the way past the current text character.) We shall discuss how to precompute this table later; fortunately, the calculations are quite simple, and we will see that they require only $O(m)$ steps.

At each step of the scanning process, we move either the text pointer or the pattern, and each of these can move at most n times. Thus at most $2n$ steps need to be performed, after the *next* table has been set up. Of course the pattern itself doesn't really move; we can do the necessary operations simply by maintaining the pointer variable j.

2. Programming the Algorithm

The pattern-matching process has the general form

```
⟨Place pattern at left⟩;
while ⟨pattern not fully matched⟩ and ⟨text not exhausted⟩ do
  begin while ⟨pattern character differs from
                              current text character⟩ do
    ⟨Shift pattern appropriately⟩;
  ⟨Advance to next character of text⟩;
  end.
```

For convenience, let us assume that the input text is present in an array *text* $[1\,..\,n]$, and that the pattern appears in *pattern* $[1\,..\,m]$. We shall also assume that $m > 0$; in other words, the pattern must be nonempty. Let k and j be integer variables such that *text* $[k]$ denotes the current text character and *pattern* $[j]$ denotes the corresponding pattern character; the pattern is essentially aligned with positions $p + 1$ through $p + m$ of the text, where $k = p + j$. Then the program above takes the following simple form:

```
j := k := 1;
while j ≤ m and k ≤ n do
  begin while j > 0 and text[k] ≠ pattern[j] do
    j := next[j];
  k := k + 1;  j := j + 1;
  end.
```

(The **and** operation here is the "conditional and," which does not evaluate the relation '*text* $[k] \neq$ *pattern* $[j]$' unless $j > 0$.) If $j > m$ at the conclusion of the program, the leftmost match has been found in positions $k - m$ through $k - 1$; but if $j \leq m$, the text has been exhausted.

The program has a curious feature, namely that the inner loop operation '$j :=$ *next* $[j]$' is performed no more often than the outer loop operation '$k := k + 1$'; in fact, the inner loop is usually performed somewhat *less* often, since the pattern generally moves right less frequently than the text pointer does.

To prove rigorously that the program above is correct, we may use the following invariant relation: "Let $p = k - j$ be the position in the text just preceding the first character of the pattern, in our assumed alignment. Then we have $\mathit{text}[p+i] = \mathit{pattern}[i]$ for $1 \le i < j$ (that is, we have matched the first $j-1$ characters of the pattern, if $j > 0$); but for each t with $0 \le t < p$ we have $\mathit{text}[t+i] \ne \mathit{pattern}[i]$ for some i with $1 \le i \le m$ (that is, there is no possible match of the entire pattern to the left of p)."

The program will of course be correct only if we can compute the *next* table so that the stated relation remains invariant when we perform the operation '$j := \mathit{next}[j]$'. Let us look at that computation now. When the program changes j to $\mathit{next}[j]$ we know that $j > 0$, and that the last j characters of the input up to and including $x = \mathit{text}[k]$ were

$$\mathit{pattern}[1] \ldots \mathit{pattern}[j-1]\, x$$

where $x \ne \mathit{pattern}[j]$. What we want is to find the least amount of shift for which these characters can possibly match the shifted pattern; in other words, we want $\mathit{next}[j]$ to be the largest i less than j such that the last i characters of the input were

$$\mathit{pattern}[1] \ldots \mathit{pattern}[i-1]\, x$$

and $\mathit{pattern}[i] \ne \mathit{pattern}[j]$. (If no such i exists, we let $\mathit{next}[j] = 0$.) With this definition of $\mathit{next}[j]$ it is easy to verify that

$$\mathit{text}[t+1] \ldots \mathit{text}[k] \ne \mathit{pattern}[1] \ldots \mathit{pattern}[k-t]$$

for $k - j \le t < k - \mathit{next}[j]$. Hence the stated relation is indeed invariant, and our program is indeed correct.

Now we must face up to the problem that we have been postponing, the task of calculating $\mathit{next}[j]$ in the first place. This problem would be easier if we didn't require $\mathit{pattern}[i] \ne \mathit{pattern}[j]$ in the definition of $\mathit{next}[j]$, so we shall consider the easier problem first. Let $f[j]$ be the largest i less than j such that

$$\mathit{pattern}[1] \ldots \mathit{pattern}[i-1] \;=\; \mathit{pattern}[j-i+1] \ldots \mathit{pattern}[j-1].$$

This condition holds vacuously for $i = 1$, so we always have $f[j] \ge 1$ when $j > 1$. By convention we let $f[1] = 0$. The pattern used in the example of §1 has the following f table:

j =	1	2	3	4	5	6	7	8	9	10
$\mathit{pattern}[j]$ =	a	b	c	a	b	c	a	c	a	b
$f[j]$ =	0	1	1	1	2	3	4	5	1	2

If $f[j] > 0$ and $\mathit{pattern}[j] = \mathit{pattern}[f[j]]$ then $f[j+1] = f[j] + 1$; but if not, we can use essentially the same pattern-matching algorithm as above to compute $f[j+1]$, with $text = pattern$! (Notice the similarity of the $f[j]$ problem to the invariant condition of the matching algorithm. Our program calculates the largest j less than or equal to k such that $pattern[1] \ldots pattern[j-1] = text[k-j+1] \ldots text[k-1]$, so we can transfer the previous technology to the present problem.)

The following program will compute $f[j+1]$, assuming that $f[j]$ and $next[1], \ldots, next[j-1]$ have already been calculated:

```
t := f[j];
while t > 0 and pattern[j] ≠ pattern[t] do
   t := next[t];
f[j + 1] := t + 1;
```

The correctness of this program is demonstrated as before; we can imagine two copies of the pattern, one sliding to the right with respect to the other. For example, suppose we have established that $f[8] = 5$ in the case above; let us consider the computation of $f[9]$:

```
      a b c a b c a c a b
  a b c a b c a c a b
                ↑
```

Since $pattern[8] \neq b$, we shift the upper copy to the right, knowing that the most recently scanned characters of the lower copy were $abcax$ for $x \neq b$. The *next* table tells us to shift right four places, obtaining

```
              a b c a b c a c a b
  a b c a b c a c a b
                ↑
```

and again there is no match. The next shift makes $t = 0$, so $f[9] = 1$.

Once we understand how to compute $f[j]$, it is only a short step to the computation of $next[j]$. A comparison of the definitions shows that, for $j > 1$,

$$next[j] = \begin{cases} f[j], & \text{if } pattern[j] \neq pattern[f[j]]; \\ next[f[j]], & \text{if } pattern[j] = pattern[f[j]]. \end{cases}$$

Therefore we can compute the *next* table as follows, without ever storing the values of $f[j]$ in memory.

```
j := 1;  t := 0;  next[1] := 0;
while j < m do
   begin comment t = f[j];
      while t > 0 and pattern[j] ≠ pattern[t] do
         t := next[t];
      t := t + 1;  j := j + 1;
```

```
      if pattern[j] = pattern[t] then next[j] := next[t]
      else next[j] := t;
   end.
```

This program takes $O(m)$ units of time, for the same reason that the matching program takes $O(n)$: The operation '$t := next[t]$' in the innermost loop always shifts the upper copy of the pattern to the right, so it is performed a total of m times at most. (A slightly different way to prove that the running time is bounded by a constant times m is to observe that the variable t starts at 0 and it is increased, $m-1$ times, by 1; furthermore its value remains nonnegative. Therefore the operation '$t := next[t]$', which always decreases t, can be performed at most $m-1$ times.)

To summarize what we have said so far: Strings of text can be scanned efficiently by making use of two ideas. We can precompute "shifts," specifying how to move the given pattern with respect to a text when a mismatch occurs at the pattern's jth character; and this precomputation of "shifts" can be performed efficiently by using the same principle, shifting the pattern against itself.

3. Gaining Efficiency

We have presented the pattern-matching algorithm in a form that is rather easily proved correct; but as so often happens, this form is not very efficient. In fact, the algorithm as presented above would probably not be competitive with the naïve algorithm on realistic data, even though the naïve algorithm has a worst-case time of order m times n instead of m plus n, because the chance of this worst case is rather slim. A well-implemented form of the new algorithm should, however, go noticeably faster, because there is no backing up after a partial match.

It is not difficult to see the source of inefficiency in the new algorithm as presented in §2: When the alphabet of characters is large, we will rarely have a partial match, and the program will waste a lot of time discovering rather awkwardly that $text[k] \neq pattern[1]$ for $k = 1, 2, 3, \ldots$. When $j = 1$ and $text[k] \neq pattern[1]$, the algorithm sets $j := next[1]$, then discovers that $j = 0$, then increases k by 1, then sets j to 1 again, then tests whether or not $1 \leq m$, and later it tests whether or not $1 > 0$. Clearly we would be much better off making $j = 1$ into a special case.

The algorithm also spends unnecessary time testing whether $j > m$ or $k > n$. A fully matched pattern can be accounted for by setting $pattern[m+1] :=$ '@' for some impossible character @ that will never be matched, and by initializing $next[m+1] := -1$; then a test for $j < 0$ can be inserted into a less frequently executed part of the code. Similarly we can for example set $text[n+1]$:= '⊥' (another impossible character) and $text[n+2] := pattern[1]$, so that the test for $k > n$ needn't be made very often. (See [17] for a discussion of such more or less mechanical transformations on programs.)

The following form of the algorithm incorporates these refinements.

```
    a := pattern[1]; pattern[m+1] := '@'; next[m+1] := −1;
    k := 1; text[n+1] := '⊥'; text[n+2] := a;
get started: j := 1;
quick test: while text[k] ≠ a do k := k + 1;
end check: if k > n then go to input exhausted;
char matched: j := j + 1; k := k + 1;
loop: comment j > 0;
    if text[k] = pattern[j] then go to char matched;
    j := next[j];
    if j = 1 then go to quick test;
    if j = 0 then
        begin k := k + 1; go to get started;
        end;
    if j > 0 then go to loop;
    comment text[k − m] through text[k − 1] have been matched;
```

This program will usually run faster than the naïve algorithm; the worst case occurs when trying to find the pattern ab in a long string of a's. Similar ideas can be used to speed up the program that prepares the *next* table.

In a text editor the patterns are usually short, so that it is most efficient to translate the pattern directly into machine-language code that implicitly contains the *next* table (see Thompson [24] and Gosper [3, Hack 179]). For example, the pattern in §1 could be compiled into the machine-language equivalent of

```
L0: k := k + 1;
L1: if text[k] ≠ 'a' then go to L0;
    k := k + 1; if k > n then go to input exhausted;
L2: if text[k] ≠ 'b' then go to L1;
    k := k + 1;
L3: if text[k] ≠ 'c' then go to L1;
    k := k + 1;
L4: if text[k] ≠ 'a' then go to L0;
    k := k + 1;
L5: if text[k] ≠ 'b' then go to L1;
```

```
        k := k + 1;
L6:  if text[k] ≠ 'c' then go to L1;
        k := k + 1;
L7:  if text[k] ≠ 'a' then go to L0;
        k := k + 1;
L8:  if text[k] ≠ 'c' then go to L5;
        k := k + 1;
L9:  if text[k] ≠ 'a' then go to L0;
        k := k + 1;
L10: if text[k] ≠ 'b' then go to L1;
        k := k + 1;
```

This expanded version will be slightly faster, since it essentially makes a special case for *all* values of j.

Curiously, people often think that the new algorithm will be slower than the naïve one, even though it does less work. Since the new algorithm is conceptually hard to understand at first, by comparison with other algorithms of the same length, we feel somehow that a computer will have conceptual difficulties too — we expect the machine to run more slowly when it gets to such subtle instructions!

4. Extensions

So far our programs have only been concerned with finding the leftmost match. However, it is easy to see how to modify the routine so that all matches are found in turn: We can calculate the *next* table for the extended pattern of length $m+1$ using $pattern[m+1] :=$ '@', and then we set $resume := next[m+1]$ before setting $next[m+1] := -1$. After finding a match and doing whatever action is desired to process that match, we can say

$j := resume$; **go to** loop;

this sequence will restart things properly. (We assume that *text* has not changed in the meantime. Note that *resume* cannot be zero.)

Another approach would be to leave $next[m+1]$ untouched in the extended pattern, never resetting it to -1. Instead, we can define integer arrays $head[1\,..\,m]$ and $link[1\,..\,n]$, initially zero, and insert the following code at label 'char matched':

$link[k] := head[j]$; $head[j] := k$;

The test '**if** $j > 0$ **then**' is also removed from the program. This extension forms linked lists for $1 \le j \le m$ of all places where the first j characters of the pattern (but no more than j) are matched in the input.

Still another straightforward modification will find the longest initial match of the pattern, namely the maximum j such that the substring $pattern[1] \ldots pattern[j]$ occurs in $text$.

In practice, the text characters are often packed into words, with say b characters per word, and the machine architecture often makes it inconvenient to access individual characters. When efficiency for large n is important on such machines, one alternative is to carry out b independent searches, one for each possible alignment of the pattern's first character in the word. These searches can treat *entire words* as "supercharacters," with appropriate masking, instead of working with individual characters and unpacking them. Since the algorithm we have described does not depend on the size of the alphabet, it is well suited to this and similar alternatives.

Sometimes we want to match two or more patterns in sequence, finding an occurrence of the first one followed by the second, etc. This problem is easily handled by consecutive searches, and the total running time will be of order n plus the sum of the individual pattern lengths.

We might also want to match two or more patterns in parallel, stopping as soon as any one of them is fully matched. A search of this kind could be done with multiple *next* and *pattern* tables, with one j pointer for each; but that would make the running time kn plus the sum of the pattern lengths, when there are k patterns. Hopcroft and Karp have observed (unpublished) that our pattern-matching algorithm can be extended so that the running time for simultaneous searches is proportional simply to n, plus the alphabet size times the sum of the pattern lengths. The patterns are combined into a "trie" whose nodes represent all of the initial substrings of one or more patterns, and whose branches specify the appropriate successor node as a function of the next character in the input text. For example, if there are four patterns $\{abcab, ababc, bcac, bbc\}$, the trie is shown in Fig. 1.

Such a trie can be constructed efficiently by generalizing the idea we used to calculate $next[j]$. Details and further refinements have been discussed by Aho and Corasick [1], who discovered the algorithm independently; see also Gosper [3, Hack 179].

Notice that this trie algorithm depends on the alphabet size. Such dependence is inherent, if we wish to keep the coefficient of n independent of k, since for example the k patterns might each consist of a single unique character of the alphabet.

It is interesting to compare this approach to what happens when the LR(0) parsing algorithm [16] is applied to the regular context-free grammar $S \to aS \mid bS \mid cS \mid abcab \mid ababc \mid bcac \mid bbc$.

node	substring	**if** a	**if** b	**if** c
0		1	7	0
1	a	1	2	0
2	ab	3	8	5
3	aba	1	4	0
4	$abab$	3	8	$ababc$
5	abc	6	7	0
6	$abca$	1	$abcab$	$bcac$
7	b	1	8	9
8	bb	1	8	bbc
9	bc	10	7	0
10	bca	1	2	$bcac$

FIGURE 1. Searching for any of four different patterns.

5. Theoretical Considerations

If the input file is being read in "real time," we might object to long delays between consecutive inputs. In this section we shall prove that the number of times '$j := next[j]$' is performed, before k is advanced, is bounded by a function of the approximate form $\log_\phi m$, where $\phi = (1+\sqrt{5})/2 \approx 1.618\ldots$ is the golden ratio, and that this bound is best possible. We shall use lowercase Latin letters to represent characters and lowercase Greek letters α, β, ... to represent strings, with ϵ the empty string and $|\alpha|$ the length of α. Thus $|a| = 1$ for all characters a; $|\alpha\beta| = |\alpha| + |\beta|$; and $|\epsilon| = 0$. We also write $\alpha[k]$ for the kth character of α, when $1 \le k \le |\alpha|$.

As a warmup for our theoretical discussion, let us consider the *Fibonacci strings* [14, exercise 1.2.8–36], which turn out to be especially pathological patterns for the algorithm above. The definition of Fibonacci strings is

$$\phi_1 = b, \quad \phi_2 = a; \quad \phi_n = \phi_{n-1}\phi_{n-2} \quad \text{for } n \ge 3. \tag{1}$$

For example, $\phi_3 = ab$, $\phi_4 = aba$, $\phi_5 = abaab$. It follows that the length $|\phi_n|$ is the nth Fibonacci number F_n, and that ϕ_n consists of the first F_n characters of an infinite string ϕ_∞ when $n \ge 2$.

Consider the pattern ϕ_8, which has the functions $f[j]$ and $next[j]$ shown in Table 1.

TABLE 1. Auxiliary functions for a Fibonacci string

$j =$	1	2	3	4	5	6	7	8	9	10	11	12	13	14	15	16	17	18	19	20	21
$pattern[j] =$	a	b	a	a	b	a	b	a	a	b	a	a	b	a	b	a	a	b	a	b	a
$f[j] =$	0	1	1	2	2	3	4	3	4	5	6	7	5	6	7	8	9	10	11	12	8
$next[j] =$	0	1	0	2	1	0	4	0	2	1	0	7	1	0	4	0	2	1	0	12	0

If we extend this pattern to ϕ_∞, we obtain infinite sequences $f[j]$ and $next[j]$ having the same general nature. It is possible to prove by induction that

$$f[j] = j - F_{k-1} \quad \text{for } F_k \le j < F_{k+1}, \tag{2}$$

because of the following remarkable near-commutative property of Fibonacci strings:

$$\phi_n\phi_{n+1} = c(\phi_{n+1}\phi_n) \quad \text{for } n \ge 1, \tag{3}$$

where $c(\alpha)$ denotes interchanging the two rightmost characters of α. For example, $\phi_6 = abaab \cdot aba$ and $c(\phi_6) = aba \cdot abaab$. Equation (3) is obvious when $n = 1$; and for $n > 1$ we have $c(\phi_n\phi_{n+1}) = \phi_n c(\phi_{n+1}) = \phi_n c(\phi_n\phi_{n-1}) = \phi_n\phi_{n-1}\phi_n = \phi_{n+1}\phi_n$ by induction; hence $\phi_n\phi_{n+1} = c(c(\phi_n\phi_{n+1})) = c(\phi_{n+1}\phi_n)$.

Equation (3) implies that the pattern ϕ_∞ has

$$next[F_k - 1] = F_{k-1} - 1 \quad \text{for } k \ge 3. \tag{4}$$

Therefore if we have a mismatch when $j = F_8 - 1 = 20$, our algorithm might set $j := next[j]$ for the successive values 20, 12, 7, 4, 2, 1 of j. Since F_k is $\phi^k/\sqrt{5}$ rounded to the nearest integer, it is possible to have up to $\sim \log_\phi m$ consecutive iterations of the '$j := next[j]$' loop.

We shall now show that Fibonacci strings actually yield the worst case, namely that $\log_\phi m$ is also an upper bound. First let us consider the concept of *periodicity* in strings. We say that p is a *period* of α if

$$\alpha[i] = \alpha[i + p] \quad \text{for } 1 \le i \le |\alpha| - p. \tag{5}$$

It is easy to see that p is a period of α if and only if

$$\alpha = (\alpha_1\alpha_2)^k\alpha_1 \tag{6}$$

for some $k \ge 0$, where $|\alpha_1\alpha_2| = p$ and $\alpha_2 \ne \epsilon$. Equivalently, p is a period of α if and only if

$$\alpha\theta_1 = \theta_2\alpha \tag{7}$$

for some θ_1 and θ_2 with $|\theta_1| = |\theta_2| = p$. Condition (6) implies (7) with $\theta_1 = \alpha_2\alpha_1$ and $\theta_2 = \alpha_1\alpha_2$. Condition (7) implies (6), for we define $k = \lfloor|\alpha|/p\rfloor$ and observe that, if $k > 0$, then $\alpha = \theta_2\beta$ implies $\beta\theta_1 = \theta_2\beta$ and $\lfloor|\beta|/p\rfloor = k - 1$; hence, reasoning inductively, $\alpha = \theta_2^k\alpha_1$ for some α_1 with $|\alpha_1| < p$, and $\alpha_1\theta_1 = \theta_2\alpha_1$. Writing $\theta_2 = \alpha_1\alpha_2$ yields (6).

The relevance of periodicity to our algorithm is clear once we consider what it means to shift a pattern. If *pattern*$[1] \ldots$ *pattern*$[j-1] = \alpha$ ends with *pattern*$[1] \ldots$ *pattern*$[i-1] = \beta$, we have

$$\alpha = \beta\theta_1 = \theta_2\beta \tag{8}$$

where $|\theta_1| = |\theta_2| = j - i$, so the amount of shift $j - i$ is a period of α.

The construction of $i =$ *next*$[j]$ in our algorithm implies further that *pattern*$[i]$, which is the first character of θ_1, is unequal to *pattern*$[j]$. Let us assume that β itself is subsequently shifted leaving a residue γ, so that

$$\beta = \gamma\psi_1 = \psi_2\gamma \tag{9}$$

where the first character of ψ_1 differs from that of θ_1. We shall now prove that

$$|\alpha| > |\beta| + |\gamma|. \tag{10}$$

If $|\beta| + |\gamma| \geq |\alpha|$, there is an overlap of $d = |\beta| + |\gamma| - |\alpha|$ characters between the occurrences of β and γ in $\beta\theta_1 = \alpha = \theta_2\psi_2\gamma$; hence the first character of θ_1 is $\gamma[d+1]$. Similarly there is an overlap of d characters between the occurrences of β and γ in $\theta_2\beta = \alpha = \gamma\psi_1\theta_1$; hence the first character of ψ_1 is $\beta[d+1]$. But those characters are distinct, so we obtain $\gamma[d+1] \neq \beta[d+1]$, contradicting (9). This contradiction establishes (10), and leads directly to the announced result:

Theorem. *The number of consecutive times that '$j :=$ next$[j]$' is performed, while one text character is being scanned, is at most $1 + \log_\phi m$.*

Proof. Let L_r be the length of the shortest string α as in the discussion above such that a sequence of r consecutive shifts is possible. Then $L_1 = 0$, $L_2 = 1$, and we have $|\beta| \geq L_{r-1}$, $|\gamma| \geq L_{r-2}$ in (10); hence $L_r \geq F_{r+1} - 1$ by induction on r. Now if r shifts occur we have $m \geq F_{r+1} \geq \phi^{r-1}$. □

The algorithm of §2 would run correctly in linear time even if $f[j]$ were used instead of *next*$[j]$, but the analog of the theorem above would then be false. For example, the pattern a^m leads to $f[j] = j - 1$ for $1 \leq j \leq m$. Therefore if we matched a^m to the text $a^{m-1}b\alpha$, using $f[j]$ instead of *next*$[j]$, the mismatch *text*$[m] \neq$ *pattern*$[m]$ would be followed by m occurrences of '$j := f[j]$' and $m - 1$ redundant comparisons of *text*$[m]$ with *pattern*$[j]$, before k is advanced to $m + 1$.

The subject of periods in strings has several interesting algebraic properties, but a reader who is not mathematically inclined may skip to §6 since the following material is primarily an elaboration of some additional structure related to the theorem above.

Lemma 1. *If p and q are periods of α, and $p+q \leq |\alpha|+\gcd(p,q)$, then $\gcd(p,q)$ is a period of α.*

Proof. Let $d = \gcd(p,q)$, and assume without loss of generality that $d < p < q = p + r$. We have $\alpha[i] = \alpha[i+p]$ for $1 \leq i \leq |\alpha| - p$ and $\alpha[i] = \alpha[i+q]$ for $1 \leq i \leq |\alpha| - q$; hence $\alpha[i+r] = \alpha[i+q] = \alpha[i]$ for $1 + r \leq i + r \leq |\alpha| - p$. Equivalently,

$$\alpha[i] = \alpha[i+r] \quad \text{for } 1 \leq i \leq |\alpha| - q.$$

Furthermore $\alpha = \beta\theta_1 = \theta_2\beta$ where $|\theta_1| = p$, and it follows that p and r are periods of β, where $p + r \leq |\beta| + d = |\beta| + \gcd(p,r)$. By induction, d is a period of β. Since $|\beta| = |\alpha| - p \geq q - d \geq q - r = p = |\theta_1|$, the strings θ_1 and θ_2 (which have the respective forms $\beta_2\beta_1$ and $\beta_1\beta_2$ by (6) and (7)) are substrings of β; so they also have d as a period. The string $\alpha = (\beta_1\beta_2)^{k+1}\beta_1$ must now have d as a period, since any characters d positions apart are contained within $\beta_1\beta_2$ or $\beta_2\beta_1$. □

The result of Lemma 1 but with the stronger hypothesis $p+q \leq |\alpha|$ was proved by Lyndon and Schützenberger in connection with a problem about free groups [19, Lemma 4]. The weaker hypothesis in Lemma 1 turns out to give the best possible bound: If $\gcd(p,q) < p < q$ we can find a string of length $p + q - \gcd(p,q) - 1$ for which $\gcd(p,q)$ is *not* a period. In order to see why this is so, consider first the example in Fig. 2 showing the most general strings of lengths 15 through 25 having both 11 and 15 as periods. (The strings are "most general" in the sense that any two character positions that can be different *are* different.)

abcdefghijkabcd
abcdafghijkabcda
abcdabghijkabcdab
abcdabchijkabcdabc
abcdabcdijkabcdabcd
abcdabcdajkabcdabcda
abcdabcdabkabcdabcdab
abcdabcdabcabcdabcdabc
abcaabcaabcabcaabcaabca
aacaaacaaacaacaaacaaacaa
aaaaaaaaaaaaaaaaaaaaaaaaa

FIGURE 2. Strings that have periods 11 and 15.

Notice that the number of degrees of freedom (the number of distinct symbols) decreases by 1 at each step. It is not difficult to prove that the number cannot decrease by *more* than 1 as we go from $|\alpha| = n-1$ to $|\alpha| = n$, since the only new relations are $\alpha[n] = \alpha[n-q] = \alpha[n-p]$; we decrease the number of distinct symbols by one if and only if positions $n-q$ and $n-p$ contain distinct symbols in the most general string of length $n-1$. The lemma tells us that we are left with at most $\gcd(p,q)$ symbols when the length reaches $p+q-\gcd(p,q)$; on the other hand we always have exactly p symbols when the length is q. Therefore each of the $p-\gcd(p,q)$ steps *must* decrease the number of symbols by 1, and the most general string of length $p+q-\gcd(p,q)-1$ must have exactly $\gcd(p,q)+1$ distinct symbols. In other words, the lemma gives the best possible bound.

When p and q are relatively prime, the strings of length $p+q-2$ on two symbols, having both p and q as periods, satisfy a number of remarkable properties, generalizing what we have observed earlier about Fibonacci strings. Since the properties of these pathological patterns may prove useful in other investigations, we shall summarize them in the following lemma.

Lemma 2. *Let the strings $\sigma(m,n)$ of length n be defined for all relatively prime pairs of integers $n \geq m \geq 0$ as follows:*

$$\sigma(0,1) = a, \quad \sigma(1,1) = b, \quad \sigma(1,2) = ab;$$

$$\left.\begin{aligned} \sigma(m,m+n) &= \sigma(n \bmod m, m)\sigma(m,n) \\ \sigma(n,m+n) &= \sigma(m,n)\sigma(n \bmod m, m) \end{aligned}\right\} \quad \text{if } 0 < m < n. \tag{11}$$

These strings satisfy the following properties:

i) $\sigma(m, qm+r)\sigma(m-r, m) = \sigma(r,m)\sigma(m, qm+r)$, *for* $m > 2$;

ii) $\sigma(m,n)$ *has period* m, *for* $m > 1$;

iii) $c(\sigma(m,n)) = \sigma(n-m, n)$, *for* $n > 2$.

(The function $c(\alpha)$ was defined in connection with (3) above.)

Proof. We have, for $0 < m < n$ and $q \geq 2$,

$$\begin{aligned} \sigma(m+n, q(m+n)+m) &= \sigma(m,m+n)\sigma(m+n,(q-1)(m+n)+m); \\ \sigma(m+n, q(m+n)+n) &= \sigma(n,m+n)\sigma(m+n,(q-1)(m+n)+n); \\ \sigma(m+n, 2m+n) &= \sigma(m,m+n)\sigma(n \bmod m, m); \\ \sigma(m+n, m+2n) &= \sigma(n,m+n)\sigma(m,n). \end{aligned}$$

Hence, if $\theta_1 = \sigma(n \bmod m, m)$ and $\theta_2 = \sigma(m,n)$ and $q \geq 1$,

$$\begin{aligned} \sigma(m+n, q(m+n)+m) &= (\theta_1\theta_2)^q\theta_1, \\ \sigma(m+n, q(m+n)+n) &= (\theta_2\theta_1)^q\theta_2. \end{aligned} \tag{12}$$

It follows that

$$\sigma(m+n, q(m+n)+m)\sigma(n, m+n) = \sigma(m, m+n)\sigma(m+n, q(m+n)+m),$$

$$\sigma(m+n, q(m+n)+n)\sigma(m, m+n) = \sigma(n, m+n)\sigma(m+n, q(m+n)+n),$$

and these formulas combine to prove (i). Property (ii) also follows immediately from (12), except for the case $m = 2$, $n = 2q+1$, $\sigma(2, 2q+1) = (ab)^q a$, which can be verified directly. Finally, it suffices to verify property (iii) for $0 < m < \frac{1}{2}n$, since $c(c(\alpha)) = \alpha$; we must show that

$$c(\sigma(m, m+n)) = \sigma(m, n)\sigma(n \bmod m, m) \quad \text{for } 0 < m < n.$$

When $m \le 2$ this property is easily checked, and when $m > 2$ it is equivalent by induction to

$$\sigma(m, m+n) = \sigma(m, n)\sigma(m - (n \bmod m), m) \quad \text{for } 2 < m < n.$$

Set $r = n \bmod m$, $q = \lfloor n/m \rfloor$, and apply property (i). □

By properties (ii) and (iii) of this lemma, $\sigma(p, p+q)$ minus its last two characters is the string of length $p+q-2$ having periods p and q. Note that Fibonacci strings are just a very special case, since $\phi_n = \sigma(F_{n-1}, F_n)$. Another property of the σ strings appears in [15]. A completely different proof of Lemma 1 and its optimality, and a completely different definition of $\sigma(m, n)$, were given by Fine and Wilf in 1965 [7]. These strings have a long history going back at least to the astronomer Johann Bernoulli in 1772; see [25, §2.13] and [21].

If α is any string, let $P(\alpha)$ be its shortest period. Lemma 1 implies that all periods q must either be multiples of $P(\alpha)$ or strictly greater than $|\alpha| - P(\alpha) + \gcd(q, P(\alpha))$. This is a rather strong condition in terms of the pattern-matching algorithm, because of the following result.

Lemma 3. *Let* $\alpha = \mathit{pattern}[1] \ldots \mathit{pattern}[j-1]$ *and let* $a = \mathit{pattern}[j]$. *In the pattern-matching algorithm, we have* $f[j] = j - P(\alpha)$ *and* $\mathit{next}[j] = j - q$, *where* q *is the smallest period of* α *that is not a period of* αa. *(If no such period exists,* $\mathit{next}[j] = 0$.*) If* $P(\alpha)$ *divides* $P(\alpha a)$ *and* $P(\alpha a) < j$, *then* $P(\alpha) = P(\alpha a)$. *If* $P(\alpha)$ *does not divide* $P(\alpha a)$ *or if* $P(\alpha a) = j$, *then* $q = P(\alpha)$.

Proof. The characterizations of $f[j]$ and $\mathit{next}[j]$ follow immediately from the definitions. Since every period of αa is a period of α, the only nonobvious statement is that $P(\alpha) = P(\alpha a)$ whenever $P(\alpha)$ divides $P(\alpha a)$ and $P(\alpha a) \neq j$. Let $P(\alpha) = p$ and $P(\alpha a) = mp$; then the (mp)th character from the right of α is a, as is the $(mp - p)$th, ..., as is the pth; hence p is a period of αa. □

Lemma 3 shows that the $j := next[j]$ loop will almost always terminate quickly. If $P(\alpha) = P(\alpha a)$, then q must not be a multiple of $P(\alpha)$; hence by Lemma 1, $P(\alpha)+q \geq j+1$. On the other hand $q > P(\alpha)$; hence $q > \frac{1}{2}j$ and $next[j] < \frac{1}{2}j$. In the other case $q = P(\alpha)$, we had better not have q too small, since q will be a period in the residual pattern after shifting, and $next[next[j]]$ will be $< q$. To keep the loop running it is necessary for new small periods to keep popping up, relatively prime to the previous periods.

6. Palindromes

One of the most outstanding unsolved questions in the theory of computational complexity is the problem of how long it takes to determine whether or not a given string of length n belongs to a given context-free language. For many years the best upper bound for this problem was $O(n^3)$ in a general context-free language as $n \to \infty$; L. G. Valiant has recently lowered this to $O(n^{\log_2 7})$. On the other hand, the problem isn't known to require more than order n units of time for any particular language. This big gap between $\Omega(n)$ and $O(n^{2.81})$ deserves to be closed, and hardly anyone believes that the final answer will be $O(n)$.

Let Σ be a finite alphabet, let Σ^* denote the strings over Σ, and let

$$P = \{\alpha\alpha^R \mid \alpha \in \Sigma^*\}.$$

Here α^R denotes the reversal of α, i.e., $(a_1a_2 \ldots a_n)^R = a_n \ldots a_2a_1$. Each string π in P is a *palindrome* of even length, and conversely every even palindrome over Σ is in P. At one time it was popularly believed that the language P^* of "even palindromes starred," namely the set of *palstars* $\pi_1 \ldots \pi_t$ where each π_i is in P, would be impossible to recognize in $O(|\pi_1 \ldots \pi_t|)$ steps on a random access computer.

It isn't especially easy to spot members of this language. For example, *aabbabba* is a palstar, but its decomposition into even palindromes might not be immediately apparent; and the reader might need several minutes to decide whether or not

$$baabbabbaababbaabbabbabaabbabbabbabbaabababbabbaab$$

is in P^*. We shall prove, however, that palstars can be recognized in $O(n)$ units of time, by using their algebraic properties.

Let us say that a nonempty palstar is *prime* if it cannot be written as the product of two nonempty palstars. A prime palstar must be an even palindrome $\alpha\alpha^R$, but the converse does not hold. By repeated decomposition, it is easy to see that every palstar β is expressible as a product $\beta_1 \ldots \beta_t$ of prime palstars, for some $t \geq 0$; what is less obvious is that such a decomposition into prime factors is unique. This "fundamental theorem of palstars" is an immediate consequence of the following basic property.

Lemma 1. *A prime palstar cannot begin with another prime palstar.*

Proof. Let $\alpha\alpha^R$ be a prime palstar such that $\alpha\alpha^R = \beta\beta^R\gamma$ for some nonempty even palindrome $\beta\beta^R$ and some $\gamma \neq \epsilon$. Furthermore, let $\beta\beta^R$ have minimum length among all such counterexamples. If $|\beta\beta^R| > |\alpha|$, then $\alpha\alpha^R = \beta\beta^R\gamma = \alpha\delta\gamma$ for some $\delta \neq \epsilon$; hence $\alpha^R = \delta\gamma$, and $\beta\beta^R = (\beta\beta^R)^R = (\alpha\delta)^R = \delta^R\alpha^R = \delta^R\delta\gamma$, contradicting the minimality of $|\beta\beta^R|$. Therefore $|\beta\beta^R| \leq |\alpha|$; hence $\alpha = \beta\beta^R\delta$ for some δ, and $\beta\beta^R\gamma = \alpha\alpha^R = \beta\beta^R\delta\delta^R\beta\beta^R$. But this implies that γ is the palstar $\delta\delta^R\beta\beta^R$, contradicting the primality of $\alpha\alpha^R$. □

Corollary (Left cancellation property). *If $\alpha\beta$ and α are palstars, so is β.*

Proof. Let $\alpha = \alpha_1 \ldots \alpha_r$ and $\alpha\beta = \beta_1 \ldots \beta_s$ be prime factorizations of α and $\alpha\beta$. If $\alpha_1 \ldots \alpha_r = \beta_1 \ldots \beta_r$, then $\beta = \beta_{r+1} \ldots \beta_s$ is a palstar. Otherwise let j be minimal with $\alpha_j \neq \beta_j$; then α_j begins with β_j or vice versa, contradicting Lemma 1. □

Lemma 2. *If α is a string of length n, we can determine the length of the longest even palindrome $\beta \in P$ such that $\alpha = \beta\gamma$, in $O(n)$ steps.*

Proof. Apply the pattern-matching algorithm with $pattern = \alpha$ and $text = \alpha^R$. When $k = n + 1$ the algorithm will stop with j maximal such that $pattern[1] \ldots pattern[j-1] = text[n+2-j] \ldots text[n]$. Now perform the following iteration:

while $j \geq 3$ **and** j even **do** $j := f[j]$.

By the theory developed in §3, this iteration terminates with $j \geq 3$ if and only if α begins with a nonempty even palindrome, and $j - 1$ will be the length of the largest such palindrome. (Note that $f[j]$ must be used here instead of $next[j]$; consider, for example, the case $\alpha = aabaab$. But the pattern-matching process takes $O(n)$ time even when $f[j]$ is used.) □

Theorem. *Let L be any language such that L^* has the left cancellation property and such that, given any string α of length n, we can find a nonempty $\beta \in L$ such that α begins with β or we can prove that no such*

β exists, in $O(n)$ steps. Then we can determine in $O(n)$ time whether or not a given string is in L^*.

Proof. Suppose that the time required to test a string of length $m > 0$ for nonempty prefixes in L is $\leq Km$. Given a string α of length $n > 0$, we begin by testing its initial subsequences of lengths 1, 2, 4, ..., 2^k, ..., and finally α itself, until either finding a prefix in L or establishing that α has no such prefix. In the latter case, α is not in L^*, and we have consumed at most $(K+K_1)+(2K+K_1)+(4K+K_1)+\cdots+(|\alpha|K+K_1) < (3K+K_1)n$ units of time, for some constant K_1. But if we find a nonempty prefix $\beta \in L$ where $\alpha = \beta\gamma$, we have used at most $(4K+K_1)|\beta|$ units of time so far. By the left cancellation property, $\alpha \in L^*$ if and only if $\gamma \in L^*$. Since $|\gamma| = n - |\beta|$ we can therefore prove by induction that at most $(4K+K_1)n$ units of time are needed to decide membership in L^*, when $n > 0$. □

Corollary. *P^* can be recognized in $O(n)$ time.* □

Notice that the related language

$$P_1^* = \{\pi \in \Sigma^* \mid \pi = \pi^R \text{ and } |\pi| \geq 2\}^*$$

cannot be handled by the techniques above, since it contains both *aaabbb* and *aaabbbba*; the fundamental theorem of palstars fails with a vengeance. It is an open problem whether or not P_1^* can be recognized in $O(n)$ time, although we suspect that it can be done.[1] Once the reader has disposed of this problem, he or she is urged to tackle another language that has recently been introduced by S. A. Greibach [11], since the latter language is known to be as hard as possible; no context-free language can be harder to recognize except by a constant factor.

7. Historical Remarks

The pattern-matching algorithm of this paper was discovered in a rather interesting way. One of the authors (J. H. Morris) was implementing a text editor for the CDC 6400 computer during the summer of 1969, and since the necessary buffering was rather complicated he sought a method that would avoid backing up the text file. Using concepts of finite automata theory as a model, he devised an algorithm equivalent to the method presented above, although his original form of presentation made it unclear that the running time was $O(m+n)$. Indeed, it turned

---

[1] Zvi Galil and Joel Seiferas have resolved this conjecture affirmatively. ["A linear-time on-line recognition algorithm for 'palstar'," *Journal of the Association for Computing Machinery* **25** (1978), 102–111.]

out that Morris's routine was too complicated for other implementors of the system to understand, and he discovered several months later that gratuitous "fixes" had turned his routine into a shambles.

In a totally independent development, another author (D. E. Knuth) learned early in 1970 of S. A. Cook's surprising theorem about two-way deterministic pushdown automata [5]. According to Cook's theorem, any language recognizable by a two-way deterministic pushdown automaton, in *any* amount of time, can be recognized on a random access machine in $O(n)$ units of time. Since Daniel Chester had recently shown that the set of strings beginning with an even palindrome could be recognized by such an automaton, and since Knuth couldn't imagine how to recognize such a language in less than about n^2 steps on a conventional computer, Knuth laboriously went through all the steps of Cook's construction as applied to Chester's automaton. His plan was to "distill off" what was happening, in order to discover why the algorithm worked so efficiently. After pondering the mass of details for several hours, he finally succeeded in abstracting the mechanism that seemed to be underlying the construction, in the special case of palindromes, and he generalized it slightly to a program capable of finding the longest prefix of one given string that occurs in another.

This was the first time in Knuth's experience that automata theory had taught him how to solve a real programming problem better than he could solve it before. He showed his results to the third author (V. R. Pratt), and Pratt modified Knuth's data structure so that the running time was independent of the alphabet size. When Pratt described the resulting algorithm to Morris, the latter recognized it as his own, and was pleasantly surprised to learn of the $O(m+n)$ time bound, which he and Pratt described in a memorandum [22]. Knuth was chagrined to learn that Morris had already discovered the algorithm, *without* knowing Cook's theorem; but the theory of finite-state machines had been of use to Morris too, in his initial conceptualization of the algorithm, so it was still legitimate to conclude that abstract automata theory had actually been helpful in this concrete practical problem.

The idea of scanning a string without backing up while looking for a pattern, in the case of a two-letter alphabet, is implicit in the early work of Gilbert [10] dealing with comma-free codes. It also is essentially a special case of Knuth's LR(0) parsing algorithm [16] when applied to the grammar

$$S \to aS, \qquad \text{for each } a \text{ in the alphabet,}$$
$$S \to \alpha$$

where α is the pattern. Diethelm and Roizen [6] independently discovered the idea in 1971. Gilbert and Knuth did not discuss the preprocessing to build the *next* table, since they were mainly concerned with other problems, and the preprocessing algorithm given by Diethelm and Roizen was of order m^2. In the case of a binary (two-letter) alphabet, Diethelm and Roizen observed that the algorithm of §3 can be improved further: We can go immediately to 'end check' after '$j := next[j]$' in this case if $next[j] > 0$.

A conjecture by R. L. Rivest led Pratt to discover the $\log_\phi m$ upper bound on pattern movements between successive input characters, and Knuth showed that this was best possible by observing that Fibonacci strings have the curious properties proved in §5. Zvi Galil has observed that a real-time algorithm can be obtained by letting the text pointer move ahead in an appropriate manner while the j pointer is moving down [9].

In his lectures at Berkeley, S. A. Cook had proved that P^* was recognizable in $O(n \log n)$ steps on a random access machine, and Pratt improved this bound to $O(n)$ using a preliminary form of the ideas in §6. The slightly more refined theory in the present version of §6 is joint work of Knuth and Pratt. Manacher [20] found another way to recognize palstars in linear time, and Galil [9] showed how to improve this result to real time. See also the independent work of Slisenko in Russia [23].

It seemed at first that there might be a way to find the *longest common substring* of two given strings, in time $O(m + n)$; but the algorithm of this paper does not readily support any such extension, and Knuth conjectured in 1970 that such efficiency would be impossible to achieve. An algorithm due to Karp, Miller, and Rosenberg [13] solved the problem in $O((m + n)\log(m + n))$ steps, and this tended to support the conjecture (at least in the mind of its originator). However, Peter Weiner has recently developed a technique for solving the longest common substring problem in $O(m+n)$ units of time with a fixed alphabet, using tree structures in a remarkable new way [26]. Furthermore, Weiner's algorithm has the following interesting consequence, pointed out by E. McCreight: A text file can be processed (in linear time) so that it is possible to determine exactly how much of a pattern is necessary to identify a position in the text uniquely; as the pattern is being typed in, the system can interrupt as soon as it "knows" what the rest of the pattern must be! Unfortunately the time and space requirements for Weiner's algorithm grow with increasing alphabet size.

If we consider the problem of scanning finite-state languages in general, it is known [2, §9.2] that the language defined by any regular expression of length m is recognizable in $O(mn)$ units of time. When the regular expression has the form

$$\Sigma^*(\alpha_{1,1}+\cdots+\alpha_{1,s(1)})\Sigma^*(\alpha_{2,1}+\cdots+\alpha_{2,s(2)})\Sigma^*\ldots\Sigma^*(\alpha_{r,1}+\cdots+\alpha_{r,s(r)})\Sigma^*$$

the algorithm we have discussed shows that only $O(m+n)$ units of time are needed (considering Σ^* as a character of length 1 in the expression). Recent work by M. J. Fischer and M. S. Paterson [8] shows that regular expressions of the form

$$\Sigma^*\alpha_1\Sigma\alpha_2\Sigma\ldots\Sigma\alpha_r\Sigma^*,$$

that is, patterns with "don't care" symbols, can be identified in at most $O(n\log m\log\log m\log q)$ units of time, where $q=|\Sigma|$ is the alphabet size and $m=|\alpha_1\alpha_2\ldots\alpha_r|+r$. The constant of proportionality in their algorithm is extremely large, but the existence of their construction indicates that efficient new algorithms for general pattern-matching problems probably remain to be discovered.

A completely different approach to pattern matching, based on hashing, has been proposed by Malcolm C. Harrison [12]. In certain applications, especially with very large text files and short patterns, Harrison's method may be significantly faster than the character-comparing method of the present paper, on the average, although the redundancy of English makes the performance of his method unclear.

8. Postscript: Faster Pattern Matching in Strings[2]

In the spring of 1974, Robert S. Boyer and J Strother Moore and (independently) R. W. Gosper noticed that there is an even faster way to match pattern strings, by skipping more rapidly over portions of the text that cannot possibly lead to a match. Their idea (see [4]) was to look first at *text*[m], instead of *text*[1]. If we find that the character *text*[m] does not appear in the pattern at all, we can immediately shift the pattern right m places. Thus, when the alphabet size q is large, we need to inspect only about n/m characters of the text, on the average! Furthermore if *text*[m] does occur in the pattern, we can shift the pattern by the minimum amount consistent with a match.

Several interesting variations on this strategy are possible. For example, if *text*[m] does occur in the pattern, we might continue the search

[2] This postscript was added by D. E. Knuth in March 1976, because of developments that occurred after preprints of this paper [18] were distributed.

by looking at $text[m-1]$, $text[m-2]$, etc.; in a random file we will usually find a small value of r such that the substring $text[m-r] \ldots text[m]$ does not appear in the pattern, so we can shift the pattern $m-r$ places. If $r = \lfloor 2 \log_q m \rfloor$, the substring $text[m-r] \ldots text[m]$ has more than m^2 possible values, but there are only $m-r$ substrings of length $r+1$ in the pattern. Hence the probability is $O(1/m)$ that $text[m-r] \ldots text[m]$ occurs. If it doesn't, we can shift the pattern right $m-r$ places; but if it does, we can determine all matches in positions $< m-r$ in $O(m)$ steps, shifting the pattern $m-r$ places by the method of this paper. Hence the expected number of characters examined among the first $m - \lfloor 2 \log_q m \rfloor$ is $O(\log_q m)$; this proves the existence of a linear-worst-case algorithm that inspects $O(n(\log_q m)/m)$ characters, on average, in a random text. This upper bound on the average running time applies to all patterns, and there are patterns such as a^m or $(ab)^{m/2}$ for which the expected number of characters examined by the algorithm is $O(n/m)$.

Boyer and Moore have refined the skipping-by-m idea in another way. Their original algorithm may be expressed as follows using our conventions:

```
k := m;
while k ≤ n do
   begin j := m;
   while j > 0 and text[k] = pattern[j] do
      begin j := j − 1;  k := k − 1;
      end;
   if j = 0 then
      begin matchfoundat(k);
      k := k + m + 1;
      end
   else k := k + max(d[text[k]], dd[j]);
   end.
```

This program calls $matchfoundat(k)$ for all $0 \le k \le n-m$ such that $pattern[1] \ldots pattern[m] = text[k+1] \ldots text[k+m]$. There are two precomputed tables, namely

$$d[a] = \min\{s \mid s = m \text{ or } (0 \le s < m \text{ and } pattern[m-s] = a)\}$$

for each of the q possible characters a, and

$$dd[j] = \min\{s + m - j \mid s \ge 1 \text{ and } ((s \ge i \text{ or } pattern[i-s] = pattern[i]) \text{ for } j < i \le m)\},$$

for $1 \le j \le m$.

The d table can clearly be set up in $O(q+m)$ steps; and the dd table can be precomputed in $O(m)$ steps using a technique analogous to the method in §2 above, as we shall see. The Boyer–Moore paper [4] contains further exposition of the algorithm, including suggestions for highly efficient implementation, and gives both theoretical and empirical analyses. In the remainder of this section we shall show how the methods above can be used to resolve some of the problems left open in [4].

First let us improve the original Boyer–Moore algorithm slightly by using dd' instead of dd, where

$$\begin{aligned} dd'[j] = \min\bigl\{s+m-j \bigm| s \ge 1 \text{ and } \\ (s \ge j \text{ or } \mathit{pattern}[j-s] \ne \mathit{pattern}[j]) \text{ and } \\ ((s \ge i \text{ or } \mathit{pattern}[i-s] = \mathit{pattern}[i]) \text{ for } j < i \le m)\bigr\}. \end{aligned}$$

(This change is analogous to using $next[j]$ instead of $f[j]$ in §2. Boyer and Moore [4] credit the improvement in this case to Ben Kuipers, but they do not discuss how to determine dd' efficiently.) The following program[3] shows how the dd' table can be precomputed in $O(m)$ steps; for purposes of comparison, the program also shows how to compute dd, which actually turns out to require slightly *more* operations than dd':

```
for k := 1 step 1 until m do dd[k] := dd'[k] := m + m − k;
j := m;  t := m + 1;
while j > 0 do
   begin f[j] := t;
   while t ≤ m and pattern[j] ≠ pattern[t] do
      begin dd'[t] := min(dd'[t], m − j);  t := f[t];
      end;
   t := t − 1;  j := j − 1;  dd[t] := min(dd[t], m − j);
   end;
for k := 1 step 1 until t do
   begin dd[k] := min(dd[k], m + t − k);
   dd'[k] := min(dd'[k], m + t − k);
   end.
```

In practice one would, of course, compute only dd', suppressing all references to dd. The example in Table 2 illustrates most of the subtleties of this algorithm.

To prove correctness, one may show first that $f[j]$ is analogous to the $f[j]$ in §2, but with the pattern considered from right to left instead

[3] This program is flawed; see the addendum below for a correct version.

TABLE 2. Auxiliary functions for the Boyer–Moore algorithm

j	=	1	2	3	4	5	6	7	8	9	10	11
$pattern[j]$	=	b	a	d	b	a	c	b	a	c	b	a
$f[j]$	=	10	11	6	7	8	9	10	11	11	11	12
$dd[j]$	=	19	18	17	16	15	8	7	6	5	4	1
$dd'[j]$	=	19	18	17	16	15	8	13	12	8	12	1

of from left to right; namely $f[m] = m+1$, and for $1 \le j < m$ we have

$$f[j] = \min\{i \mid j < i \le m \text{ and } pattern[i+1] \ldots pattern[m] = pattern[j+1] \ldots pattern[m+j-i]\}.$$

Furthermore the final value of t corresponds to $f[0]$ in this definition; $m-t$ is the maximum overlap of the pattern on itself. The correctness of $dd[j]$ and $dd'[j]$ for all j now follows without much difficulty, by showing that the minimum value of s in the definition of $dd[j_0]$ or $dd'[j_0]$ is discovered by the algorithm when $(t, j) = (j_0, j_0 - s)$.

The Boyer–Moore algorithm and its variants can have curiously anomalous behavior in unusual circumstances. For example, the method discovers more quickly that the pattern $aaaaaaacb$ does not appear in the text $(ab)^n$ if it suppresses the d heuristic entirely, by setting $d[t]$ to $-\infty$ for all t. Likewise, dd actually turns out to be better than dd' when matching $a^{15}bcbabab$ in $(baabab)^n$, for large n.

Boyer and Moore showed that their algorithm has quadratic behavior in the worst case; the running time can be essentially proportional to pattern length times text length, for example when the pattern $ca(ba)^m$ is being sought in the text $(x^{2m}aa(ba)^m)^n$. They observed that this particular example is handled in linear time when the improvement of Kuipers (dd' for dd) is made; but they left open the question of the true worst-case behavior of the improved algorithm.

There are trivial cases in which the Boyer–Moore algorithm has quadratic behavior, when matching all occurrences of the pattern, for example when matching the pattern a^m in the text a^n. But we are probably willing to accept such behavior when there are so many matches; the crucial issue is how long the algorithm takes in the worst case to scan over a text that does *not* contain the pattern at all. By extending the techniques of §5, it is possible to show that the modified Boyer–Moore algorithm is *linear* in such a situation:

Theorem. *If the algorithm above is used with dd' replacing dd, and if the text does not contain any occurrences of the pattern, the total number of characters successfully matched is at most $6n$.*

Proof. An execution of the algorithm consists of a series of *stages*, in which m_k characters are matched and then the pattern is shifted s_k places, for $k = 1, 2, \ldots$. We want to show that $\sum m_k \le 6n$; the proof is based on breaking this cost into three parts, two of which are trivially $O(n)$ and the third of which is less obviously so.

Let $m'_k = m_k - 2s_k$ if $m_k > 2s_k$; otherwise let $m'_k = 0$. When $m'_k > 0$, we will say that the leftmost m'_k text characters matched during the kth stage have been "tapped." It suffices to prove that the algorithm taps characters at most $4n$ times, since $\sum m_k \le \sum m'_k + 2\sum s_k$ and $\sum s_k \le n$. But it is possible for some characters of the text to be tapped roughly $\log m$ times, so we need to argue carefully that $\sum m'_k \le 4n$.

Suppose the rightmost m''_k of the m'_k text characters tapped during the kth stage are matched again during some later stage, but the leftmost $m'_k - m''_k$ are being matched for the last time. Clearly $\sum(m'_k - m''_k) \le n$, so it remains to show that $\sum m''_k \le 3n$.

Let p_k be the amount by which the pattern would shift after the kth stage if the $d[a]$ heuristic were not present ($d[a] = -\infty$); then $p_k \le s_k$, and p_k is a period of the string matched at stage k.

Consider a value of k such that $m''_k > 0$, and suppose that the text characters matched during the kth stage form the string $\alpha = \alpha_1\alpha_2$ where $|\alpha| = m_k$ and $|\alpha_2| = m''_k + 2s_k$; hence the text characters in α_1 are matched for the last time. Since the pattern does not occur in the text, it must end with $x\alpha$ and the text scanned so far must end with $z\alpha$, where $x \ne z$. At this point the algorithm will shift the pattern right s_k positions and will enter stage $k+1$. We distinguish two cases: (i) The pattern length m exceeds $m_k + p_k$. Then the pattern can be written $\theta\beta\alpha$ where $|\beta| = p_k$; the last character of β is x and the last character of θ is $y \ne x$, by definition of dd'. Otherwise (ii) $m \le m_k + p_k$; the pattern then has the form $\beta\alpha$, where $|\beta| \le p_k \le s_k$. By definition of m''_k and the assumption that the pattern does not occur in the text, we have $|\beta\alpha| > s_k + |\alpha_2|$, thus $|\beta| > s_k - |\alpha_1|$. In both cases (i) and (ii), p_k is a period of $\beta\alpha$.

Now consider the first subsequent stage k' during which the *leftmost* of the m''_k text characters tapped during stage k is matched again; we shall write $k \to k'$ when the stages are in this relation. Suppose the mismatch occurs this time when text character z' fails to match pattern character x'. If z' occurs in the text within α_1, regarding α as fixed in its stage k position, then x' cannot be within $\beta\alpha$ where $\beta\alpha$ now occurs in the stage k' position of the pattern, since p_k is a period of $\beta\alpha$ and the character p_k positions to the right of x' is a z' (it matches a z' in the text). Thus x' now appears within θ. On the other hand, if z' occurs to the left of α, we must have $|\alpha_1| = 0$, since the characters of α_1

are never matched again. In either event, case (ii) above proves to be impossible. Hence case (i) always occurs when $m''_k > 0$, and x' always appears within θ.

To complete the argument, we shall show that $\sum_{k \to k'} m''_k$, for all fixed k', is at most $3s_{k'}$. Let $p' = p_{k'}$ and let α' denote the pattern matched at stage k'. Let $k_1 < \cdots < k_r$ be the values of k such that $k \to k'$. If $|\alpha'| + p' \le m$, let $\beta'\alpha'$ be the rightmost $p' + |\alpha'|$ characters of the pattern. Otherwise let α'' be the leftmost $|\alpha'| + p' - m$ characters of α'; and let $\beta'\alpha'$ be α'' followed by the pattern. Notice that in both cases α' is an initial substring of $\beta'\alpha'$ and $|\beta'| = p'$. In both cases, the actions of the algorithm during stages $k_1 + 1$ through k' are completely known if we are given the pattern and β', and if we know z' and the place within β' where stage $k_1 + 1$ starts matching. This follows from the fact that β' by itself determines the text, so that if we match the pattern against the string $z'\beta'\beta'\beta' \ldots$ (starting at the specified place for stage $k_1 + 1$) until the algorithm first tries to match z' we will know the length of α'. (If $|\alpha'| < p'$ then β' begins with α' and this statement holds trivially; otherwise, α' begins with β' and has period p'; hence $\beta'\beta'\beta' \ldots$ begins with α'.) The algorithm cannot begin two different stages at exactly the same position within β', for then it would loop indefinitely, contradicting the fact that it does terminate. This property will be our key tool for proving the desired result.

Let the text strings matched during stages $k_1, \ldots, k_r$ be $\alpha_1, \ldots, \alpha_r$, and let their periods determined as in case (i) be $p_1, \ldots, p_r$, respectively; we have $p_j < \frac{1}{2}|\alpha_j|$ for $1 \le j \le r$. Suppose that during stage k_j the mismatch of $x_j \ne z_j$ implies that the pattern ends with $y_j\beta_j\alpha_j$, where $|\beta_j| = p_j$. We shall prove that $|\alpha_1| + \cdots + |\alpha_r| \le 3p'$. First let us prove that $|\alpha_j| < p'$ for all j: We have observed that x' always occurs within θ_j; hence $y_j\beta_j\alpha_j$ occurs as a rightmost substring of $x'\alpha'$. If $|\alpha_j| \ge p'$ then $p_j + p' \le |\beta_j\alpha_j|$; hence the character p_j positions to the right of y_j in $x'\alpha'$ is x_j, as is the character $p_j + p'$ positions to the right of y_j. But the character p' positions to the right of y_j in $x'\alpha'$ is y_j, since p' is a period of $x'\alpha'$; hence the character $p' + p_j$ positions to the right of y_j is also y_j, contradicting $x_j \ne y_j$.

Since $|\alpha_j| < p'$, each string α_j for $j \ge 2$ appears somewhere within β', when β' is regarded as a cyclic string, joined end-for-end. (It follows from the definition of $k \to k'$ that $z_j\alpha_j$ is a substring of α' for $j \ge 2$.) We shall prove that the rightmost halves of these strings, namely the rightmost $\lceil \frac{1}{2}|\alpha_j| \rceil$ characters as they appear in β', are disjoint. This implies that $\frac{1}{2}|\alpha_2| + \cdots + \frac{1}{2}|\alpha_r| \le p'$, and the proof will be complete (because $|\alpha_1| \le p'$).

Suppose therefore that the right half of the appearance of α_i overlaps the right half of the appearance of α_j within β', for some $2 \le i < j < k'$, where the rightmost character of α_i is within α_j. This means that the algorithm at stage k_i begins to match characters starting within α_j more than p_j characters to the right of z_j where $z_j\alpha_j$ appears in β', when the text α' is treated modulo p'. (Recall that $p_j < \frac{1}{2}|\alpha_j|$.) The pattern ends with $x_j\alpha_j$, and p_j is a period of $x_j\alpha_j$. The algorithm must work correctly when the text equals the pattern; so there must come a stage, before shifting the pattern to the right of the appearance of α_j, where the algorithm scans left until hitting z_j. At this point, call it stage k'', there must be a mismatch of $z_j \ne x_j$, since p_j or more characters have been matched. (The character p_j positions to the right of z_j is x_j, by periodicity.) Hence $k'' < k'$; and it follows that $k'' = k_i$. (If $k'' > k_i$, we have $z_i\alpha_i$, entirely contained within α'', but then $k_i \to k'$ implies that $k'' = k'$.) Now $k'' = k_i$ implies that $z_j = z_i$ and $x_j = x_i$. We shall obtain a contradiction by showing that the algorithm "synchronizes" its stage $k_i + 1$ behavior with its stage $k_j + 1$ behavior, modulo p', causing an infinite loop as remarked above. The main point is that the dd' table will specify shifting the pattern p_j steps, so that y_j is brought into the position corresponding to z_j, in stage k_i as well as in stage k_j. (Any lesser shift brings an x_j into position p_j spaces to the right of z_j; hence it puts $y_i = x_j$ into the position corresponding to z_j, by periodicity, contradicting $x_i \ne y_i$.) The amount of shift depends on the maximum of the d and dd' entries, and the d entry will be chosen (in either k_i or k_j) if and only if z_j is not a character of β_j; but in this case, the d entry will also specify the same shift both for stage k_i and stage k_j. ☐

The constant 6 in the theorem above is probably much too large, and the proof given seems to be much too long; the reader is invited to improve the theorem in either or both respects. An interesting example of the rather complex behavior possible with this algorithm occurs when the pattern is $b\psi_r$ and the text is $\psi_r a \psi_r$ for large r, where

$$\psi_0 = a, \qquad \psi_{n+1} = \psi_n\psi_n b\psi_n.$$

Corollary. *The worst-case running time of the Boyer–Moore algorithm with dd' replacing dd is $O(n+rm)$ character comparisons, if the pattern occurs r times in the text.*

Proof. Let $T(n,r)$ be the worst-case running time as a function of n and r, when m is fixed. The theorem implies that $T(n,0) \le 7n$, counting the mismatched characters as well as the matched ones. Furthermore, if

$r > 0$ and if the first appearance of the pattern ends at position n_0 we have $T(n, r) \le 7(n_0 - 1) + m + T(n - n_0 + m - 1, r - 1)$. It follows that $T(n, r) \le 7n + 8rm - 14r$. □

When the Boyer–Moore algorithm implicitly shifts the pattern to the right it forgets all that it "knows" about characters already matched; this is why the linearity theorem is not trivial. A more complex algorithm can be envisaged, with a finite number of states corresponding to which text characters are known to match the pattern in its current position; when in state q we fetch the character $x := text[k - t[q]]$, then we set $k := k + s[q, x]$ and go to state $q'[q, x]$. For example, consider the pattern *abacbaba*, and the specification of t, s, and q' in Table 3; exactly 41 distinguishable states can arise. An asterisk ($*$) in that table shows where the pattern has been fully matched.

The number of states in this generalization of the Boyer–Moore algorithm can be rather large, as the example shows, but the patterns that occur most often in practice probably do not imply many states. The number of states is always less than 2^m, and perhaps a much smaller upper bound is possible; it is unclear which patterns of a given length lead to the most states, and it does not seem obvious that this maximum number of states is exponential in m.

If the characters of the pattern are distinct, say $a_1 a_2 \ldots a_m$, this generalization of the Boyer–Moore algorithm leads to exactly $\frac{1}{2}(m^2 + m)$ states (namely, all states $\bullet \ldots \bullet a_k \bullet \ldots \bullet a_{j+1} \ldots a_m$ for $0 \le k < j \le m$, with a_k suppressed if $k = 0$). By merging several of these states we obtain the following simple algorithm, which uses a table $c[x]$ where $c[a_j] = m - j$ and $c[x] = -1$ when $x \notin \{a_1, \ldots, a_m\}$. The algorithm works only when all pattern characters are distinct, but it improves slightly on the Boyer–Moore technique in this important special case:

```
j := k := m;
while k ≤ n do
   begin i := c[text[k]];
      if i = 0 then
         begin for i := 1 step 1 until m − 1 do
            if text[k − i] ≠ pattern[m − i] then go to no match;
         matchfoundat(k − m);
      no match: j := m;
         end
      else if i + j ≥ m then j := i else j := m;
      k := k + j;
   end.
```

TABLE 3. States of a generalized Boyer–Moore algorithm

			$s[q,x]$, $q'[q,x]$			
state q	known characters	$t[q]$	$x = a$	$x = b$	$x = c$	other x
0	••••••••	0	0, 1	1, 8	4, 9	8, 0
1	•••••••a	1	7, 10	0, 2	7, 10	7, 10
2	••••••ba	2	0, 3	7, 10	2, 11	7, 10
3	•••••aba	3	5, 12	0, 4	5, 12	5, 12
4	••••baba	4	5, 12	5, 12	0, 5	5, 12
5	•••cbaba	5	0, 6	5, 12	5, 12	5, 12
6	••acbaba	6	5, 12	0, 7	5, 12	5, 12
7	•bacbaba	7	*5, 12	5, 12	5, 12	5, 12
8	••••••b•	0	0, 2	8, 0	8, 0	8, 0
9	•••c••••	0	0, 13	6, 14	4, 9	8, 0
10	a•••••••	0	0, 15	1, 8	4, 9	8, 0
11	•••cba••	0	0, 16	6, 14	8, 0	8, 0
12	aba•••••	0	0, 17	3, 18	4, 9	8, 0
13	•••c•••a	1	7, 10	0, 19	7, 10	7, 10
14	•b••••••	0	0, 20	3, 18	4, 9	8, 0
15	a••••••a	1	7, 10	0, 21	7, 10	7, 10
16	•••cba•a	1	7, 10	0, 5	7, 10	7, 10
17	aba••••a	1	7, 10	0, 22	7, 10	7, 10
18	••••b•••	0	0, 23	3, 24	8, 0	8, 0
19	•••c••ba	2	0, 25	7, 10	7, 10	7, 10
20	•b•••••a	1	7, 10	0, 26	7, 10	7, 10
21	a•••••ba	2	0, 27	7, 10	2, 11	7, 10
22	aba•••ba	2	0, 28	7, 10	2, 29	7, 10
23	••••b••a	1	7, 10	0, 30	7, 10	7, 10
24	•b••b•••	0	0, 31	3, 24	8, 0	8, 0
25	•••c•aba	3	5, 12	0, 5	5, 12	5, 12
26	•b••••ba	2	0, 32	7, 10	2, 11	7, 10
27	a••••aba	3	5, 12	0, 33	5, 12	5, 12
28	aba••aba	3	5, 12	0, 34	5, 12	5, 12
29	a••cba••	0	0, 35	6, 14	8, 0	8, 0
30	••••b•ba	2	0, 4	7, 10	7, 10	7, 10
31	•b••b••a	1	7, 10	0, 36	7, 10	7, 10
32	•b•••aba	3	5, 12	0, 37	5, 12	5, 12
33	a•••baba	4	5, 12	5, 12	0, 38	5, 12
34	aba•baba	4	5, 12	5, 12	*5, 12	5, 12
35	a••cba•a	1	7, 10	0, 38	7, 10	7, 10
36	•b••b•ba	2	0, 37	7, 10	7, 10	7, 10
37	•b••baba	4	5, 12	5, 12	0, 39	5, 12
38	a••cbaba	5	0, 40	5, 12	5, 12	5, 12
39	•b•cbaba	5	0, 7	5, 12	5, 12	5, 12
40	a•acbaba	6	5, 12	*5, 12	5, 12	5, 12

Let us close this section by making a preliminary investigation into the question of "fastest" pattern matching in strings, or *optimum* algorithms. What algorithm minimizes the number of text characters examined, over all conceivable algorithms for the problem we have been considering? In order to make this question nontrivial, we shall ask for the minimum *average* number of characters examined when finding *all* occurrences of the pattern in the text, where the average is taken uniformly with respect to strings of length n over a given alphabet. (The minimum worst-case number of characters examined is of no interest, since it is between $n - m$ and n for all patterns[4]; therefore we ask for the minimum average number. It might be argued that the minimum average number, taken over random strings, is of little interest, since people rarely search in random strings; they usually search for patterns that actually appear. However, the random-string model is a reasonable approximation when we consider those stretches of text that do not contain the pattern, and the algorithm obviously must examine every character in those places where the pattern does occur.)

The case of patterns of length 2 can be solved exactly; it is somewhat surprising to find that the analysis is not completely trivial even in this highly restricted case. Consider first the pattern ab where $a \neq b$. Let q be the alphabet size, $q \geq 2$. Let $f(n)$ denote the minimum average number of characters examined by an algorithm that finds all occurrences of the pattern in a random text of length n, and let $g(n)$ denote the minimum average number of characters examined in a random text of length $n+1$ that is known to begin with a, not counting the examination of the known first character. These functions can be computed by the following recurrence relations:

$$
\begin{aligned}
&f(0) = f(1) = g(0) = 0, \qquad g(1) = 1;\\
&f(n) = 1 + \min_{1\leq k\leq n}\Bigl(\frac{1}{q}(f(k-1) + g(n-k)) + \frac{1}{q}(g(k-1) + f(n-k))\\
&\qquad\qquad\qquad\qquad\qquad + \Bigl(1 - \frac{2}{q}\Bigr)(f(k-1) + f(n-k))\Bigr);\\
&g(n) = 1 + \frac{1}{q}g(n-1) + \Bigl(1 - \frac{1}{q}\Bigr)f(n-1), \qquad n \geq 2.
\end{aligned}
$$

[4] This observation is clear when we must find all occurrences of the pattern; R. L. Rivest has recently proved it also for algorithms that stop after finding one occurrence. ["On the worst-case behavior of string-searching algorithms," *SIAM Journal on Computing* **6** (1977), 669–674.]

The recurrence for f follows by considering which character is examined first; the recurrence for g follows from the fact that the second character must be examined in any case, so it can be examined first without loss of efficiency. It can be shown that the minimum is always assumed for $k = 2$; hence $f(n) = g(n-1) + 1/q$ for $n \geq 2$, and we obtain the closed form solution

$$\begin{aligned} f(n) &= \frac{n(q^2+q-1)}{q(2q-1)} - \frac{(q-1)(q^2+2q-1)}{q(2q-1)^2} + \frac{(1-q)^n}{q^{n-3}(q-1)(2q-1)^2}, \\ g(n) &= \frac{n(q^2+q-1)}{q(2q-1)} + \frac{(q-1)(q^2-3q+1)}{q(2q-1)^2} - \frac{(1-q)^n}{q^{n-2}(2q-1)^2}, \qquad n \geq 1. \end{aligned}$$

(To prove that these functions satisfy the stated recurrences reduces to showing that the minimum of

$$\left(\frac{1-q}{q}\right)^{k-1} + \left(\frac{1-q}{q}\right)^{n-k}$$

for $1 \leq k \leq n$ occurs for $k = 2$, whenever $n \geq 2$ and $q \geq 2$.)

If the pattern is aa, the recurrence for f changes to

$$f(n) = 1 + \min_{1\leq k\leq n}\left(\frac{1}{q}(g(k-1)+g(n-k)) + \left(1-\frac{1}{q}\right)(f(k-1)+f(n-k))\right),$$

when $n \geq 2$; but this is actually no change!

Hence the following is an optimum algorithm for all patterns of length 2, in the sense of minimum average text characters inspected to find all matches in a random string:

```
k := 2;
while k ≤ n do
  begin c := text[k];
  if c = pattern[2] and text[k − 1] = pattern[1] then
    matchfoundat(k − 2);
  while c = pattern[1] do
    begin k := k + 1; c := text[k];
    if c = pattern[2] then matchfoundat(k − 2);
    end;
  k := k + 2;
  end.
```

For patterns of length 3 the recurrence relations become more complex; they depend on more than simply the length of the strings and local knowledge about characters at the boundaries. The determination of an optimum strategy in this case remains an open problem.

The algorithm sketched at the beginning of this section shows that an average of $O(n(\log m)/m)$ bit inspections suffices for any pattern of length m over a binary alphabet. Clearly $\lfloor n/m \rfloor$ is a lower bound, since the algorithm must inspect at least one bit in any block of m consecutive bits. The pattern a^m can be handled with $O(n/m)$ bit inspections on the average; but it seems reasonable to conjecture that patterns of length m exist for arbitrarily large m, such that an average of at least $cn(\log m)/m$ bits must be inspected for all large n. Here c denotes a positive constant, independent of m and n.

Acknowledgment

Robert S. Boyer and J Strother Moore suggested many important improvements to early drafts of this postscript, especially in connection with errors in the author's first attempts at proving the linearity theorem.

Donald E. Knuth's work was supported in part by the National Science Foundation under grant GJ 36473X and by the Office of Naval Research under contract NR 044-402. James H. Morris, Jr.'s work was supported in part by the National Science Foundation under grant GP 7635 at the University of California, Berkeley. Vaughan R. Pratt's work was supported in part by the National Science Foundation under grant GP 6945 at the University of California, Berkeley, and under grant GJ 992 at Stanford University.

References

[1] Alfred V. Aho and Margaret J. Corasick, "Efficient string matching: An aid to bibliographic search," *Communications of the ACM* **18** (1975), 333–340.

[2] Alfred V. Aho, John E. Hopcroft, and Jeffrey D. Ullman, *The Design and Analysis of Computer Algorithms* (Reading, Massachusetts: Addison–Wesley, 1974).

[3] M. Beeler, R. W. Gosper, and R. Schroeppel, *HAKMEM*, Artificial Intelligence Memo No. 239 (Cambridge, Massachusetts: Massachusetts Institute of Technology, 29 February 1972), 95+10 pages.

[4] Robert S. Boyer and J Strother Moore, "A fast string searching algorithm," (Menlo Park, California: Stanford Research Institute, and Palo Alto, California: Xerox Palo Alto Research Center, 29 December 1975), 35 pages. Subsequently published in *Communications of the ACM* **20** (1977), 762–772; **22** (1979), 679–680.

[5] S. A. Cook, "Linear time simulation of deterministic two-way pushdown automata," *Information Processing 71*, Proceedings of IFIP Congress 71 (Amsterdam: North-Holland, 1972), 75–80.

[6] Pascal Diethelm and Peter Roizen, "An efficient linear search for a pattern in a string," unpublished manuscript (Geneva, Switzerland: World Health Organization, April 1972), 3 pages.

[7] N. J. Fine and H. S. Wilf, "Uniqueness theorems for periodic functions," *Proceedings of the American Mathematical Society* **16** (1965), 109–114.

[8] Michael J. Fischer and Michael S. Paterson, "String matching and other products," *SIAM–AMS Proceedings* **7** (Providence, Rhode Island: American Mathematical Society, 1974), 113–125.

[9] Zvi Galil, "On converting on-line algorithms into real-time and on real-time algorithms for string-matching and palindrome recognition," *SIGACT News* **7**, 4 (November–December 1975), 26–30.

[10] E. N. Gilbert, "Synchronization of binary messages," *IRE Transactions on Information Theory* **IT-6** (1960), 470–477.

[11] Sheila A. Greibach, "The hardest context-free language," *SIAM Journal on Computing* **2** (1973), 304–310.

[12] Malcolm C. Harrison, "Implementation of the substring test by hashing," *Communications of the ACM* **14** (1971), 777–779.

[13] Richard M. Karp, Raymond E. Miller, and Arnold L. Rosenberg, "Rapid identification of repeated patterns in strings, trees, and arrays," *ACM Symposium on Theory of Computing* **4** (1972), 125–136.

[14] Donald E. Knuth, *Fundamental Algorithms*, Volume 1 of *The Art of Computer Programming* (Reading, Massachusetts: Addison–Wesley, 1968).

[15] Donald E. Knuth, "Sequences with precisely $k+1$ k-blocks," Solution to problem E2307, *American Mathematical Monthly* **79** (1972), 773–774.

[16] Donald E. Knuth, "On the translation of languages from left to right," *Information and Control* **8** (1965), 607–639. [Reprinted with revisions as Chapter 15 of *Selected Papers on Computer Languages*, CSLI Lecture Notes 139 (Stanford, California: Center for the Study of Language and Information, 2003), 327–375.]

[17] Donald E. Knuth, "Structured programming with **go to** statements," *Computing Surveys* **6** (1974), 261–301. [Reprinted with

revisions as Chapter 2 of *Literate Programming*, CSLI Lecture Notes 27 (Stanford, California: Center for the Study of Language and Information, 1992), 17–89.]

[18] Donald E. Knuth, James H. Morris, Jr., and Vaughan R. Pratt, "Fast pattern matching in strings," Computer Science Technical Report STAN-CS-74-440 (Stanford University, 1974), i + 32 pages.

[19] R. C. Lyndon and M. P. Schützenberger, "The equation $a^M = b^N c^P$ in a free group," *Michigan Mathematical Journal* **9** (1962), 289–298.

[20] Glenn Manacher, "A new linear-time 'on-line' algorithm for finding the smallest initial palindrome of a string," *Journal of the Association for Computing Machinery* **22** (1975), 346–351.

[21] A. Markoff, "Sur une question de Jean Bernoulli," *Mathematische Annalen* **19** (1882), 27–36.

[22] J. H. Morris, Jr. and Vaughan R. Pratt, "A linear pattern-matching algorithm," Computing Center Technical Report 40 (Berkeley, California: University of California, 1970), 6 pages.

[23] А. О. Слисенко, «Распознавание предиката симметрии многоголовчатыми машинами тьюринга со входом», *Труды математического института имени В. А. Стеклова* **129** (1973), 30–202, 267. English translation: A. O. Slisenko, "Recognizing a symmetry predicate by multihead Turing machines with input," *Proceedings of the Steklov Institute of Mathematics* **129** (1976), 25–208.

[24] Ken Thompson, "Regular expression search algorithm," *Communications of the ACM* **11** (1968), 419–422.

[25] B. A. Venkov, *Elementary Number Theory* (Groningen, The Netherlands: Wolters-Noordhoff, 1970). [Translation of a Russian book originally published in 1937.]

[26] Peter Weiner, "Linear pattern matching algorithms," *IEEE Symposium on Switching and Automata Theory* **14** (1973), 1–11.

Addendum

This paper has spawned an enormous number of sequels, so the following remarks are confined to works that are closely related to issues that were discussed explicitly above.

Kurt Mehlhorn wrote me in October 1977 to point out that my preprocessing program for the Boyer–Moore algorithm with pattern aaa gives $dd'[2] = 4$, while the correct answer is 3. He explained how to cure the problem for all patterns, by adding eight more lines to the program.

Ole-Johan Dahl visited Stanford during autumn quarter 1978, and used his expertise in algorithm verification to devise a new program that is not only simpler than my original, it is also correct:

```
for k := 1 step 1 until m do dd[k] := dd'[k] := 0;
j := m;  t := m+1;
while j > 0 do
   begin f[j] := t;
   while t ≤ m and pattern[j] ≠ pattern[t] do
      begin if dd'[t] = 0 then dd'[t] := m − j;
      t := f[t];
      end;
   t := t − 1;  j := j − 1;
   if dd[t] = 0 then dd[t] := m − j;
   end;
for k := 1 step 1 until t do if dd[k] = 0 then dd[k] := m+t−k;
for k := 1 step 1 until m do
   begin if t < k then t := f[t];
   if dd'[k] = 0 then dd'[k] := m+t−k;
   end;
```

Let $R(s)$ be the statement that

$$pattern[s{+}1] \ldots pattern[m] = pattern[1] \ldots pattern[m{-}s].$$

Dahl's invariant relation for the final loop on k in his program was

$$(t \geq k-1) \text{ and } R(t) \text{ and } (\text{if } k-1 \leq s < t \text{ then not } R(s)).$$

Actually, he noted that this invariant fails when $k = 1$ and $s = 0$; but he pointed out that he could have set $f[0] := t$ and $t := 0$ just before the loop, making the invariant true, and causing the algorithm to set $t := f[0]$ immediately!

If the statement '$dd'[0] := m + t$' is inserted just before Dahl's final loop on k, Zvi Galil noted that the Boyer–Moore algorithm can be streamlined further by changing its final lines to

```
      if j = 0 then matchfoundat(k);
      k := k + max(d[text[k]], dd'[j]);
      end;
   end.
```

["On improving the worst case running time of the Boyer-Moore string matching algorithm," *Communications of the ACM* **22** (1979), 505–508.

We do need to assume that *text*[0] is a valid character.] Galil also showed how to make the algorithm run in worst-case time $O(n)$ even when the pattern appears more than $2n/m$ times, because the pattern has period $f[0] \leq m/2$ in such cases.

I'm not sure why neither Mehlhorn nor Dahl submitted their observations for publication at the time. Shortly afterwards a more complicated correction to my faulty algorithm was published independently by Wojciech Rytter, "A correct preprocessing algorithm for Boyer–Moore string-searching," *SIAM Journal on Computing* **9** (1980), 509–512. Mehlhorn's correction was eventually published by G. de V. Smit, "A comparison of three string matching algorithms," *Software — Practice & Experience* **12** (1982), 57–66, page 62; corrected by Jalel Mzali and Jean-Jacques Thiel, *Software — Practice & Experience* **13** (1983), 1095.

My theorem that the dd'-style Boyer–Moore algorithm makes at most $7n$ character comparisons when the pattern isn't present was improved to an upper bound of $4n$ by Leo J. Guibas and Andrew M. Odlyzko ["A new proof of the linearity of the Boyer-Moore string searching algorithm," *SIAM Journal on Computing* **9** (1980), 672–682]. They conjectured that $2n$ might in fact be provable. But many years later, Richard Cole ["Tight bounds on the complexity of the Boyer–Moore string matching algorithm," *SIAM Journal on Computing* **23** (1994), 1075–1091] finally succeeded in taming this difficult problem by proving both upper and lower worst-case bounds of $3n$, plus lower-order terms.

The maximum number of states in an extended Boyer–Moore automaton such as Table 3 is still unknown. Guibas and Odlyzko (unpublished) have found examples in a 3-letter alphabet that require $\Omega(m^3)$ states. Is exponential growth possible?

The conjecture at the end of the paper, that $\Theta(n(\log m)/m)$ bits must be inspected on the average by any algorithm that finds all occurrences of certain m-bit patterns in an n-bit string, was resolved by Andrew Chi-Chih Yao ["The complexity of pattern matching for a random string," *SIAM Journal on Computing* **8** (1979), 368–387]. He proved, in fact, that the analogous result for q-letter alphabets holds for almost all patterns.

My analysis of optimum algorithms for patterns of length 2, in the sense of minimum expected number of character inspections when the text is uniformly random, has not yet been extended to patterns of length 3, as far as I know. Michael Paterson observes that the problem of searching optimally for the pattern *aaa* is rather like a one-dimensional version of the game known as "Battleship."

I learned in 2012 that Yuri Matiyasevich had anticipated the linear-time pattern matching and pattern preprocessing algorithms of this paper, in the special case of a binary alphabet, already in 1969. He presented them as constructions for a Turing machine with a two-dimensional working memory in the paper "Real-time recognition of the inclusion relation," *Journal of Soviet Mathematics* **1** (1973), 64–70, which is a translation of his original Russian article in *Записки Научных Семинаров Ленинградского Отделения Математического Института имени В. А. Стеклова* **20** (1971), 104–114].

Chapter 10

Addition Machines

[Written with Robert W. Floyd. Originally published in SIAM Journal on Computing **19** *(1990), 329–340.]*

It is possible to compute $\gcd(x,y)$ *efficiently with only* $O(\log xy)$ *additions and subtractions, when three arithmetic registers are available but not when there are only two. Several other functions, such as* $x^y \bmod z$*, are also efficiently computable in a small number of registers, using only addition, subtraction, and comparison.*

An addition machine is a computing device with a finite number of registers, limited to the following six types of operations:

read x	{input to register x}
$x \leftarrow y$	{copy register y to register x}
$x \leftarrow x + y$	{add register y to register x}
$x \leftarrow x - y$	{subtract register y from register x}
if $x \geq y$	{compare register x to register y}
write x	{output from register x}

The register contents are assumed to belong to a given set A, which is an additive subgroup of the real numbers. If A is the set of all integers, we say the device is an *integer addition machine*; if A is the set of all real numbers, we say the device is a *real addition machine.*

We will consider how efficiently an integer addition machine can do operations such as multiplication, division, greatest common divisor, exponentiation, and sorting. We will also show that any addition machine with at least six registers can compute the ternary operation $x\lfloor y/z\rfloor$ with reasonable efficiency, given $x, y, z \in A$ with $z \neq 0$.

Remainders

As a first example, consider the calculation of

$$x \bmod y = \begin{cases} x - y\lfloor x/y\rfloor\,, & \text{if } y \neq 0; \\ x\,, & \text{if } y = 0. \end{cases}$$

This binary operation is well defined on any additive subgroup A of the reals, and we can easily compute it on an addition machine as follows:

```
P1: read x; read y; z ← z − z;
    if y ≥ z then
       if z ≥ y then {y = 0, do nothing}
       else if x ≥ z then while x ≥ y do x ← x − y
          else repeat x ← x + y until x ≥ z
    else if z ≥ x then while y ≥ x do x ← x − y
          else repeat x ← x + y until z ≥ x;
    write x.
```

(There is implicitly a finite-state control. A pidgin Pascal program such as this one is easily converted to other formalisms, as in [4].)

Program *P1* handles all sign combinations of x and y; therefore it is rather messy. In the special case where $x \geq 0$ and $y > 0$, a much simpler program applies:

```
P2: read x; read y;   {assume that x ≥ 0 and y > 0}
    while x ≥ y do x ← x − y;
    write x.
```

Any program for this special case can be converted to a program of comparable efficiency for the general case by using the identities

$$-x = (x - x) - x;$$

$$(-x) \bmod (-y) = -(x \bmod y);$$

$$(-x) \bmod y = \begin{cases} y - (x \bmod y), & \text{if } x \bmod y \neq 0; \\ 0, & \text{if } x \bmod y = 0. \end{cases}$$

General programs for multiplication, division, and gcd can, similarly, be constructed from algorithms that assume positive operands. We shall therefore restrict consideration to positive cases in the algorithms below.

Program *P2* performs $\lfloor x/y \rfloor$ subtractions. Can we do better? Yes; here, for example, is a program that uses a doubling procedure to subtract larger multiples of y:

```
P3: read x; read y;   {assume that x ≥ 0 and y > 0}
    while x ≥ y do
       begin z ← y;
       repeat w ← z; z ← z + z; until not x ≥ z;
       x ← x − w;
       end;
    write x.
```

This program repeatedly subtracts $2^k y$ from x, where $k = \lfloor \log_2(x/y) \rfloor$; thus, it implicitly computes the binary representation of $\lfloor x/y \rfloor$, from left to right. The total running time of *P3* is bounded by $O\big(\log(x/y)\big)^2$, which is considerably smaller than $\lfloor x/y \rfloor$ when $\lfloor x/y \rfloor$ is large.

Further improvement, to a running time that is only $O\big(\log(x/y)\big)$ instead of $O\big(\log(x/y)\big)^2$, appears at first sight to be impossible, because an addition machine has only finitely many registers and it cannot divide by 2. Therefore the numbers y, $2y$, $4y$, $8y$, $\ldots$ must all apparently be computed again and again if we want to use a trick based on doubling.

A Fibonacci Method

Remainders can, however, be computed with the desired efficiency $O\big(\log(x/y)\big)$ if we implicitly use the Fibonacci representation of $\lfloor x/y \rfloor$ instead of the binary representation. Define Fibonacci numbers as usual by

$$F_0 = 0\,; \quad F_1 = 1\,; \quad F_n = F_{n-1} + F_{n-2}\,, \qquad \text{for } n \ge 2\,.$$

Every nonnegative integer n can be uniquely represented [9] in the form

$$n = F_{l_1} + F_{l_2} + \cdots + F_{l_t}\,, \qquad l_1 \gg l_2 \gg \cdots \gg l_t \gg 0\,,$$

where $t \ge 0$ and $l \gg l'$ means that $l - l' \ge 2$. If $n > 0$, this representation can be found by choosing l_1 such that

$$F_{l_1} \le n < F_{l_1+1}\,,$$

so that $n - F_{l_1} < F_{l_1+1} - F_{l_1} = F_{l_1-1}$, and by writing

$$n = F_{l_1} + (\text{Fibonacci representation of } n - F_{l_1})\,.$$

We shall let

$$\lambda n = l_1 \qquad \text{and} \qquad \nu n = t$$

denote respectively the index of the leading term and the number of terms, in the Fibonacci representation of n. By convention, $\lambda 0 = 1$.

Fibonacci numbers are well suited to addition machines because we can go from the pair $\langle F_l, F_{l+1} \rangle$ up to the next pair $\langle F_{l+1}, F_{l+2} \rangle$ with a single addition, or down to the previous pair $\langle F_{l-1}, F_l \rangle$ with a single subtraction. Furthermore, Fibonacci numbers grow exponentially, about 69% as fast as powers of 2. They have been used as power-of-2 analogs in a variety of algorithms (see, for instance, "Fibonacci numbers" in the index to [2]), and in Matijasevič's solution to Hilbert's tenth problem [6].

If we let two registers of an addition machine contain the pair of numbers $\langle yF_l, yF_{l+1}\rangle$, where l is an implicit parameter, it is easy to implement the operations

$$l \leftarrow 1\,, \qquad l \leftarrow l+1\,, \qquad l \leftarrow l-1$$

and to test the conditions

$$x \ge yF_l\,, \qquad x < yF_{l+1}\,, \qquad l = 1\,.$$

Therefore we can compute $x \bmod y$ efficiently by implementing the following procedure:

```
read x;  read y;   {assume that x ≥ 0 and y > 0}
if x ≥ y then
   begin l ← 1;
   repeat l ← l + 1 until x < yF_{l+1};
   repeat if x ≥ yF_l then x ← x − yF_l;
      l ← l − 1;
   until l = 1;
   end;
write x.
```

The first **repeat** loop increases l until we have

$$yF_l \le x < yF_{l+1}\,,$$

that is, until $l = \lambda n$, where $n = \lfloor x/y \rfloor$. The second loop decreases l while subtracting

$$yF_{l_1} + yF_{l_2} + \cdots + yF_{l_t} = yn$$

from x according to the Fibonacci representation of n. The final result, $x - ny = x \bmod y$, has been computed with

$$2\lambda n - 2 + \nu n = O\bigl(\log(x/y)\bigr)$$

additions and subtractions altogether.

Here is the same program expressed directly in terms of additions and subtractions, using only three registers:

```
P4: read x;  read y;   {assume that x ≥ 0 and y > 0}
    if x ≥ y then
       begin z ← y;
```

```
        repeat ⟨y, z⟩ ← ⟨z, y + z⟩ until not x ≥ z;
        {at this point x ≥ y still holds}
        repeat if x ≥ y then x ← x − y;
           ⟨y, z⟩ ← ⟨z − y, y⟩;
        until y ≥ z;
        end;
     write x.
```

The multiple assignment '$\langle y, z\rangle \leftarrow \langle z, y+z\rangle$' is an abbreviation for the operation "set $y \leftarrow y+z$ and interchange the roles of registers y and z in the subsequent program"; the assignment '$\langle y, z\rangle \leftarrow \langle z-y, y\rangle$' is similar. By making two copies of this program code, in one of which the variables y and z are interchanged, we can jump from one copy to the other and obtain a legitimate addition-machine program. (See [3, Example 7].)

A formal proof of correctness for program *P4* would establish the invariant relation

$$\exists l \geq 1\,(y = y_0 F_l \text{ and } z = y_0 F_{l+1})$$

in the case $x_0 \geq y_0$, where x_0 and y_0 are the initial values of x and y.

Multiplication and Division

We can use essentially the same idea to compute the ternary operation $x\lfloor y/z \rfloor$ efficiently on any addition machine. This time we accumulate multiples of x as we discover the Fibonacci representation of $\lfloor y/z \rfloor$:

```
     read x; read y; read z;    {assume that y ≥ 0 and z > 0}
     w ← 0;
     if y ≥ z then
        begin l ← 1;
        repeat l ← l + 1 until y < zF_{l+1};
        repeat if y ≥ zF_l then ⟨w, y⟩ ← ⟨w + xF_l, y − zF_l⟩;
           l ← l − 1;
        until l = 1;
        end;
     write w.
```

The actual addition-machine code requires six registers, because we need Fibonacci multiples of x as well as z:

```
P5:  read x; read y; read z;    {assume that y ≥ 0 and z > 0}
     w ← w − w;
     if y ≥ z then
```

```
    begin u ← x;  v ← z;
    repeat ⟨u, x⟩ ← ⟨x, u + x⟩;  ⟨v, z⟩ ← ⟨z, v + z⟩;
    until not y ≥ z;    {y ≥ v still holds}
    repeat if y ≥ v then ⟨w, y⟩ ← ⟨w + u, y − v⟩;
      ⟨u, x⟩ ← ⟨x − u, u⟩;  ⟨v, z⟩ ← ⟨z − v, v⟩;
    until v ≥ z;
    end;
  write w.
```

The key invariant relations, in the case $y_0 \geq z_0$, are now

$$\exists\, l \geq 1\,(u = x_0 F_l\,,\ x = x_0 F_{l+1}\,,\ v = z_0 F_l\,,\ z = z_0 F_{l+1})\,;$$
$$\exists\, n \geq 0\,(w = x_0 n\,,\ y = y_0 - z_0 n)\,.$$

If we suppress x, u, and w from this program, the **repeat** statements act on $\langle y, v, z\rangle$ exactly as the **repeat** statements in our previous program act on $\langle x, y, z\rangle$. Therefore, if $y_0 \geq z_0$, we have

$$y = y_0 \bmod z_0 = y_0 - z_0 \lfloor y_0/z_0 \rfloor$$

after the **repeat** statements in the new program. Hence $w = x_0 \lfloor y_0/z_0 \rfloor$ as desired. The total number of additions and subtractions is

$$4\lambda n - 3 + 2\nu n = O\bigl(\log(y_0/z_0)\bigr)\,,$$

where $n = \lfloor y_0/z_0 \rfloor$.

An integer addition machine can make use of the constant 1 by reading that constant into a separate, dedicated register. Then we can specialize the ternary algorithm by setting $z \leftarrow 1$ (for multiplication) or $x \leftarrow 1$ (for division). Thus we can compute the product xy in $O\bigl(\log \min(|x|, |y|)\bigr)$ operations, and the quotient $\lfloor y/z \rfloor$ in $O(\log |y/z|)$ operations, using only addition, subtraction, and comparison of integers. (Multiplication and division clearly cannot be done unless such constants are used, since any function $f(x, y, \ldots)$ computed by an addition machine that inputs the sequence of values $\langle x, y, \ldots\rangle$ must satisfy $f(\alpha x, \alpha y, \ldots) = \alpha f(x, y, \ldots)$ for all $\alpha > 0$.)

Greatest Common Divisors

Euclid's algorithm for the greatest common divisor of two positive integers x and y can be formulated as follows:

```
read x;  read y;    {assume that x > 0 and y ≥ 0}
while y > 0 do ⟨x, y⟩ ← ⟨y, x mod y⟩;
write x.
```

The **while** loop preserves the invariant relation $\gcd(x, y) = \gcd(x_0, y_0)$. After the first iteration, we have $x > y \geq 0$; the successive values of x are strictly decreasing and positive, so the algorithm must terminate.

We can therefore use our method for computing $x \bmod y$ to calculate $\gcd(x, y)$ on an integer addition machine:

```
P6: read x; read y;   {assume that x > 0 and y ≥ 0}
    z ← y; z ← z + z;
    while not y ≥ z do   {equivalently, y > 0, since z = 2y}
      begin while x ≥ z do ⟨y, z⟩ ← ⟨z, y + z⟩;
      repeat if x ≥ y then x ← x − y;
        ⟨y, z⟩ ← ⟨z − y, y⟩;
      until y ≥ z;
      ⟨x, y⟩ ← ⟨y, x⟩; z ← y; z ← z + z;
      end;
    write x.
```

(Here the operation $\langle x, y\rangle \leftarrow \langle y, x\rangle$ should not really be performed; it means that the roles of registers x and y should be interchanged. The implementation jumps between six copies of this program, one for each permutation of the register names x, y, z.)

This algorithm will compute $\gcd(x, y)$ correctly on a general addition machine, whenever the ratio y/x is rational. Otherwise it will loop forever.

The total number of operations performed by program *P6* is

$$T(x, y) = f(q_1) + f(q_2) + \cdots + f(q_m) + 6\,,$$

when $y > 0$, where $\langle q_1, q_2, \ldots, q_m\rangle$ is the sequence of quotients $\lfloor x/y\rfloor$ in the respective iterations of Euclid's algorithm, and where $f(q)$ counts the number of operations in one iteration of the outermost **while** loop. If $q = 0$ (this case can occur only on the first iteration), we have one assignment, one addition, one subtraction, and four comparisons; so $f(0) = 7$. If $q > 0$ we have one assignment, $\lambda q - 1$ additions, $\lambda q + \nu q - 1$ subtractions, and $3\lambda q - 2$ comparisons; so

$$f(q) = 5\lambda q + \nu q - 3$$

in that case. We have $f(1) = 8$, $f(2) = 13$, and, in general, $f(F_l) = 5l - 2$ for all $l \geq 2$.

This three-register algorithm for greatest common divisor turns out to be quite efficient, even though it uses only addition, subtraction, and comparison. Indeed, the numbers in the registers never exceed $2\max(x, y)$, where x and y are the given inputs, and we can obtain rather precise bounds on the running time.

Theorem 1. *Let $N = \max(x,y)/\gcd(x,y)$. The number of operations $T(x,y)$ performed by program P6 satisfies*

$$3\log_\phi N + \alpha \le T(x,y) \le 13.5\log_\phi N + \beta\,,$$

for some constants α and β, where $\phi = (1+\sqrt{5}\,)/2$.

Proof. We can assume that $x > y$; then all the q's are positive. If $F_l \le q < F_{l+1}$ we have $\lambda q = l$ and $1 \le \nu q \le l/2$, hence

$$5l - 2 \le f(q) \le 5.5l - 3\,.$$

Furthermore we have $\phi^{l-2} \le F_l \le \phi^{l-1}$, hence

$$5\log_\phi(q+1) - 2 \le f(q) \le 5.5\log_\phi q + 8\,.$$

Summing over all values $q_1, \ldots, q_m$ gives

$$5\log_\phi\bigl((q_1{+}1)\ \ldots\ (q_m{+}1)\bigr) - 2m \le T(x,y) - 6 \le 5.5\log_\phi(q_1 \ldots q_m) + 8m\,.$$

Now let the values occurring in Euclid's algorithm be $x_0, x_1, \ldots, x_m, x_{m+1}$, where $x_0 = x$, $x_1 = y$, $x_{j+1} = x_{j-1} \bmod x_j$, $x_m = \gcd(x,y)$, and $x_{m+1} = 0$. Then $q_j = \lfloor x_{j-1}/x_j \rfloor$ for $1 \le j \le m$, and we have

$$q_1 q_2 \ldots q_m \le \frac{x_0}{x_1}\,\frac{x_1}{x_2}\ \ldots\ \frac{x_{m-1}}{x_m} < (q_1+1)(q_2+1)\ldots(q_m+1)\,.$$

The product $(x_0/x_1)(x_1/x_2)\ \ldots\ (x_{m-1}/x_m) = x_0/x_m$ is just what we have called N. Furthermore we have $m \le \log_\phi N$ by a well-known theorem of Lamé [1, Theorem 4.5.3F]. □

When the inputs $\langle x, y\rangle$ to program P6 are consecutive Fibonacci numbers $\langle F_m, F_{m+1}\rangle$ with $m \ge 2$, we have $q_1 = 0$, $q_2 = \cdots = q_{m-1} = 1$, $q_m = 2$, and the total running time is

$$T(F_m, F_{m+1}) = 7 + 8(m-2) + 13 + 6 = 8m + 10\,.$$

This appears to be the worst case, in the sense that it seems to maximize $T(x,y)$ over all pairs $\langle x,y\rangle$ with $\max(x,y) \le F_{m+1}$. Computations for small n support this conjecture, which (if true) would imply that the upper bound in Theorem 1 could be improved to $8\log_\phi N + \beta$.

Stacks and Sorting

Euclid's algorithm defines a one-to-one correspondence between pairs of relatively prime positive integers $\langle x, y\rangle$ with $x > y$ and sequences of positive integers $\langle q_1, \ldots, q_m\rangle$ where each $q_j \geq 1$ and $q_m \geq 2$. We can push a new integer q onto the front of such a sequence by setting $\langle x, y\rangle \leftarrow \langle qx + y, x\rangle$; we can pop $q_1 = \lfloor x/y \rfloor$ from the front by setting $\langle x, y\rangle \leftarrow \langle y, x \bmod y\rangle$.

Therefore an integer addition machine can represent a stack of arbitrary depth in two of its registers. The operation of pushing or popping a positive integer q can be done with $O(\log q)$ operations, using a few auxiliary registers.

Here, for example, is the outline of an integer addition program that reads a sequence of positive integers followed by zero and writes out those positive integers in reverse order:

$\langle x, y\rangle \leftarrow \langle 2, 1\rangle$; {the empty stack}
repeat read q;
  if $q \geq 1$ **then** $\langle x, y\rangle \leftarrow \langle qx + y, x\rangle$;
until not $q \geq 1$;
repeat $\langle q, x, y\rangle \leftarrow \langle \lfloor x/y \rfloor, y, x \bmod y\rangle$;
  if $y \geq 1$ **then write** q;
until not $y \geq 1$.

This program uses the algorithms for multiplication and division shown earlier. The total running time to reverse the input $\langle q_1, q_2, \ldots, q_m, 0\rangle$ is $O(m + \log q_1 q_2 \ldots q_m)$.

We can sort a given list of positive integers $\langle q_1, q_2, \ldots, q_m\rangle$ in a similar way, using the classical algorithms for merge sorting with three or more magnetic tapes that can be "read backwards" [2, Section 5.4.4]. The basic operations required are essentially those of a stack; so we can sort in $O\big((m + \log q_1 q_2 \ldots q_m) \log m\big)$ steps if there are at least 12 registers.

Exponentiation mod *z*

We can now show that an integer addition machine is able to compute

$$x^y \bmod z$$

in $O\big((\log y)(\log z) + \log(x/z)\big)$ operations. The basic idea is simple: We first form the numbers

$$x_l = x^{F_l} \bmod z$$

for $2 \leq l \leq \lambda y$; this requires one multiplication mod z for each new value of l, once $x_2 = x \bmod z$ has been found in $O\big(\log(x/z)\big)$ operations.

Then we use the Fibonacci representation of y to compute $x^y \bmod z$ with $\nu y - 1$ further multiplications mod z. For example, $x^{11} \bmod z$ is computed by successively forming the powers

$$x^1, \quad x^2, \quad x^3, \quad x^5, \quad x^8, \quad x^{8+3}$$

modulo z.

There is, however, a difficulty in carrying out this plan with only finitely many registers, since the method we have used to discover the Fibonacci representation of y determines the relevant terms F_l in reverse order from the way we need to calculate the relevant factors x_l.

One solution is to push the numbers $x_2, x_3, \ldots, x_{\lambda y}$ onto a simulated stack as they are being computed. Then we can pop them off in the desired order as we discover the Fibonacci representation of y. Each stack operation takes $O(\log z)$ time, since each x_l is less than z; hence the stacking and unstacking requires only $O\bigl((\log y)(\log z)\bigr)$ operations, and the overall running time changes by at most a constant factor.

But the stacking operation forms extremely large integers, having $\Theta\bigl((\log y)(\log z)\bigr)$ bits, so it is not a practical solution if we are concerned with the size of the numbers being added and subtracted as well as the number of additions and subtractions. An algorithm that needs only $O\bigl((\log y)(\log z)\bigr)$ additions and subtractions of integers that never get much larger than z would be far more useful in practice.

We can obtain such an algorithm if we first compute the "Fibonacci reflection" of y, namely the number

$$y^R = F_{2+\lambda y - l_1} + F_{2+\lambda y - l_2} + \cdots + F_{2+\lambda y - l_t}$$

when y has the Fibonacci representation

$$y = F_{l_1} + F_{l_2} + \cdots + F_{l_t}\,.$$

Then we can use the Fibonacci representation of y^R to determine the relevant factors x_l as we compute them; no stack is needed.

Here is a program that computes y^R, assuming that $y > 0$ and that both y and the constant 1 have already been read into registers named y and 1.

$u \leftarrow 1;\;\; v \leftarrow 1;\quad \{u = F_l,\; v = F_{l+1},\; l = 1\}$
repeat $\langle u, v\rangle \leftarrow \langle v, u+v\rangle$ **until not** $y \geq v$;
  $\{\text{now } u = F_l,\; v = F_{l+1},\; l = \lambda y\}$
$r \leftarrow 1;\;\; s \leftarrow 1;\;\; t \leftarrow t - t;\;\; w \leftarrow y;$

```
repeat if w ≥ u then
      begin w ← w − u;  t ← t + s;
      end;
   ⟨u, v⟩ ← ⟨v − u, u⟩;  ⟨r, s⟩ ← ⟨s, r + s⟩;    {l ← l − 1}
until u ≥ v.
```

Throughout this program we have $u = F_l$ and $v = F_{l+1}$, where l begins at 1, rises to λy, and returns to 1. During the second **repeat** statement we have also

$$r = F_{1+\lambda y - l}\,, \qquad s = F_{2+\lambda y - l}\,, \qquad t = (y - w)^R\,.$$

The program terminates with $l = 1$ and $w = 0$; hence we have

$$r = F_{\lambda y}\,, \qquad s = F_{\lambda y + 1}\,, \qquad t = y^R\,.$$

The full program for $x^y \bmod z$ can now be written as follows, using routines described earlier:

```
read x;  read y;  read z;      {assume that x ≥ 0, y > 0, z > 0}
⟨r, s, t⟩ ← ⟨F_λy, F_λy+1, y^R⟩;
x ← x mod z;  w ← x;  u ← 1;    {x = x_l, w = x_l+1, l = 1}
repeat if t ≥ r then
      begin t ← t − r;  u ← (uw) mod z;
      end;
   ⟨r, s⟩ ← ⟨s − r, r⟩;  ⟨x, w⟩ ← ⟨w, (xw) mod z⟩;    {l ← l + 1}
until r ≥ s;
write u.
```

The invariant relations

$$x = x_l\,, \qquad w = x_{l+1}\,, \qquad r = F_{1+\lambda y - l}\,, \qquad s = F_{2+\lambda y - l}$$

are maintained in the final **repeat** loop as l increases from 1 to λy.

For example, if $y = 11 = 8 + 3 = F_6 + F_4$, we have $\lambda y = 6$ and $y^R = F_2 + F_4 = 1 + 3 = 4$. Hence $r = 8$, $s = 13$, $t = 4$, $u = 1$, and $x = w = x_0 \bmod z$ at the beginning of the final **repeat**. The registers will then contain the following respective values at the moments when the final **until** statement is encountered:

r	s	t	u	x	w
5	8	4	1	$x_0 \bmod z$	$x_0^2 \bmod z$
3	5	4	1	$x_0^2 \bmod z$	$x_0^3 \bmod z$
2	3	1	$x_0^3 \bmod z$	$x_0^3 \bmod z$	$x_0^5 \bmod z$
1	2	1	$x_0^3 \bmod z$	$x_0^5 \bmod z$	$x_0^8 \bmod z$
1	1	0	$x_0^{11} \bmod z$	$x_0^8 \bmod z$	$x_0^{13} \bmod z$

The statement '$u \leftarrow (uw) \bmod z$' can be implemented by first forming uw and then taking the remainder mod z, using the multiplication and division algorithms presented earlier. But we can do better by changing the multiplication algorithm so that the quantities being added together for the final product are maintained modulo z: We simply change appropriate operations of the form $\alpha \leftarrow \alpha + \beta$ to the sequence

$\alpha \leftarrow \alpha + \beta$;
if $\alpha \geq z$ **then** $\alpha \leftarrow \alpha - z$.

Then the register contents never get large. In fact, if x and y are initially nonnegative and less than z, all numbers in the algorithm will be nonnegative and less than $2z$. We have proved the following result:

Theorem 2. *If* $0 \leq x < z \leq 2^{n-1}$ *and* $0 \leq y < z \leq 2^{n-1}$, *the quantity* $x^y \bmod z$ *can be computed from* x, y, *and* z *with* $O\bigl((\log y)(\log z)\bigr)$ *additions and subtractions of integers in the interval* $[0\,..\,2^n)$, *on a machine with finitely many registers.* □

Indeed, the constant implied by this O is reasonably small. The algorithm just sketched may therefore find practical application in the design of special-purpose hardware for $x^y \bmod z$, which is the fundamental operation required by the RSA scheme of encoding and decoding messages [7].

Lower Bounds

Some of the algorithms presented above can be shown to be optimal, up to a constant factor. For example, we obviously need $\Omega\bigl(\log \min(x,y)\bigr)$ additions to compute the product xy; we cannot compute any number larger than $2^k \max(x,y)$ with k additions, and if $2^k < \min(x,y)$ this is less than $\min(x,y)\max(x,y) = xy$.

Logarithmic time is also necessary for division and gcd, even if we extend addition machines to *addition-multiplication machines* (which can perform multiplication as well as addition in one step). An elegant proof of this lower bound was given by L. J. Stockmeyer in an unpublished report [8]. We reproduce his proof here for completeness.

Theorem 3 (Stockmeyer). *An integer addition-multiplication machine requires* $\Omega(\log x)$ *arithmetic operations to compute* $\lfloor x/2 \rfloor$, $x \bmod 2$, *or* $\gcd(x,2)$, *for infinitely many* x.

Proof. If we can compute $\lfloor x/2 \rfloor$ or $\gcd(x,2)$ in t steps, we can compute $x \bmod 2 = x - 2\lfloor x/2 \rfloor = 2 - \gcd(x,2)$ in at most $t+2$ steps. So it suffices to prove that $x \bmod 2$ requires $\Omega(\log x)$ steps.

Any computation of an integer addition-multiplication machine on a given input x forms polynomials in x and compares polynomial values. A t-step computation defines at most 2^t different *computation paths*, depending on the results of **if** tests. For convenience we assume that each statement of the form '**write** w' is changed to

if $0 \geq w$ **then write** w **else write** w.

Then a program that computes $x \bmod 2$ must take a different path when x is changed to $x + 1$.

Each computation path is defined by a sequence of polynomial tests

$$q_1(x) : 0, \qquad q_2(x) : 0, \quad \ldots, \quad q_s(x) : 0$$

made at times $t_1 < t_2 < \cdots < t_s \leq t$. (Different paths have different polynomials in general, although $q_1(x)$ will be the same on each path.) If $q_j(x)$ corresponds to a test at time t_j, the degree of $q_j(x)$ is at most $2^{t_j - 1}$. Therefore the sum of the degrees of the $q_j(x)$ is less than 2^t. Therefore the total number of roots of all the polynomials $q_j(x)$, taken over all computation paths of length t, is less than 2^{2t}.

Let m be the least integer $\geq 2^{2t}$ such that none of the polynomials described in the previous paragraph has a root in the closed interval $[m \mathinner{.\,.} m+1]$. Each root can exclude at most two values of m; therefore $m \leq 2^{2t} + 2^{2t+1}$. By definition, the addition-multiplication program takes the same computation path when it is applied to $x = m$ and to $x = m + 1$; therefore it does not compute $x \bmod 2$ on both of these values. Therefore there is an integer x_t in the interval $[2^{2t} \mathinner{.\,.} 2^{2t+2})$ such that the value $x_t \bmod 2$ has not been computed at time t on any of the computation paths. Therefore there are infinitely many x for which the time to compute $x \bmod 2$ is $\Omega(\log x)$.

So far we have counted both arithmetic operations and conditional tests as steps of the computation. This also gives a lower bound on the number of arithmetic operations, since we can assume without loss of generality that no computation path makes more than $\binom{k}{2}$ consecutive conditional tests when there are k registers. □

Notice that Stockmeyer's argument establishes the lower bound $\Omega(\log x)$ on the total computation time even if the number of registers is unbounded, and even if the programs are allowed to introduce arbitrary constants. A straightforward generalization of the proof shows that an integer addition-multiplication machine needs $\Omega\bigl(\log(x/y)\bigr)$ steps to compute $x \bmod y$, uniformly for all $y > 0$ and for infinitely many x when y is given. However, the argument does not apply to machines with unbounded registers and indirect addressing; for this case Stockmeyer [8] used a more complex argument to obtain the lower bound $\Omega(\log x/\log\log x)$. It is still unknown whether indirect addressing can be exploited to do better than $O(\log x)$. When integer division is allowed, as well as addition and multiplication, the bound $\Omega\bigl(\log\log\log\min(x,y)\bigr)$ on arithmetic operations needed to compute $\gcd(x,y)$ has been proved by Mansour, Schieber, and Tiwari [5].

Our efficient constructions have all been for addition machines that contain at least three registers. The following theorem shows that 2-register addition machines cannot do much:

Theorem 4. *Any algorithm that computes* $\gcd(x,y)$ *on an integer addition machine with only two registers needs at least* $n-1$ *operations to compute* $\gcd(n,1)$.

Lemma. *Consider a graph on unordered pairs* $\{x,y\}$ *of nonnegative integers, where* $\{x,y\}$ *is adjacent to* $\{x,x+y\}$, $\{x+y,y\}$, $\{|x-y|,y\}$, *and* $\{x,|x-y|\}$. *The shortest path from* $\{n,1\}$ *to* $\{1,1\}$ *in this graph has length* $n-1$, *for all* $n\ge 1$.

Proof of the lemma (by Tomás Feder). Consider the following four operations on unordered pairs $\{x,y\}$:

$\underline{A}$. Replace $\min(x,y)$ by $x+y$.

$\overline{A}$. Replace $\max(x,y)$ by $x+y$.

$\underline{S}$. Replace $\min(x,y)$ by $\max(x,y)-\min(x,y)$.

$\overline{S}$. Replace $\max(x,y)$ by $\max(x,y)-\min(x,y)$.

Then $\underline{A}\,\overline{S}=\overline{A}\,\overline{S}=\underline{S}\,\underline{S}=$ identity and $\underline{S}\,\overline{S}=\overline{S}$. Furthermore $\underline{S}$ is either $\overline{S}\,\underline{A}$ or $\overline{S}\,\overline{A}$, hence $\underline{A}\,\underline{S}$ and $\overline{A}\,\underline{S}$ are either $\underline{A}$ or $\overline{A}$. Any minimal sequence of operations must therefore begin with S's and end with A's; in fact, the sequence of S's must be either $\overline{S}^k$ or $\overline{S}^k\underline{S}$. But $\overline{S}^k$ applied to $\{n,1\}$ yields $\{n-k,1\}$, for $k<n$; and A's do not decrease anything. Therefore the shortest path is $\overline{S}^{n-1}$. □

Proof of the theorem. As in the proof of Theorem 3, the sequence of **if** tests made by an addition machine defines a computation path, dependent on the inputs. We say that the test '**if** $x\ge y$' is *critical* if it is performed at a moment when the contents of registers x and y happen to be identical.

Let M be a 2-register addition machine that produces the output $M(a,b)$ when applied to inputs $\langle a,b\rangle$. We assume that a and b are initially present in the two registers; therefore the computation path corresponding to $\langle a, b\rangle$ will be the computation path corresponding to $\langle ma, mb\rangle$ for all integers $m \geq 1$.

Every computation path defines constants α and β such that $M(a,b) = \alpha a + \beta b$ for all $\langle a, b\rangle$ leading to this path. If M never encounters a critical test when applied to $\langle a, b\rangle$, it will follow the same path on inputs $\langle am, bm\rangle$ and $\langle am+1, bm\rangle$ for all sufficiently large values of m. Therefore we will have

$$M(am+1, bm) = M(am, bm) + \alpha$$

for all large m; and M cannot be a valid program for computing the gcd. We have proved that every 2-register gcd program must make a critical test before it produces an output.

Next we show that every 2-register gcd machine must make a critical test before it uses any instruction of the form $x \leftarrow x - x$ or $x \leftarrow x + x$. Suppose M performs such an instruction when it is applied to inputs $\langle a, b\rangle$; these inputs determine a computation path defining constants α and β such that the other register, y, contains $\alpha a + \beta b$ when $x \leftarrow x - x$ or $x \leftarrow x + x$ is performed. If no critical tests have occurred, the same computation path will be followed when the inputs are $\langle a^2bm+1, ab^2m\rangle$ and $\langle a^2bm, ab^2m+1\rangle$, for all sufficiently large m. But

$$\gcd(a^2bm+1, ab^2m) = \gcd(a^2bm, ab^2m+1) = 1;$$

hence register y must contain an odd value when M is applied to either $\langle a^2bm+1,\ ab^2m\rangle$ or $\langle a^2bm,\ ab^2m+1\rangle$. (If y is even when x is being set to $x - x$ or $x + x$, both registers will contain an even value; hence M cannot subsequently output '1'.) Hence $\alpha(a^2bm+1) + \beta(ab^2m)$ and $\alpha(a^2bm) + \beta(ab^2m+1)$ are odd, for all sufficiently large m; hence α and β are both odd. But $\gcd(2a^2bm+1, 2ab^2m+1)$ is odd, and the inputs $\langle 2a^2bm+1, 2ab^2m+1\rangle$ follow the same path as $\langle a, b\rangle$ for all large m; hence $\alpha(2a^2bm+1) + \beta(2ab^2m+1)$ must be odd, a contradiction.

Therefore every 2-register gcd machine must make a critical test, before which it has performed only operations of the forms $x \leftarrow x \pm y$, $y \leftarrow y \pm x$. Such operations correspond to the transformations considered in the lemma.

Suppose M is applied to the inputs $\langle n, 1\rangle$. When the first critical test occurs, we have $x = y$; and $\gcd(x,y) = \gcd(n,1) = 1$, because $\gcd(x,y)$ is preserved by all of the operations $x \leftarrow x \pm y$ or $y \leftarrow y \pm x$ that have been performed so far. Thus $x = y = \pm 1$; the algorithm must have followed a path from $\{n, 1\}$ to $\{1, 1\}$ in the sense of the lemma. So the algorithm must have performed at least $n - 1$ operations before reaching the first critical test. $\square$

Further Restrictions

A "minimalist" definition of addition machines would eliminate the copy operation $x \leftarrow y$, because that operation can be achieved by

$$x \leftarrow x - x\,; \qquad x \leftarrow x + y\,.$$

We can also simplify the **if** tests, allowing only the one-register form '**if** $x \geq 0$', because a general two-register comparison '**if** $x \geq y$ **then** α **else** β' can be replaced by

$x \leftarrow x - y$;
if $x \geq 0$ **then begin** $x \leftarrow x + y$; α **end**
else begin $x \leftarrow x + y$; β **end**.

Similarly, we can do away with addition, if we add a new register ξ, because $x \leftarrow x + y$ can be achieved by three subtractions:

$$\xi \leftarrow \xi - \xi\,; \qquad \xi \leftarrow \xi - y\,; \qquad x \leftarrow x - \xi\,.$$

Addition cannot be eliminated without increasing the number of registers, in general. For we can prove that the operation $x_1 \leftarrow x_1 + x_2$ cannot be achieved by any sequence of operations of the forms $x_i \leftarrow x_i - x_j$, for $1 \leq i, j \leq r$, without losing information. The proof can be formulated in matrix theory as follows:

Let E_{ij} be the matrix that is all zeros except for a 1 in row i and column j. We want to show that the matrix $I + E_{12}$ cannot be obtained as a product of matrices of the form $I - E_{ij}$. Clearly we cannot use the matrices $I - E_{jj}$, whose determinant is zero; so we must use $I - E_{ij}$ with $i \neq j$. But the inverse of $I - E_{ij}$ is $I + E_{ij}$, when $i \neq j$. So if

$$I + E_{12} = (I - E_{i_1 j_1}) \ldots (I - E_{i_m j_m})$$

we have, taking inverses,

$$I - E_{12} = (I + E_{i_m j_m}) \ldots (I + E_{i_1 j_1})\,,$$

which is patently absurd because the right side contains no negative coefficients.

Open Questions

1) Can the upper bound in Theorem 1 be replaced by $8 \log_\phi N + \beta$?

2) Can an integer addition machine with only 5 registers compute x^2 in $O(\log x)$ operations? Can it compute the quotient $\lfloor y/z \rfloor$ in $O(\log(y/z))$ operations?

3) Can an integer addition machine compute the quantity $x^y \bmod z$ in $o\big((\log y)(\log z)\big)$ operations, given $0 \le x < z$ and $0 \le y < z$?

4) Can an integer addition machine sort an arbitrary sequence of positive integers $\langle q_1, q_2, \ldots, q_m \rangle$ in $o\big((m + \log q_1 q_2 \ \ldots \ q_m) \log m\big)$ steps?

5) Can the powers of 2 in the binary representation of x be computed and output by an integer addition machine in $o(\log x)^2$ steps? For example, if $x = 13$, the program should output the numbers 8, 4, 1 in some order.

6) Is there an efficient algorithm to determine whether a given $r \times r$ matrix of integers is representable as a product of matrices of the form $I + E_{ij}$?

Acknowledgment

We wish to thank Baruch Schieber for calling our attention to Stockmeyer's paper [8].

This research was supported in part by the National Science Foundation under grant CCR-86-10181, and by Office of Naval Research contract N00014-87-K-0502.

References

[1] Donald E. Knuth, *Seminumerical Algorithms*, Volume 2 of *The Art of Computer Programming* (Reading, Massachusetts: Addison–Wesley, 1969).

[2] Donald E. Knuth, *Sorting and Searching*, Volume 3 of *The Art of Computer Programming* (Reading, Massachusetts: Addison–Wesley, 1973).

[3] Donald E. Knuth, "Structured programming with **go to** statements," *Computing Surveys* **6** (1974), 261–301. Reprinted with revisions as Chapter 2 of *Literate Programming*, CSLI Lecture Notes 27 (Stanford, California: Center for the Study of Language and Information, 1992), 17–89.

[4] Donald E. Knuth and Richard H. Bigelow, "Programming languages for automata," *Journal of the Association for Computing Machinery* **14** (1967), 615–635. [Reprinted as Chapter 12 of *Selected Papers on Computer Languages*, CSLI Lecture Notes 139 (Stanford, California: Center for the Study of Language and Information, 2003), 237–262.]

[5] Yishay Mansour, Baruch Schieber, and Prasoon Tiwari, "Lower bounds for integer greatest common divisor computations," *IEEE Symposium on Foundations of Computer Science* **29** (1988), 54–63.

[6] Ю. В. Матиясевич, «Диофантовость перечислимых множеств», *Доклады Академия Наук СССР* **191** (1970), 279–282. English translation: Ju. V. Matijasevič, "Enumerable sets are diophantine," *Soviet Mathematics Doklady* **11** (1970), 354–358.

[7] R. L. Rivest, A. Shamir, and L. Adleman, "A method for obtaining digital signatures and public-key cryptosystems," *Communications of the ACM* **21** (1978), 120–126.

[8] Larry J. Stockmeyer, "Arithmetic versus Boolean operations in idealized register machines," IBM Thomas J. Watson Research Center report RC 5954 (Yorktown Heights, New York: International Business Machines Corporation, 21 April 1976), 11 pages.

[9] E. Zeckendorf, "Représentation des nombres naturels par une somme de nombres de Fibonacci ou de nombres de Lucas," *Bulletin de la Société Royale des Sciences de Liège* **41** (1972), 179–182.

Addendum

To the best of my knowledge, none of the six open problems have yet captured the attention of any researchers who are able to solve them.

Chapter 11

A Simple Program Whose Proof Isn't

[Originally published in Beauty Is Our Business: A Birthday Salute to Edsger W. Dijkstra, edited by W. H. J. Feijen, A. J. M. van Gasteren, D. Gries, and J. Misra (New York: Springer-Verlag, 1990), 233–242.]

As I was writing the TeX program, I needed to construct subroutines for many small tasks. The solution to one of those problems turned out to be especially interesting — to me at least — because it was a very short and simple piece of code, yet I could see no easy way to demonstrate its correctness by conventional methods.

My purpose in this note is to exhibit that subroutine with a sketch of the best proof I know, hoping that others with more experience in formal methods will agree that the algorithm is interesting, and will help me figure out what I should have done.

1. Converting Decimal Fractions to Fixed-Point Binary

To warm up, let me present another short program whose proof is quite easy. TeX works internally with integer multiples of 2^{-16}, but its input language uses decimal notation. Therefore TeX needs a routine to translate a given decimal fraction

$$.d_1 d_2 \ldots d_k$$

to the nearest representable binary fraction. Here $k \geq 1$, and each digit d_j is an integer in the range $0 \leq d_j < 10$, for $1 \leq j \leq k$. The problem is to find the nearest integer multiple of 2^{-16}; in other words, we want to round the quantity

$$2^{16} \sum_{j=1}^{k} d_j / 10^j$$

to the nearest integer n. If two integers are equally near this quantity, we will let n be the larger; thus

$$n = \left\lfloor 2^{16} \sum_{j=1}^{k} d_j/10^j + 1/2 \right\rfloor.$$

Notice that the smallest possible value of n is 0; this occurs if and only if the input is strictly less than

$$.00000762939453125,$$

the decimal representation of 2^{-17}. The largest possible value of n is 65536; this occurs if and only if the input is greater than or equal to

$$.99999237060546875,$$

the decimal representation of $1 - 2^{-17}$. Since the output value n is a nonnegative integer in a limited range, and since the input values d_j are small nonnegative integers, it is desirable to compute n with integer arithmetic in such a way that the intermediate results stay reasonably small.

The total number of input digits, k, can be arbitrarily large. Therefore we cannot solve the problem by simply computing the integer

$$N = 10^k \left(2^{16} \sum_{j=1}^{k} d_j/10^j + 1/2 \right)$$

and then letting n be the quotient $\lfloor N/10^k \rfloor$; the values of N and 10^k may be too large for our computer's hardware.

We can, however, note that the values of d_j for $j > 17$ have absolutely no effect on the answer n. Suppose k is at least 17. Then we can use the well-known law

$$\left\lfloor \frac{x+a}{b} \right\rfloor = \left\lfloor \frac{\lfloor x \rfloor + a}{b} \right\rfloor,$$

which holds for all integers a and b with $b > 0$, to prove that

$$n = \left\lfloor \frac{10^{17} \sum_{j=1}^{k} d_j/10^j + 5^{17}}{2 \cdot 5^{17}} \right\rfloor = \left\lfloor \frac{10^{17} \sum_{j=1}^{17} d_j/10^j + 5^{17}}{2 \cdot 5^{17}} \right\rfloor$$
$$= \left\lfloor 2^{16} \sum_{j=1}^{17} d_j/10^j + 1/2 \right\rfloor.$$

(The stated law of floors within floors is exercise 1.2.4–35 in [3].) Therefore TeX need only maintain an array capable of holding up to 17 digits; all digits after the 17th may be discarded.

The decimal representation of 2^{-17}, given above, shows that 17 digits are not only sufficient; they are sometimes also necessary to determine the correct value of n. We can always ignore d_{18}, but the value of d_{17} might matter.

It is convenient to compute n by writing

$$n = \left\lfloor \frac{m_0 + 1}{2} \right\rfloor, \quad m_l = \left\lfloor 2^{17} \sum_{j=l+1}^{k} d_j / 10^{j-l} \right\rfloor.$$

The intermediate values m_l obey a simple recurrence,

$$m_k = 0; \quad m_{l-1} = \lfloor (2^{17} d_l + m_l)/10 \rfloor.$$

Therefore the following program does the desired conversion to fixed-binary fractions.

```
P1:  l := min(k, 17);  m := 0;
     repeat m := (131072 * d[l] + m) div 10;
        l := l − 1;
     until l = 0;
     n := (m + 1) div 2.
```

(The proof is easy: We have $m = m_l$ at the beginning of the **repeat** loop, assuming that $k \le 17$; and we have shown that it is legitimate to replace k by 17 if k is larger.)

Notice that the intermediate values computed by program *P1* are nonnegative, and they never exceed 1310720.

2. Converting the Other Way

Now let's consider the inverse problem: Given an integer n, which we shall assume is in the range

$$0 < n < 2^{16},$$

find a decimal fraction

$$.d_1 d_2 \ldots d_k$$

that approximates $2^{-16} n$ so closely that our previous algorithm for converting decimal fractions will reproduce n exactly.

This problem, like the other, has a simple solution: We can insist that $k = 5$, and we can let $d_1 d_2 d_3 d_4 d_5$ be the decimal digits of the integer

$$D = \left\lfloor 10^5 \frac{n}{2^{16}} + \frac{1}{2} \right\rfloor.$$

Then $D/10^5$ must reproduce n, for the conversion algorithm finds a number n' such that

$$\left|\frac{D}{10^5} - \frac{n'}{2^{16}}\right| \le 2^{-17}.$$

We also have

$$\left|D - 10^5 \frac{n}{2^{16}}\right| \le \frac{1}{2},$$

by definition; hence

$$\begin{aligned} |n - n'| &\le |n - 2^{16}D/10^5| + |n' - 2^{16}D/10^5| \\ &\le 2^{15}/10^5 + 1/2 = .82768 < 1 \end{aligned}$$

and n must equal n'.

The original implementation of TEX took $k = 5$ as suggested above. But that choice turned out to be unsatisfactory, because a user who asked for a .4-point rule was told that TEX had actually typeset a rule of .39999 points. TEX's response was reasonably honest, but the extra detail was distracting and unnecessarily messy.

Therefore it is preferable to find a solution to the inverse conversion problem such that k is as small as possible. We seek a *shortest* decimal fraction that will reproduce the given value of n.

Moreover, there often are two decimal fractions of the same length that both yield the desired value. (For example, both .00001 and .00002 yield $n = 1$, because both 0.65536 and 1.31072 round to 1.) In such cases it is desirable to choose the decimal fraction that is closest to $n/2^{16}$ (namely .00002 when $n = 1$).

It turns out that there is a simple program to compute such shortest decimal fractions. But — as I said in the introduction — I don't know of an equally simple way to derive that program or to prove it correct. Here is the program that I came up with.

```
P2:  j := 0;  s := 10 * n + 5;  t := 10;
     repeat if t > 65536 then s := s + 32768 − (t div 2);
        j := j + 1;  d[j] := s div 65536;
        s := 10 * (s mod 65536);  t := 10 * t;
     until s ≤ t;
     k := j.
```

Why does this work? Everybody knows Dijkstra's famous dictum that testing can reveal the presence of errors but not their absence [1]. However, a program like this, with only finitely many inputs, is a counterexample! Suppose we test it for all 65535 values of n, and suppose the resulting fractions $.d_1 \ldots d_k$ all reproduce the original value when converted back. Then we need only verify that none of the shorter fractions or neighboring fractions of equal length are better; this testing will prove the program correct.

But testing is still not a good way to guarantee correctness, because it gives us no insight into generalizations. Therefore we seek a proof that is comprehensible and educational.

Even more, we seek a proof that reflects the ideas used to create the program, rather than a proof that was concocted ex post facto. The program didn't emerge by itself from a vacuum, nor did I simply try all possible short programs until I found one that worked.

3. Germs of a Proof

The invariant relationship that I had in mind when I wrote the program in the previous section — the relation that explains the intrinsic meaning of the variables s and t in connection with other data of the problem — is not easy for me to formalize. Perhaps I am too close to the program, too incapable of analyzing my own thought processes. I was thinking (I think) of the set of all possible continuations of the digits already determined.

In other words, if $d_1 \ldots d_j$ have already been computed, I was thinking of the set of all strings of decimal digits $d_{j+1} \ldots d_k$ such that

$$.d_1 \ldots d_j d_{j+1} \ldots d_k$$

would produce the number n when processed by the first program above. I wanted this set to be nonempty; and I wanted to stop if the empty string was in the set. Moreover, I wanted to make sure that my program would satisfy the optimality property (namely that it would round to the nearest of two equal-length possibilities). The latter condition is harder to deal with, so let's think first about the former one.

In the following proof I will use Lyle Ramshaw's convention, inspired by Pascal syntax, that $[a \mathinner{.\,.} b)$ denotes the set of real numbers x in the range

$$a \le x < b.$$

Similar notations apply to open intervals $(a \mathinner{.\,.} b)$ and to closed intervals $[a \mathinner{.\,.} b]$; in this way we avoid conflict with the many other mathematical interpretations of the notations (a, b) and $[a, b]$. (A similar convention, but with three dots instead of two, was first proposed by Hoare [2].)

The set of digit strings $d_{j+1} \ldots d_k$ that produce a given result n, when preceded by $.d_1 \ldots d_j$, can be characterized as a set of decimal fractions

$$.d_{j+1} \ldots d_k$$

that lie in some half-open interval $[\alpha \mathinner{.\,.} \beta)$. Initially we have $j = 0$, and this interval is

$$\Bigl[2^{-16}\Bigl(n - \frac{1}{2}\Bigr) \mathinner{.\,.} 2^{-16}\Bigl(n + \frac{1}{2}\Bigr)\Bigr).$$

In general if the interval is $[\alpha \mathinner{.\,.} \beta)$, the empty string is in the set of possible continuations $d_{j+1} \ldots d_k$ if and only if $\alpha \le 0$ and $\beta > 0$. Furthermore, the permissible values of d_{j+1}, if we wish to increase j and retain a nonempty set of continuations, are those decimal digits d such that the interval

$$[d/10 \mathinner{.\,.} (d+1)/10)$$

has a nonempty intersection with $[\alpha \mathinner{.\,.} \beta)$. This means that the conditions

$$0 \le d \le 9, \quad d/10 < \beta, \quad \text{and} \quad \alpha < (d+1)/10$$

are necessary and sufficient for d to be an acceptable choice of d_{j+1} (if we ignore complications of optimality).

Suppose we set $j := j + 1$ and $d[j] := d$. Then the new interval $[\alpha' \mathinner{.\,.} \beta')$ replacing $[\alpha \mathinner{.\,.} \beta)$ should be such that

$$\frac{d + \alpha'}{10} = \alpha, \qquad \frac{d + \beta'}{10} = \beta.$$

In other words, the new interval should be

$$[10\alpha - d \mathinner{.\,.} 10\beta - d).$$

This analysis has used real numbers, but we want to stick to integers in the calculation. Let us therefore represent α and β implicitly by the integer variables s and t, where

$$10\alpha = 2^{-16}(s - t); \quad 10\beta = 2^{-16}s.$$

The initial values of s and t, corresponding to the initial α and β values $2^{-16}(n - \frac{1}{2})$ and $2^{-16}(n + \frac{1}{2})$, are therefore

$$s = 10n + 5; \quad t = 10.$$

The conditions for an admissible d translate into

$$0 \le d \le 9, \quad s > 2^{16}d, \quad \text{and} \quad s - t < 2^{16}d + 2^{16}.$$

When j is increased by 1 and the next output digit is set to d, we want to set

$$s := 10(s - 2^{16}d), \quad t := 10t.$$

These manipulations on s and t change $[\alpha\,..\,\beta)$ into $[\alpha'\,..\,\beta')$ as discussed above.

From such considerations we can prove that the following program does almost what we want.

```
P3:  j := 0;  s := 10 * n + 5;  t := 10;
     repeat j := j + 1;  d[j] := s div 65536;
        s := 10 * (s mod 65536);  t := 10 * t;
     until s ≤ t;
     k := j.
```

This program preserves the desired invariant relation between variables, namely that a string $d_{j+1} \ldots d_k$ of decimal digits will have the property that $.d_1 \ldots d_j d_{j+1} \ldots d_k$ converts to $n/2^{16}$ if and only if

$$2^{-16}\frac{s-t}{10} \le .d_{j+1} \ldots d_k < 2^{-16}\frac{s}{10}.$$

We can verify this as follows: First we note that the conditions

$$s > t, \quad t = 10^{j+1}, \quad 0 \le s \le 655355$$

hold on each entry to the body of the **repeat** loop. Hence $10^{j+1} < 655355$ whenever the **repeat** loop begins, and we must have $j \le 5$ when the program terminates.

It is not difficult to verify that s is an odd multiple of 2^j, just before our new program increases j, because j never reaches 16. Therefore s is never a multiple of $2^{16} = 65536$, and the digit $d = s$ **div** 65536 always satisfies the conditions

$$0 \le d \le 9 \quad \text{and} \quad s - t - 2^{16} < 2^{16}d < s$$

that we have derived for admissibility of d as a digit.

This argument establishes that the decimal fraction $.d_1 \ldots d_k$ computed by program *P3* will be converted back to n by *P1*. Furthermore, the value of k will be at most 5.

The value of k produced by *P3* is actually minimum; no shorter fraction than $.d_1 \ldots d_k$ will reproduce n. To prove this, we observe that the algorithm always chooses the largest possible digit d; if two or more values of d satisfy

$$s - t - 2^{16} < 2^{16} d < s,$$

then s **div** 65536 is the largest one. Hence, any fraction $.d'_1 \ldots d'_{k'}$ with $k' < k$ that program *P1* converts to n has $d'_j < d_j$ for some j. This implies that

$$.d_1 \ldots d_k - .d'_1 \ldots d'_{k'} \geq 10^{-k'}.$$

But we have

$$\left| .d'_1 \ldots d'_{k'} - \frac{n}{2^{16}} \right| \leq 2^{-17} \quad \text{and} \quad \left| .d_1 \ldots d_k - \frac{n}{2^{16}} \right| \leq 2^{-17};$$

hence

$$10^{-k'} \leq 2^{-17} + 2^{-17} = 2^{-16}$$

and $k' \geq 5 \geq k$, a contradiction.

Therefore *P3* is almost a solution to our problem. The only remaining task is to find an approximation $.d_1 \ldots d_k$ of minimum length that is as close as possible to $n/2^{16}$. If the output of *P3* is not the best approximation then the only better ones are at least 10^{-k} less than $.d_1 \ldots d_k$, and the argument just given implies that two distinct approximations of length k are possible only when $k \geq 5$. Therefore *P3* gives the correct answer whenever it finds an approximation of length 4 or less.

In all other cases, program *P3* eventually begins its **repeat** loop with $j = 4$ and $t = 10^5$. We want to modify the calculation on the final round so that the final digit d_5 will be "best possible" among the available choices. This means that we want to compute the quantity

$$\frac{n}{2^{16}} \text{ rounded to 5 decimal places} = \frac{\lfloor 10^5 n/2^{16} + 1/2 \rfloor}{10^5}$$

whenever there is no suitable approximation with fewer than 5 decimal places.

4. Completion of the Proof

If we change the penultimate line of *P3* to '**until** *false*', so that the program loops forever, our interpretation of the variable s implies that the resulting program will compute the decimal expansion $.d_1 d_2 d_3 \ldots$ of $(n + 1/2)/2^{16}$. (This expansion is infinite, although d_j will be zero for all $j \geq 18$.) Stating this another way, we can prove without difficulty that the variables $d_1, \ldots, d_j$, and s of program *P3* obey the invariant relation

$$\frac{s}{10^{j+1}} + 2^{16}\sum_{i=1}^{j} d_i 10^{-i} = n + \frac{1}{2}.$$

Now let's return to program *P2*, which is identical to *P3* except that the conditional instruction

$$\textbf{if } t > 65536 \textbf{ then } s := s + 32768 - (t \textbf{ div } 2)$$

has been inserted at the very beginning of the **repeat** loop. This instruction takes effect only when $j = 4$, namely when $t = 10^5$, because $t = 10^{j+1}$ and the loop is never entered when $j > 4$.

When $j = 4$ at the beginning of the loop, we are about to compute d_5. And we know that $d_5 > 0$, because we have proved that *P3* computes a shortest possible decimal expansion; therefore we have $s > 65536$. Therefore the new value of s in program *P2*,

$$s' = s + 32768 - (t \textbf{ div } 2),$$

is nonnegative, but less than 655355. Therefore the corresponding quantity $d_5 = s' \textbf{ div } 65536$ will be a digit in the interval $(0\,..\,9]$.

But by the invariant between d_1, d_2, d_3, d_4, and s, we know that

$$\begin{aligned} s' + 2^{16}\sum_{i=1}^{4} d_i 10^{5-i} &= s + 2^{15} - \frac{10^5}{2} + 2^{16}\sum_{i=1}^{4} d_i 10^{5-i} \\ &= 10^5 n + 2^{15}. \end{aligned}$$

Therefore

$$\left\lfloor \frac{10^5 n}{2^{16}} + \frac{1}{2} \right\rfloor = \left\lfloor \frac{s'}{2^{16}} + \sum_{i=1}^{4} d_i 10^{5-i} \right\rfloor = \sum_{i=1}^{5} d_i 10^{5-i};$$

in other words, $.d_1 d_2 d_3 d_4 d_5$ is equal to $n/2^{16}$ rounded to five decimal places, as desired.

We have shown that *P2* either computes the (unique) shortest decimal approximation $.d_1 \ldots d_k$ of length $k < 5$, or it computes the best approximation of length 5 (when there is no shorter one). This completes the proof that *P2* is correct.

So. Is there a better program, or a better proof, or a better way to solve the problem?

5. Closing Remarks

We have based our discussion on the particular case of 16-bit fixed-point binary fractions. But an examination of the proof shows that the ideas are quite general: If m is any positive number for which we seek the shortest-and-best decimal representation of n/m, given $0 < n < m$, we can use the following modification of program *P2*.

```
P4:  j := 0;  s := 10 * n + 4;  t := 10;
     repeat if t > m then s := s + 1 + (m div 2) − (t div 2);
        j := j + 1;  d[j] := s div m;
        s := 10 * (s mod m) + 9;  t := 10 * t;
     until s < t;
     k := j.
```

The resulting fraction $.d_1 \ldots d_k$ is an optimum representation subject to the conditions

$$\frac{n-1/2}{m} \le .d_1 \ldots d_k < \frac{n+1/2}{m}.$$

(If equality is impossible in shortest representations, a program more like *P2* will also be valid.)

When the denominator is not a power of 2, the converse problem (corresponding to *P1*) becomes much more interesting, especially if m is divisible by at least one prime number other than 2 or 5. Then the digits d_j can be relevant for arbitrarily large j. The reader may enjoy trying to construct an algorithm that reads a decimal input fraction $.d_1 \ldots d_k$ "on line" from an input tape and computes the unique value of n that satisfies the inequalities above, given m. It is desirable to limit the amount of auxiliary memory to $O(\log m)$ bits, not counting the digits on the tape. The input digits must be read once only, from left to right.

An algorithm similar to program *P3* was published in 1959 by Donald Taranto [6]. A generalization of Taranto's method, due to Guy L. Steele Jr. and Jon L White, appears in the answer to exercise 4.4–3 of [4]. Programs *P1* and *P2* appear in sections 102 and 103, respectively, of the program for TEX [5].

After writing this paper I decided to optimize *P2* in TEX by changing '$+\,32768 - (t$ **div** $2)$' to '$-\,17232$'.

Acknowledgments. I wish to thank David Gries for helping me locate reference [1]. The preparation of this note was supported in part by the National Science Foundation under grant CCR-86-10181. *(End of acknowledgments.)*

References

[1] Edsger W. Dijkstra, comment in *Software Engineering Techniques*, edited by J. N. Buxton and B. Randell (Brussels: NATO Science Committee, 1970), 21. Reprinted in *Software Engineering: Concepts and Techniques*, edited by Peter Naur, Brian Randell, and J. N. Buxton (New York: Petrocelli, 1976), 159.

[2] C. A. R. Hoare. "A note on the **for** statement," *BIT* **12** (1972), 334–341.

[3] Donald E. Knuth, *Fundamental Algorithms*, Volume 1 of *The Art of Computer Programming* (Reading, Massachusetts: Addison–Wesley, 1968).

[4] Donald E. Knuth, *Seminumerical Algorithms*, Volume 2 of *The Art of Computer Programming* (Reading, Massachusetts: Addison–Wesley, 1969).

[5] Donald E. Knuth, *TEX: The Program*, Volume B of *Computers & Typesetting* (Reading, Massachusetts: Addison–Wesley, 1986).

[6] Donald Taranto, "Binary conversion, with fixed decimal precision, of a decimal fraction," *Communications of the ACM* **2**, 7 (July 1959), 27.

Addendum

Formal developments of program *P2* were subsequently presented by David Gries ["Binary to decimal, one more time," in *Beauty Is Our Business* (New York: Springer-Verlag, 1990), 141–148]; by R. S. Bird ["Functional pearls: Two greedy algorithms," *Journal of Functional Programming* **2** (1992), 237–244]; and by Jinyun Xue and Ruth Davis ["A derivation and proof of Knuth's binary to decimal conversion program," *Software — Concepts and Tools* **18** (1997), 149–156].

For general procedures that convert accurately between floating-point binary and floating-point decimal numbers according to IEEE standard conventions, see Donald E. Knuth, *MMIXware: A RISC Computer for the Third Millennium, Lecture Notes in Computer Science* **1750** (Berlin: Springer-Verlag, 1999), 86–100.

Chapter 12

Verification of Link-Level Protocols

[Originally published in BIT **21** *(1981), 31–36.]*

Stein Krogdahl [1] has given an interesting demonstration of the partial correctness of a "protocol skeleton," by which the validity of the essential aspects of a large variety of data transmission schemes can be demonstrated. This note presents a simpler way to obtain the same results, by first establishing the validity of a less efficient skeleton and then "optimizing" the algorithms. The present approach, which was introduced for a particular protocol by N. V. Stenning [2], also solves a wider class of problems that do not require first-in-first-out transmissions.

1. Introduction

Alice wants to send messages $M_0\,M_1\,M_2\,\ldots$ to Bill over noisy transmission lines. They decide to handle the problem in the following way: Alice keeps a local variable

A = the number of consecutive messages that Alice is sure Bill has received and stored;

Bill keeps a local variable

B = the number of consecutive messages that Bill is sure he has received and stored.

Initially $A = B = 0$; we ignore problems of termination, since they can be dealt with as in [1]. Alice does two types of operations:

A1. Send message M_j, where j is an integer, $A \le j < A + k$.

A2. Receive an acknowledgment 'b' and set $A \leftarrow b$.

Bill also does two types of operations:

B1. Send an acknowledgment 'B'.

B2. Receive message M_j, and optionally store it; then set B to any value $b \ge B$ such that messages $M_0\,M_1\,\ldots\,M_{b-1}$ have been received and stored.

Here k is a constant representing the size of some internal buffer storage maintained by Alice. We shall assume as in [1] that the "send" operation either inserts an item at the rear of a queue, or it causes nothing at all to happen; the latter event accounts for transmission errors, since a garbled message or a garbled acknowledgment will be treated as if it has not arrived at all. (Items that do reach the queue, however, are transmitted faithfully.) According to this convention, the sender does not know whether the sent item has been put into the queue or not. The "receive" operation is performed only when the queue is nonempty; in such a case the receiver reads and deletes the item at the front of the queue.

Thus there are two queues, one containing messages from Alice and the other containing acknowledgments from Bill. The only essential difference between the conventions above and those of [1] is that we assume as in [2] that each message M_j in the message queue specifies its own integer index j, and each acknowledgment in the acknowledgment queue specifies an integer b, where j and b can be arbitrarily large. After this simple but unrealistic model has been examined, it will be clear that only a limited amount of information about j and b need actually be sent.

The particular order in which Alice and Bill decide to perform operations A1, A2, B1, and B2 is immaterial to us, and so are the particular choices of optional actions in steps A1 and B2. Our goal is to derive facts about *any* scheme that is based on these four operations; it is in this sense that we are studying a "protocol skeleton" for a large class of conceivable protocols. The facts we shall derive are expressed in terms of relations that remain *invariant* under all four of the basic operations A1, A2, B1, B2.

2. Invariants

The first invariant relation we shall prove can be stated as follows:

Lemma 1. *Let the contents of the acknowledgment queue be*

$$b_1 \ldots b_r$$

from the front to the rear, where $r \ge 0$. *Then*

$$A \le b_1 \le \cdots \le b_r \le B.$$

Proof. This condition holds initially, when $A = B = r = 0$, and it is unaffected by operation A1. Operation A2 is performed only when $r > 0$, and it replaces $(A, b_1, \ldots, b_r)$ by $(b_1, b_2, \ldots, b_r)$; operation B1 either does nothing or replaces $(b_1, \ldots, b_r, B)$ by $(b_1, \ldots, b_r, B, B)$; and operation B2 has no effect except possibly to increase B. Thus the stated relation remains invariant. □

As a corollary of Lemma 1, we conclude that variable A never decreases during the course of a computation, since it changes only during A2. Notice that the invariant in Lemma 1 expresses a *joint property* of the entire communication system. Although Alice does not know the value of B and Bill does not know the value of A, and although neither knows the contents of the queue, they can be sure that the unknown quantities satisfy the invariant relation. The introduction of system-wide invariants like this is one of the main features of Krogdahl's treatment.

Lemma 2. *Let the contents of the message queue be*

$$M_{j_1} \ldots M_{j_r}$$

from the front to the rear, where $r \geq 0$, and let $j_{\max}$ be the maximum index of any message that has ever been removed from the message queue. (If nothing has ever been removed, let $j_{\max} = 0$.) Let $j_0 = j_{\max}$ and $j_{r+1} = A$. Then

$$j_i < j_{i'} + k \quad \text{for } 0 \leq i < i' \leq r+1\,.$$

Proof. Initially $r = 0$, so we have $i = 0$, $i' = 1$, and $0 = j_0 < j_1 + k = A + k = k$. Operation A1 either does nothing or replaces $(j_1, \ldots, j_r)$ by $(j_1, \ldots, j_r, j)$ for some j, where $A \leq j < A + k$; this leaves the stated relation invariant (we must consider two new cases, namely $j_i = j$ and $j_{i'} = j$). Operation A2 does not decrease A, as we have already observed; and operation B1 changes nothing. Operation B2 is performed only when $r > 0$, and it replaces $(j_{\max}, j_1, \ldots, j_r)$ by $\bigl(\max(j_{\max}, j_1), j_2, \ldots, j_r\bigr)$; again the relation remains invariant. □

3. Consequences

The comparatively simple invariants proved in Lemmas 1 and 2 lead immediately to our main result:

Theorem. *If M_j is in the message queue, we have*

$$B - k \leq j < B + k\,.$$

If b is in the acknowledgment queue, we have

$$A \leq b \leq A + k\,.$$

Proof. We know from Lemma 2 that $j < A + k$ and from Lemma 1 that $A \leq B$, hence $j < B + k$. Furthermore $B - 1 \leq j_{\max}$, where $j_{\max}$ is defined in Lemma 2, because messages $M_0\, M_1\, \ldots\, M_{B-1}$ have all been removed from the message queue; hence $B-1 < j+k$ and $B-1 < A+k$ by Lemma 2. This completes the proof, since $A \le b \le B$ by Lemma 1. □

The theorem tells us that only a limited amount of information about j needs to appear in the message queue, and only a limited amount about b needs to appear in the acknowledgment queue. Let us consider b first: If m_1 is any fixed integer $> k$, it suffices to send the remainder $B \bmod m_1$ instead of the arbitrarily large integer B in step B1, since Alice will be able to construct the full acknowledgment 'b' from the remainder $b \bmod m_1$ received in step A2, given the fact that $A \le b \le A+k$. Indeed, the operation $A \leftarrow b$ is simply replaced by

$$A \leftarrow A + (b' - A) \bmod m_1$$

where $b' = b \bmod m_1$ is the acknowledgment that was received.

Let us suppose that Bill will store a message M_j that he receives in operation B2 only if $B \le j < B + l$, where l represents a fixed amount of buffer storage. There is of course no point in storing M_j when $j < B$, since all such messages have already been stored. We might as well assume that $l \le k$, because j will always be less than $B + k$. In this case it suffices to send only the remainder $j \bmod m_2$ as an identification number for M_j, instead of the full integer j, provided that $m_2 \ge k + l$. For we know that the index j received by Bill in B2 must satisfy $B - k \le j < B + k$; the values of $j \bmod m_2$ in the range $B \le j < B + l$ are distinct, and they are disjoint from the values of $j \bmod m_2$ in the range $B - k \le j < B$ or $B + l \le j < B + k$. The fact that $(B + l) \bmod m_2$ might coincide with $(B - k) \bmod m_2$ does not matter; Bill would not store such a message in either case, and he doesn't care about the precise value of j when the message isn't being stored since such messages might as well have been dropped.

Krogdahl's paper [1] essentially discusses the case $l = 1$ and $m_1 = m_2 = k + 1$ in detail; he also gives a sketch of the case $l = k$, $m_1 = m_2 = 2k$ without proof. The argument above is not only simpler and more general, it shows that the moduli $m_1 = k + 1$ and $m_2 = 2k$ are sufficient when $l = k$.

4. Generalization

Krogdahl conjectured that the theory can be extended to the case where the queues do not quite operate in a first-in-first-out manner. It is clear that we cannot avoid sending the full integer j or b when the queuing discipline allows the deletion of items in *arbitrary* order, since small values might remain in the queue until they coexist with large ones. Let us suppose, however, that if entries are inserted in the order $x_1\,x_2\,x_3\,\ldots$ and deleted in the order $x_{p(1)}\,x_{p(2)}\,x_{p(3)}\,\ldots$, then $p(1)\,p(2)\,p(3)\,\ldots$ is a permutation of the positive integers such that we have $|p(i)-i| \le q$ for all i and some parameter q. Furthermore the indices $p(i)$ must be consistent with the actual sequence of insertions and deletions made to the queue, in the sense that at least $p(i)$ elements must have been inserted at the time of the ith deletion. How does our previous analysis of the case $q = 0$ extend to this more general situation?

In the first place it is clear that the assignment at the end of operation A2 should be replaced by

$$A \leftarrow \max(A, b)$$

in this more general setting, otherwise the monotone growth of A would be destroyed.

Before considering the general protocol problem in detail, it is useful to study the general queuing discipline more carefully. If $i < i'$ and $p(i) > p(i')$, let us say that element $x_{p(i)}$ "passes" element $x_{p(i')}$, since it was inserted later but deleted earlier.

Lemma 3. *A permutation $p(1)\,p(2)\,p(3)\,\ldots$ of the positive integers satisfies the condition $p(i) \ge i - q$ for all i if and only if no element of the corresponding queuing discipline is passed by more than q other elements. It satisfies the condition $p(i) \le i + q$ for all i if and only if no element of the corresponding queuing discipline passes more than q other elements.*

Proof. If $p(i) \ge i - q$ for all i, then $p(i') > i - q$ for all $i' > i$; but p is a permutation, so exactly $i - q$ of the indices $i' \le i$ have $p(i') \le i - q$. This leaves at most q indices $i' < i$ that could have $p(i') > p(i)$; so $x_{p(i)}$ is not passed by more than q other elements. Conversely, if $p(i) < i - q$ for some i, then at most $p(i) - 1$ indices $i' < i$ have $p(i') < p(i)$, so at least $i - p(i)$ indices $i' < i$ have $p(i') > p(i)$; in other words, at least $q+1$ elements pass $x_{p(i)}$. The second half of the lemma follows from the first half, if we replace p by the inverse permutation. □

As long as we are generalizing the case $q = 0$, we might as well generalize further by supposing that the queuing discipline satisfies

$$i - q \le p(i) \le i + q'$$

for all i. Here $q = 0$ if and only if $q' = 0$, but each pair of positive integers (q, q') defines a different queuing discipline. We shall assume that the acknowledgment queue satisfies such a discipline with parameters q_1 and q'_1, while the message queue satisfies such a discipline with parameters q_2 and q'_2.

Let $b_1\, b_2\, b_3\, \ldots$ be the entries that are inserted into the acknowledgment queue, and let $j_1\, j_2\, j_3\, \ldots$ be the indices of the messages inserted into the message queue. We can prove as before that $b_1 \le b_2 \le \cdots \le b_n \le B$, after n acknowledgments have been inserted; that $j_i < A + k$ for $1 \le i \le n$, after n messages have been inserted; and that

$$j_i < j_{i'} + k \quad \text{for } 1 \le i < i' \,.$$

It follows that $A \le B \le A + k$.

We can now show that all entries b in the acknowledgment queue satisfy

$$A - q_1 k \le b \le A + k \,.$$

The upper bound is obvious, because $b \le B$. To prove the lower bound, we may suppose that $b < A$. When b was first placed into the queue, we can assume that we had $b = B \ge A$; so A must have increased since then, by being set to other entries read from the queue. Suppose that n of these other entries have "passed" b, in the sense that they were inserted after b; only the entries inserted after b can have a value $> b$. Before the first such entry was read by Alice, we had $b \ge A$; afterwards we had $b \ge A - k$, because A cannot increase by more than k during operation A2. (All entries in the queue at that time are $\le B$.) By induction we have $b \ge A - nk$ if n entries have passed b, but Lemma 3 tells us that $n \le q_1$.

Finally, we can prove that all indices j in the message queue satisfy

$$B - k - q_2 \le j < B + k \,.$$

Again the upper bound is obvious, since $j < A + k$. To prove the lower bound, suppose that n message indices have "passed" j in the queue; all other indices j' read by Bill satisfy $j' \le j + k - 1$. Therefore if Bill has received and stored messages $M_0 \ldots M_{B-1}$, we have $B - 1 \le j + k - 1 + n$, with equality only if the n messages that passed M_j were distinct messages whose index lies in the interval $[j + k \,.\,.\, j + k - 1 + n]$. By Lemma 3, we have $n \le q_2$.

It is not difficult to verify that the inequalities above on b and j are best possible, by constructing scenarios in which the extreme values occur. As before, we can conclude that it suffices to transmit only the residues $b \bmod m_1$ in the acknowledgment queue and $j \bmod m_2$ in the message queue, where m_1 and m_2 are any integers that satisfy

$$m_1 > (q_1 + 1)k\,,$$
$$m_2 \geq k + l + q_2\,;$$

we assume that Bill has a buffer for receiving up to $l \leq k$ messages whose indices lie in $\{B, B+1, \ldots, B+l-1\}$. It is curious that q_1' and q_2' do not enter into these formulas.

The protocol of Stenning [2] requires that at least one acknowledgment be transmitted per message received; in this special case the bound $m_1 \geq k + l + q_1$ is necessary and sufficient, where q_1 is the maximum number of other acknowledgments that can be sent and received between the transmission and receipt of any particular acknowledgment.

In practice, Alice is a system program that receives messages sequentially from some user, and Bill is a system program that delivers messages sequentially to another user. Therefore, as Krogdahl has observed, the variables A and B need not be explicitly maintained; only their values modulo a common multiple of m_1 and m_2 are needed.

The preparation of this paper was supported in part by National Science Foundation grants MCS-77-23738 and IST-79-21977, and by Office of Naval Research contract N00014-76-C-0330.

References

[1] Stein Krogdahl, "Verification of a class of link-level protocols," *BIT* **18** (1978), 436–448.

[2] N. V. Stenning, "A data transfer protocol," *Computer Networks* **1** (1976), 99–110.

Chapter 13

Additional Comments on a Problem in Concurrent Programming Control

[The following letter to the editor was originally published in Communications of the ACM **9** *(1966), 321–322, 878. It refers to the problem of "mutual exclusion" that arises when concurrent programs with unpredictable speeds must communicate reliably via a common memory, with no interlocks except that the operations of reading from memory or writing to memory are indivisible and cannot lead to conflict. Furthermore, each independent processor is supposed to use essentially the same algorithm, so that none has precedence over another. Interesting programs that attempted to solve this problem had previously been published by Edsger Dijkstra (Fig. 2) and by Harris Hyman (Fig. 1).]*

```
C0:  b[i] := false;
C1:  if k ≠ i then
        begin C2: if ¬b[1 − i] then go to C2;
        k := i;  go to C1;
        end;
     critical section;
     b[i] := true;
     remainder of program;  go to C0;
```

FIGURE 1. Hyman's program for computer i, which may be 0 or 1. Global variables are the integer k and the Boolean array $b[0:1]$.

EDITOR:

Professor Dijkstra's ingenious construction ["Solution of a problem in concurrent programming control," *Communications of the ACM* **8** (1965), 569] is not quite a solution to a related problem almost identical to the problem he posed there, and Mr. Hyman's "simplification" for the case of two computers [*Communications of the ACM* **9** (1966), 45] hardly works at all. I hope that by this letter I can save people some of the problems they would encounter if they were to use either of those methods.

It is easy to find a counterexample to Mr. Hyman's "solution." If initially $k = 0$ and $b[0] = b[1] =$ **true**, computer 1 may start the process by setting $b[1]$ **false**, subsequently finding $b[0]$ is **true**. Then computer 0 sets $b[0]$ **false** and finds $k = 0$, whereupon it starts to execute its critical section. But computer 1 now sets $k = 1$ and executes its critical section at the same time.

```
        begin integer j;
Li0:    b[i] := false;
Li1:    if k ≠ i then
Li2:       begin c[i] := true;
Li3:       if b[k] then k := i;
           go to Li1;
           end;
Li4:    c[i] := false;
        for j := 1 step 1 until N do
           if j ≠ i ∧ ¬c[j] then go to Li1;
        critical section;
        c[i] := true;  b[i] := true;
        remainder of the cycle, in which stopping is allowed;
        go to Li0;
        end
```

FIGURE 2. Dijkstra's program for computer i, which may be 1, 2, ..., or N. Global variables are the integer k and the Boolean arrays $b[1:N]$, $c[1:N]$.

Professor Dijkstra's solution is harder to attack; however, there is a definite possibility that one or more of the computers that want to execute their critical sections may have to wait "until eternity" while other programs are executing theirs. In other words, although Dijkstra's algorithm ensures that all computers are not *simultaneously* blocked, it is still possible that an *individual* computer will be blocked. (He decided to allow this possibility in his statement of the problem since he was interested in cases where there is comparatively low average demand for the use of critical sections; but there are certainly many applications for which the possibility of individual blocking is unacceptable.) For example, suppose time passes in discrete intervals; this assumption is valid for many computer systems and it is convenient but not strictly necessary for this example. Assume that computer 1 is looping, finding $b[k] =$ **false** and $k \neq 1$, and that it is positioned at label $L13$ in Dijkstra's program at times 0, 10, 20, 30, ... It is quite possible for the other computers to set $b[k]$ and change k at other times so that computer 1 never breaks out of the loop. For example, if $k = N = 2$, computer 2 can set $b[2]$ **true** at times $10n + 2$ and come back to $L20$ to set it **false** again at time $10n + 9$, for arbitrarily many n.

I tried out over a dozen ways to solve this problem before I found what I believe is a correct solution. The program for the ith computer $(1 \le i \le N)$, using the common store

integer array *control* $[1 : N]$; **integer** k

(initially zero) is the following:

```
      begin integer j;
L0:   control[i] := 1;
L1:   for j := k step −1 until 1, N step −1 until 1 do
         begin if j = i then go to L2;
         if control[j] ≠ 0 then go to L1;
         end;
L2:   control[i] := 2;
      for j := N step −1 until 1 do
         if (j ≠ i) ∧ (control[j] = 2) then go to L0;
L3:   k := i;
      critical section;
      k := if i = 1 then N else i − 1;
L4:   control[i] := 0;
L5:   remainder of the cycle, in which stopping is allowed;
      go to L0;
      end
```

To prove that this construction works, first observe that no two computers can be simultaneously positioned between their statements $L3$ and $L4$, for the same reason that this is true in Dijkstra's algorithm. Secondly, observe that the entire system cannot be blocked until all computers are done with their critical section computations, for if no computer after a certain point executes a critical section, the value of k will stay constant, and the first computer in the cyclic ordering (k, $k-1$, $\ldots$, 1, N, $N-1$, $\ldots$, $k+1$) that subsequently would wish to perform a critical section would meet no restraint.

Finally it is necessary to prove that no individual computer can become blocked. The proof of this is not trivial, for it can be shown that in unfavorable circumstances a computer positioned at $L1$ may have to wait as many as $2^{N-1} - 1$ turns—while other computers do critical sections — before it can get into its own critical section. If, for example, $N = 4$, suppose computer 4 is at $L1$ and computers 1, 2, 3 are at $L2$. Then

(i) computer 1 goes (at high speed) from $L2$ to $L5$;

(ii) computer 2 goes from $L2$ to $L5$, then computer 1 goes from $L5$ to $L2$;

(iii) computer 1 goes from $L2$ to $L5$;

(iv) computer 3 goes from $L2$ to $L5$, then computer 1 goes from $L5$ to $L2$, then computer 2 goes from $L5$ to $L2$;

(v), (vi), (vii) are like (i), (ii), (iii), respectively.

Meanwhile computer 4 has been unfortunate enough to miss the momentary values of k that would enable it to get through to $L2$.

To prove that computer i_0 will ultimately execute its critical section after it reaches $L1$, note that since the system does not get completely blocked, i_0 can be blocked only if there is at least one other computer j_0 that does get to execute its critical section arbitrarily often. But every time j_0 gets through from $L0$ to $L5$, with $control[i_0] \neq 0$, the value of k it encounters at $L1$ must have been set by a computer k_0 that follows i_0 and precedes j_0 in the cyclic ordering $(N, N-1, \ldots, 2, 1, N)$. Therefore some k_0 that follows i_0 and precedes j_0 in the ordering must also execute a critical section arbitrarily often. This is a contradiction if we choose j_0 to be the first successor of i_0 having this property.

Lest someone write another letter just to give the special case of the algorithm above when there are only two computers, here is the program for computer i in the simple case when computer j is the only other computer present and when k is either i or j:

```
L0:  control[i] := 1;
L1:  if k = i then go to L2;
     if control[j] ≠ 0 then go to L1;
L2:  control[i] := 2;
     if control[j] = 2 then go to L0;
L3:  k := i;  critical section;  k := j;
L4:  control[i] := 0;
L5:  remainder; go to L0;
```

When N is large or when it is variable, the algorithm above is not very efficient. Considerably more efficient methods can easily be designed if we modify the basic assumption that the only indivisible operations of the computers are single reads or writes to a store. Suppose, for example, that we have the common store

$$\textbf{integer array } Q[0:N];\ \textbf{integer } T$$

(initially zero) and assume that the three operations

```
procedure add to queue(i);
  T := Q[T] := i;
Boolean procedure head of queue(i);
  head of queue := (i = Q[0]);
procedure remove(i);
  if T = i then Q[0] := T := 0 else Q[0] := Q[i];
```

are each indivisible, hardware operations. Then an efficient solution for the ith computer, $1 \le i \le N$, is:

```
L0:  add to queue(i);
L1:  if ¬head of queue(i) then go to L1;
L3:  critical section;
L4:  remove(i);
L5:  remainder; go to L0;
```

The loop at $L1$ can be handled by interrupts; so the processor can do other work while waiting in the queue. This method is "fairer" than the others, and it can be modified to work with priorities.

Addendum

A subsequent letter by N. G. de Bruijn [*Communications of the ACM* **10** (1967), 137–138] observed that my program can be improved by replacing the lines

```
L3:  k := i;
     critical section;
     k := if i = 1 then N else i − 1;
```

by

```
L3:  critical section;
     if (control[k] = 0) ∨ (k = i) then
        k := if k = 1 then N else k − 1;
```

and requiring that the initial value of k be one of the numbers $\{1, \ldots, N\}$. He showed that no processor needs to wait more than $N(N-1)/2$ turns to execute a critical section when these changes are made.

I once believed that my solution to the starvation-free mutual-exclusion problem for two processors was as simple as possible. But Gary L. Peterson found a better way that is amazingly elegant:

$$\begin{array}{ll} L0\colon & b[i] := \textbf{true};\ \ k := i; \\ L1\colon & \textbf{if}\ b[j] \wedge (k = i)\ \textbf{then go to}\ L1; \\ & \textit{critical section};\ \ b[i] := \textbf{false}; \\ & \textit{remainder};\ \textbf{go to}\ L0; \end{array}$$

He also extended this idea to N processors; see "Myths about the mutual exclusion problem," *Information Processing Letters* **12** (1981), 115–116.

Chapter 14

Optimal Prepaging and Font Caching

[Written with David R. Fuchs. Originally published in ACM Transactions on Programming Languages and Systems **7** *(1985), 62–79.]*

An efficient algorithm is presented for the communication of letter-shape information from a high-speed computer with a large memory to a typesetting device that has a limited memory. The encoding is optimum, in the sense that the total time for typesetting is minimized, using a model that generalizes well-known "demand paging" strategies to the case when changes to the cache are allowed before the associated information is actually needed. Extensive empirical data show that good results are obtained even when difficult technical material is being typeset on a machine that can store information concerning only 100 *characters. The methods of this paper are also applicable to other hardware and software caching applications with restricted lookahead.*

1. Introduction

The purpose of this paper is to study a data-reduction problem that arises when computers are applied to phototypesetting. A page that is printed with modern typesetting equipment may be regarded as a gigantic matrix of 0s and 1s, where 0 represents a blank space and 1 represents ink. For example, the particular machine used in our experiments has approximately 19,000,000 such bits per square inch; therefore a typical page of technical text from a book like [4], which was printed on that machine, is essentially a matrix of more than 727 million bits. This data must be reduced by more than three orders of magnitude in order to be transmitted from the host computer to the typesetter at a rate of 9600 bits per second, if the page is to be finished in two minutes. Typical methods of data compression are considered excellent if they achieve a reduction factor of only 50 percent, so it is clear that special techniques are needed if high-resolution digital printing is to be efficient.

The main factor accounting for this thousand-fold reduction in the number of information bits is, of course, the fact that pages are composed from letters that have comparatively simple shapes. For example, a typical character that measures 5×10 printer's points, where there are 72.27 printer's points per inch, has a digital pattern occupying about 182,000 bits on the machine mentioned above, but this pattern can be specified satisfactorily with about 250 bytes $=$ 2000 bits.

Even with this reduction, however, there remain about 2500 characters per page, so about 5,000,000 bits still need to be transmitted. The problem would be simple if the typesetter knew all of the digital patterns for all of the letters, since then we would merely have to transmit letter codes. But typical technical texts involve a variety of different fonts and special symbols, and many typesetting machines have only a limited local memory for the storage of character patterns. Character shapes must therefore be transmitted from the host computer to the typesetter, and the only way we can compress this data is by using the fact that most characters are used again and again.

For example, suppose that the typesetter has enough memory to record the shapes of 60 characters. This is just barely enough for the letters a to z and A to Z, but we also need to deal with numerals and punctuation marks, together with italic and bold variations, and with changes in size and style. The standard industrial practice has been to solve the size-change problem by doing simple scaling operations, so that "8-point type" is obtained as an 80 percent reduction of "10-point type"; but typographers are very unhappy about this compromise, because the results were much better on the old hot-lead machines when every point size was designed separately. Fortunately, it turns out that individual lines of text hardly ever need a great variety of characters even without the compromise; therefore the typesetter can use its memory as a "cache" for 60 characters, including the 30 or so that it needs on the current line.

The typesetter might also be able to accept a few more character descriptions that will be needed on subsequent lines, at the same time as it is setting type on the current line; these new characters can replace "dead" ones in the cache, and with luck the cache will be up-to-date at all times. For example, if there is time to make five adjustments to the cache on each line, 30 new characters can be brought in when a new font is desired, if the changes begin six lines in advance. By looking ahead to see which characters need to be sent in the future, the host computer can control the typesetter's cache contents in an efficient way. The purpose of this paper is to examine suitable algorithms by which the host computer can exhibit such clairvoyant behavior, and to study how much is gained by such techniques.

Section 2 presents a theoretical model of a general cache allocation problem, and Section 3 derives an optimal allocation strategy for that model. Data structures and algorithms by which the optimum strategy can be computed with reasonable efficiency are described in Sections 4 and 5. The concluding section presents empirical results that illustrate what can be achieved.

Although this paper is oriented towards a particular application to typesetting, the reader is encouraged to speculate about how the same methods could be applied to the design of ultra-high-speed computers. One can imagine a pipelined arithmetic unit, playing a role analogous to that of the typesetter, taking orders from another computer, whose function is to preload a cache memory with numeric data, based on the knowledge of a particular algorithm's control structure. Instead of relying on the conventional architecture of a general-purpose computer, one could apply the methods of this paper to a large class of important computation-intensive algorithms whose control structure is predictable.

2. A Cache-Allocation Model

Consider an alphabet of m possible characters that might be kept in a cache that can hold at most s characters at once. We wish to implement a sequence of commands of the following three types, where $1 \le i \le m$:

$L(i)$	Lock character i in the cache.
$U(i)$	Unlock character i.
G	Get any character and place it into the cache.

Such a sequence is called a "job." Character i is said to be "bound" at a certain point of a job if more $L(i)$ commands than $U(i)$ commands have occurred before that point. We assume that the $U(i)$ command appears only when character i is bound; thus, at any point in reading through a job, we will never have seen more $U(i)$ commands than $L(i)$ commands for any i. Furthermore, we assume that there are never more than s different characters bound at any one time; a job that does not meet this requirement needs a larger cache.

Initially the cache has s empty slots. When an $L(i)$ command occurs and character i is not present in the cache, we say that there is a "fault." In case of a fault, the typesetter comes to a halt while the cache receives character i; that character either goes into an empty slot or replaces some unbound character. At the time of a G command, any character not present in the cache can, similarly, be brought into a cache slot that is not occupied by a bound character. We do not consider that a fault has occurred in the latter case, since the typesetter is still busy doing a previous line. Thus, G commands allow us to anticipate L commands so that future faults are avoided. It is also possible to "pass" a G command, leaving the cache unchanged, if passing seems more desirable than bringing in a new character.

Note that a character must be in the cache whenever it is bound, because an $L(i)$ command guarantees that character i is present and because no character can be replaced until it has become unbound.

This model is more general than the "page reference" model that is usually used to study cache behavior in a virtual memory system. The page reference model is the special case in which there are no G commands and where each $L(i)$ command is immediately followed by $U(i)$. Our model also assumes that we know the entire sequence of commands in a job before the job is begun.

In our application, the typesetting of a full line must operate in real time with no waiting for faults. Thus, a line of type containing the characters $i_1 i_2 \ldots i_r$ might be represented by the command sequence

$$L(i_1)L(i_2)\ldots L(i_r)G^k U(i_1)U(i_2)\ldots U(i_r)\,,$$

if there is time to bring into the cache as many as k characters for future lines while the line is being typeset. The actual typesetting of the line starts after the command $L(i_r)$ has been completed. This sequence of commands ensures that all characters needed on the line will be present in the cache before typesetting takes place. The L instructions for line $i+1$ are not begun until the typesetter has completed line i, so that no characters needed on the line will get overwritten before the typesetter is done with them.

The model does not specify what character is replaced at the time of a fault or of a G command. A *caching strategy* is a set of rules that governs what happens to the cache at such times. A strategy *trace* is the output of a strategy when it is fed a job; in other words, it is a list that tells us at each G and L command which character, if any, is to be brought into which cache slot.

3. An Optimum Caching Strategy

In this section we shall see that an intuitively plausible strategy for cache allocation actually minimizes the total number of faults, among all possible strategies for a given command sequence. (This generalizes Belady's well-known "MIN" method in the page reference model [1, 2, 5].) The strategy is simply this:

(1) Whenever a character is brought into the cache, place it in an empty slot, if possible; otherwise let it replace an unbound character i that never appears in a subsequent $L(i)$ command, if possible; otherwise let it replace the unbound character i in the cache whose next appearance in an $L(i)$ command is as late as possible. Since at most s characters can be bound at once, one of these three cases must always hold.

(2) Whenever a G command appears, bring in the character i, not currently in the cache, whose next appearance in an $L(i)$ command is as soon as possible, unless all unbound characters currently in the cache will be locked by L commands that occur between the current G command and that $L(i)$ command, or unless no such character i exists.

When a character is brought into the cache by rule (2), its cache slot is selected by rule (1). The case in rule (2) where no such i exists occurs when all characters needed by the rest of the job are already in the cache.

To prove that this strategy S is optimum, we compare its trace on any job to any other possible trace for the same job, and show that S's trace leads to no more faults. More precisely, let X_J be any trace for job J, and let S_J be S's trace for that job. If X_J is different from S_J, we construct a trace X'_J that has no more accumulated faults than X_J at any time, and such that X'_J agrees with S_J longer than X_J does. In other words, if X_J agrees with S_J on the first $t-1$ commands but differs from it at command number t, then X'_J will agree with S_J for at least the first t commands. Repeated application of this argument will show that S leads to the smallest possible accumulated number of faults at all times.

The construction we define makes use of a "trace subroutine," which uses a fully specified trace to complete another trace that has been specified only in part. The input to this subroutine consists of a job J, a trace W_J that implements J, and a partial trace Y_J that is defined only through the $(p-1)$st command of J. The subroutine will complete the definition of Y_J such that it is at least as good as W_J. The following conditions must hold just after command number $p-1$ for both W_J and Y_J (omitting the implied subscript 'J'):

i) There are characters w and y such that the cache for W has the form $\{w\} \cup C$ and the cache for Y has the form $\{y\} \cup C$, for some set of characters C, where $w \notin C$ and $y \notin C$. In other words, the caches are identical except for at most one element.

ii) Trace W has had at least as many faults as trace Y.

iii) If the sequence of future commands causes w to be locked before y, where w and y are the characters mentioned in condition (i), then W has already had more faults than Y.

The second condition says that Y is no worse than W. The third condition says in effect that character w cannot be a "better" thing to have in the cache than y, unless Y can afford a fault without falling behind W.

If these three conditions are satisfied, we say that "relation (w, y) holds for (W, Y, J)" at the current position in the job. The trace subroutine is called only when relation (w, y) holds for (W, Y, J) after command number $p-1$. The subroutine proceeds by figuring out what Y should do for the pth command in order to preserve these invariant conditions. In other words, if relation (w, y) holds before the pth command of J, the subroutine must define the next step of Y so that relation (w', y') holds after the pth command, for some w' and y'. The subroutine can now do the $(p+1)$st command, and so on, until Y has been defined for all of J. Since the invariants still hold, we know by (ii) that Y is no worse than W, so the subroutine does what was claimed.

Now to define Y on the pth command. We know that relation (w, y) holds. If $w = y$, so that both traces currently have the same cache contents according to condition (i), we simply let Y be the same as W on command p. Relation (w, y) still holds. (In this case the next iteration of the subroutine will have $w = y$, so the $(p+1)$st action of Y will again be defined to be the same as that of W by this rule, and so on; from this time henceforth, Y will be the same as W.)

On the other hand, if $w \neq y$, note that both w and y must currently be unbound, since w does not occur in Y's cache and y does not appear in W's. The following subcases arise in defining Y on command p:

a) If the command is $L(y)$ so that a fault occurs with trace W, suppose W replaces z by y. Trace Y has no fault, so it cannot bring a character into the cache; but after the command $L(y)$, it is easy to check that relation (w, z) holds, because condition (ii) implies that W has now had more faults than Y.

b) If the command is $L(w)$ so that a fault occurs with Y but not W, then Y replaces y in its cache by w. This replacement is legitimate because y is currently unbound. Afterwards relation (w, w) holds, since condition (iii) implies that this case can arise only if Y can afford at least one fault.

c) If the command is $L(i)$ where $i \neq w$, $i \neq y$, and $i \notin C$, a fault occurs in both traces. If W replaces w by i, then Y replaces y by i; relation (i, i) holds. Otherwise, if W replaces z by i for some $z \in C$, then Y also replaces z by i, and relation (w, y) still holds.

d) If the command is $L(i)$ where $i \in C$, or if it is $U(i)$, no fault occurs for either W or Y, and relation (w, y) remains true.

e) If the command is G and if W replaces w by v, then Y replaces y by v, and relation (v, v) holds.

f) Finally, if the command is G and if W replaces z by v for some $z \in C$, or if W does nothing, then Y likewise replaces z by v or does nothing. Relation (w, y) still holds.

This completes the definition of Y from W, except in one degenerate case: Suppose that the command is $L(y)$ and that W brings character y into an empty position in its cache. This is a variation on case (a), where Y cannot bring in a character because no fault has occurred. We can avoid this situation by assuming that the set C in condition (i) always contains $s-1$ elements (or equivalently that there are no empty positions). For we can fill each empty position with distinct dummy characters that do not appear in any commands. This convention makes the proof go through.

Now that the trace subroutine has been specified, we use it to prove the optimality of strategy S. Suppose X_J is any trace different from S_J for some job J, where the first difference occurs at the tth command in the traces. We create a trace X'_J to be the same as S_J up to and including the tth command, such that relation (x, y) holds for (X_J, X'_J, J), for some x and y. We can then call the trace subroutine to complete X'_J such that it is at least as good as X_J. Then we are able to repeat the process with S_J and X'_J, getting X''_J, which is like S_J through the $(t+1)$st command and at least as good as X'_J (and therefore at least as good as X_J), and so on. The final result is that S_J is the same as $X_J^{(n)}$ for some $n \leq \text{length}(J)$, and S_J is no worse than X_J. Since J and X are arbitrary, this proves that S is optimal. (Once again, we drop the J when it is understood.)

The only task left, therefore, is to show that if X' is defined to be the same as S through command t, then relation (x, x') holds for (X, X', J), for some x and x'. Just before command t, both X and X' have had the same number of faults and both their caches have the same contents. The tth command must either be an L command that causes a fault or a G command on which S and X did not both pass. Suppose first that command t is $L(i)$, where i is not in the cache at time t, and X replaces character j by i while S replaces character k by i. Relation (k, j) holds for (X, X', J) because rule (1) guarantees that character k is not locked before character j.

Similarly, if the tth command is G and if X passes while S replaces k by z, relation (k, z) holds for (X, X', J), since rules (1) and (2) imply that k is not locked before z. And if X replaces j by w when implementing a G command, while S passes on that G, relation (w, j) holds, since rule (2) ensures that w is not locked before j.

The only remaining case is that the tth command is G and that trace X replaces j by z while S replaces k by w. If $j = k$, relation (z, w) holds for (X, X', J) because of rule (2). On the other hand, if $j \neq k$, we have to invoke the trace subroutine twice before obtaining a trace that dominates X and agrees with S on commands 1 through t: First we let Z be a trace that replaces k by z so that Z is a mixture of X and S. At this point, relation (k, j) holds for (X, Z, J) because of rule (1). Completing Z with the trace subroutine, we obtain another trace that differs from S in the tth command. In this G command, Z replaces k by z while S and X' replace k by w; hence relation (z, w) holds for (Z, X', J).

We have now shown that it is always possible to set up X'_J to obey the invariant conditions, thus completing the proof that S is optimum.

4. Implementing the Optimum Strategy

Let m be the total number of possible characters, let s be the size of the cache, and let n be the number of commands in job J. Our goal is to design an algorithm that computes the optimal trace S_J. Job J's commands are given in arrays *op* and *char* before the algorithm begins. If the jth command is $L(i)$, $U(i)$, or G, then

$$\begin{array}{lll} & op[j] = \text{`}L\text{'}, & char[j] = i, \\ \text{or} & op[j] = \text{`}U\text{'}, & char[j] = i, \\ \text{or} & op[j] = \text{`}G\text{'}, & char[j] = \text{undefined}, \end{array}$$

respectively, for $1 \leq j \leq n$. For the present we shall pretend that we have enough memory to store all n of the commands at once.

The algorithm records the resulting trace in the *cache* and *char* arrays. If $cache[j] > 0$, step j of the trace says to bring character $char[j]$ into cache slot $cache[j]$; and when $cache[j] = 0$, then no character is to be brought into the cache during step j. Thus, if a fault occurs at the jth command, the algorithm should set $cache[j]$ to the cache position that S allocates to $char[j]$, where $1 \leq cache[j] \leq s$. If $op[j] = $ 'G' and if strategy S replaces cache position k by character c, the algorithm should set $cache[j] \leftarrow k$ and $char[j] \leftarrow c$. In other cases the algorithm should set $cache[j] \leftarrow 0$. Note that the *char* array is altered by this algorithm, but only in G commands.

Our algorithm works with two pointers p and q, where $1 \leq p \leq q \leq n+1$. Pointer p represents the current position where we are defining the trace; we say that the trace has been defined "up to time p," thinking of a clock that advances when p increases. Pointer q looks ahead to the first L command that locks a character not in the cache at time p; if no such commands exist, we have $q = n+1$. For each character i there are two values:

$$slot[i] = \begin{cases} 0, \text{ if } i \text{ is not present in the cache at time } p; \\ \text{the cache position of } i, \text{ otherwise;} \end{cases}$$

$$\begin{aligned} usage[i] = {} & (\text{the number of } L(i) \text{ instructions before command } q) \\ & - (\text{the number of } U(i) \text{ instructions before command } p). \end{aligned}$$

For each cache position k, where $1 \le k \le s$, we have

$$contents[k] = \begin{cases} i, & \text{if } slot[i] = k; \\ 0, & \text{if position } k \text{ is empty.} \end{cases}$$

Suppose that character i appears in r_i different "lock" commands, numbered $j_{i1} < j_{i2} < \cdots < j_{ir_i}$. A preliminary pass over the *char* array suffices to fill two auxiliary arrays *first*$[i]$ and *next*$[j]$, for $1 \le i \le m$ and $1 \le j \le n$, so that

$$first[i] = j_{i1},\ next[j_{i1}] = j_{i2},\ \ldots,\ next[j_{ir_i}] = n+1.$$

If $r_i = 0$, we can set *first*$[i] = n+1$; however, this value will not be looked at, so it really does not matter.

Initially $p = q = 1$, *usage*$[i]$ = *slot*$[i] = 0$ for $1 \le i \le m$, and *contents*$[k] = 0$ for $1 \le k \le s$. The initial value of *first*$[i]$ will be j_{i1} as stated above; but as the algorithm progresses, *first*$[i]$ will be updated so that it is the smallest element $\ge q$ of the set $\{j_{i1}, j_{i2}, \ldots, j_{ir_i}\}$. For convenience, we also set *first*$[0] = n+1$ and *usage*$[0] = 0$, so that 0 is essentially a character that never appears.

One more thing completes this family of data structures: There is a priority queue Q of all cache positions k such that *usage*[*contents*$[k]$] = 0; these positions are ordered by *first*[*contents*$[k]$]. Initially Q contains all positions $\{1, \ldots, s\}$ in arbitrary order. Any suitable scheme for implementing a priority queue can be used for Q. If s is small, a sorted linear list is adequate; if s is large, a method that requires at most $O(\log s)$ steps per operation might be most appropriate. Note that Q contains all cache positions whose contents will be unbound at all times between p and q inclusive, sorted in order of the first time they will be locked after time q.

The algorithm proceeds by advancing p one step at a time, first moving q as far as it can ahead of p:

```
while p ≤ n do
   begin integer i; comment the character to bring in next;
   ⟨Move q forward until reaching L(i) with i not present⟩;
   ⟨Process command p, attempting to bring in i⟩;
   p ← p + 1;
   end
```

The subalgorithm that moves q forward sets i to the character that should be brought into the cache next; this is the character not present at time p that is going to be needed soonest. If no such characters exist, the subalgorithm sets $q \leftarrow n+1$ and $i \leftarrow 0$:

```
⟨Move q forward until reaching L(i) with i not present⟩ =
   begin i ← 0;
   while q ≤ n and i = 0 do
      if op[q] ≠ 'L' then q ← q + 1
      else begin i ← char[q];
         if slot[i] > 0 then
            begin first[i] ← next[q];
            if usage[i] = 0 then delete slot[i] from Q;
            usage[i] ← usage[i] + 1;
            q ← q + 1;
            i ← 0;
            end;
         end;
   end
```

When deleting $slot[i]$ from Q, it may help to know that $slot[i]$ is at the rear of Q; i is the character that would currently be chosen last for replacement in the cache on the basis of priority, since it has the minimal value of $first[contents[k]]$.

The processing of command p has two main components, depending on whether the command is for unlocking or bringing in a character:

```
⟨Process command p, attempting to bring in i⟩ =
   begin cache[p] ← 0; comment this value may be changed later;
   if op[p] = 'U' then ⟨Unlock char[p]⟩
   else if i > 0 and (op[p] = 'G' or p = q) then
      ⟨Try to bring in i and advance q⟩;
   end
```

The first of these is a simple update to the data structures:

```
⟨Unlock char[p]⟩ =
   begin integer j; comment unlock this character;
   j ← char[p];
   usage[j] ← usage[j] − 1;
   if usage[j] = 0 then insert slot[j] into Q with key first[j];
   end
```

The other operation is the most interesting:

```
⟨Try to bring in i and advance q⟩ =
   if Q is empty then
      begin if p = q then report overflow error;
      end
   else begin integer k; comment change this cache position;
      delete k from Q with maximum first[contents[k]];
      cache[p] ← k; char[p] ← i;
      slot[contents[k]] ← 0; slot[i] ← k; contents[k] ← i;
      first[i] ← next[q]; usage[i] ← 1; q ← q + 1;
      end
```

Note that if $p = q$, we have $op[p] =$ 'L', and a fault has occurred. An overflow error is detected if $p = q$ and Q is empty, since this means that the pth command is trying to lock some character not in the cache, while s other characters are already bound.

It is straightforward to verify that the operations preserve the invariant relations we have stated for the data structures, and therefore that an optimum strategy S is being found.

The running time of this implementation is of order $m + n \log s$. If G commands occur frequently, the pointer q tends to be quite far ahead of p, so that comparatively few characters in the cache will have zero usage; Q will not contain many entries, and the running time will be essentially linear. Thus, additional G commands tend to make the algorithm faster, even though they cause it to find the optimum over a larger space of possible strategies.

5. Refinements to the Implementation

The algorithm of Section 4 can be modified in various ways to improve its efficiency and to take account of practical constraints.

In the first place, the running time will be improved if we realize that p usually increases several times before q moves. If $i > 0$, so that $op[q] =$ 'L' and $char[q] = i$ needs to be brought in, pointer q will stand still until the code ⟨Try to bring in i and advance q⟩ is actually executed. Therefore the main loop of the program can be reorganized as a loop on q, followed by a loop on p, followed by an operation that increases both p and q.

In the second place, the fact that n is large means that it is undesirable to have a separate array $next[j]$ for $1 \le j \le n$; this additional array limits the number of commands that can be accommodated. By looking at the way the algorithm uses *next*, we can see that the *next* and *char* arrays can be overlapped at the expense of a (shorter) array $second[j]$ for $1 \le j \le m$. The new conventions are as follows, if the "lock" commands following time q for character i are $j_{i1} < \cdots < j_{ir_i}$:

If $r_i = 0$: $first[i] = n+1$, $second[i] =$ undefined.
If $r_i = 1$: $first[i] = j_{i1}$, $second[i] = n+1$, $char[j_{i1}] = i$.
If $r_i = 2$: $first[i] = j_{i1}$, $second[i] = j_{i2}$, $char[j_{i1}] = i$, $char[j_{i2}] = n+1$.
If $r_i \ge 3$: $first[i] = j_{i1}$, $second[i] = j_{i2}$, $char[j_{i1}] = i$, $char[j_{i2}] = j_{i3}$, $\ldots$, $char[j_{i(r_i-1)}] = j_{ir_i}$, $char[j_{ir_i}] = n+1$.

The operation '$first[i] \leftarrow next[q]$', which appears twice in the algorithm of Section 4 at times when $r_i > 0$, is now changed to the following code:

```
begin integer j;
j ← second[i]; first[i] ← j;
if j ≤ n then
   begin second[i] ← char[j]; char[j] ← i;
   end;
end
```

In the third place, we must face the fact that jobs generally have more commands than could possibly be held in our computer's memory. Rather than having the algorithm read in an entire job, figure out the cache trace and put it into the *cache* array, we instead regard the cache allocation algorithm as a coroutine that does the caching "online" as it reads the commands. In other words, if we can store only n_0 commands in memory at once, we would like to have an algorithm that will have read n_0 commands ahead of the one it is actually implementing at any given time. When the coroutine is called on to provide the value of $cache[x]$, it will have elements x through $x + n_0 - 1$ of the *char* and *op* arrays in cyclic buffers in memory. The coroutine is supposed to figure out what to do for step x, and then it should read in command $x + n_0$, overwriting $op[x]$ and $char[x]$.

We might not discover a truly optimum trace if the lookahead is limited to n_0 future commands; but the only errors we make are to remove certain items from the cache in a different order when those items are not used at all during the next n_0 steps. If n_0 is large enough compared to the cache size, all such items will probably leave the cache anyway, even in an optimum trace, so a limited-lookahead method will usually be no worse than the optimum. Indeed, our proof of optimality in Section 3 shows that a variety of strategies will usually perform no worse than strategy S.

Implementation of the coroutine philosophy means that we need to update the *first*, *second*, and *char* arrays online instead of assuming that they have been initialized by a preliminary pass over all the commands. For this purpose we need another array $last[i]$ for $1 \le i \le m$, containing the value of j_{ir_i}, if $r_i > 0$; we leave $last[i]$ undefined if $r_i = 0$. Furthermore, some other sentinel value must be used instead of $n+1$ in the *first* and *second* arrays, since we do not know what n is. We shall use 0; the test '$j \le n$' above should therefore be changed to '$j > 0$'.

The algorithm now starts by filling up the *op* and *char* arrays with the first n_0 commands in the job; the *first*, *second*, and *last* arrays are set up to reflect those commands; and p and q are set to 1. The entire data structure must be kept up to date as p and q change. For instance, when p is incremented to 2, the algorithm should put command number n_0+1 into $op[1]$ and $char[1]$ and update the *first*, *second*, and *last* arrays to reflect this new command. Thus, the statement '$p \leftarrow p+1$' is replaced by

```
⟨Advance p⟩ =
   begin (op[p], char[p]) ← next command in the job;
   if op[p] = 'L' then
      begin integer i;  i ← char[p];
      if first[i] = 0 then
         begin first[i] ← p;  second[i] ← 0;
         ⟨If slot[i] ∈ Q, change its key⟩;
         end
      else begin char[p] ← 0;
         if second[i] = 0 then second[i] ← p
         else char[last[i]] ← p;
         end;
      last[i] ← p;
      end;
   if p = n0 then p ← 1 else p ← p + 1;
   end;
```

the statements '$q \leftarrow q + 1$' are changed to

$$\textbf{if } q = n_0 \textbf{ then } q \leftarrow 1 \textbf{ else } q \leftarrow q + 1;$$

and a few other changes to the code are required to keep q from incrementing when it gets n_0 commands ahead of p.

A "dead" character is one that, as far as we can tell from our limited lookahead, will never again be used in the job. Thus, character i is dead if and only if $usage[i] = 0$ and $first[i] = 0$. It is convenient to split Q into two separate parts: Q_0, which is simply an unordered set of all cache positions that are empty or contain dead characters, and the remaining part Q_1, which is a priority queue ordered by the nonzero key values $first[contents[k]]$. These key values are to be "circularly ordered" in the sense that we regard $x > y$ if $x < p \le q \le y$, since x is one lap ahead of y in such a case. Note that the operation ⟨If $slot[i] \in Q$, change its key⟩ simply removes $slot[i]$ from Q_0 and enters it into Q_1 with key p, which will be higher than any other key currently in Q_1. The elements of Q_0 are all regarded as having higher keys than those of Q_1.

It is a simple matter to fill in all the remaining details — namely, to take care of shutting down the input operations when all commands have been read and to terminate the coroutine when all of the cache commands have been performed.

6. Empirical Tests

The authors have used these procedures to drive an Alphatype CRS phototypesetter, producing dozens of technical books (beginning with [4]). In this application the characters in the cache have variable size, so the actual cache storage is allocated dynamically. When a new character is brought into the cache, there may already be room; on the other hand, two or more other characters might need to be removed before a hole appears that is large enough to accommodate an especially large newcomer. The number of G commands at the end of a line is not fixed, because it depends on the sizes of characters that are actually brought in.

In other words, the theoretical model studied earlier in this paper is a rather drastic simplification of the actual problem that had arisen in practice. As usual. But (as usual) the theoretical considerations provided valuable guidelines for a practical implementation; and by using an algorithm that is optimal, or near-optimal under the simplifying assumptions, the authors were able to achieve quite satisfactory results, even though those assumptions were violated.

Indeed, it would almost surely be unfeasible to develop an optimum strategy that takes account of all the details of the actual application, since the problem of optimum dynamic storage allocation is already NP-complete before we add the extra complexities of cache management (see [3], problem SR2). Instead of worrying about special schemes for dynamic allocation, the authors found that it was sufficient to replace unbound characters simply on the basis of their priority, without regard to their size or to the priorities or sizes of their neighbors.

Figure 1 shows a five-page sample text that was subjected to a variety of experiments discussed below. This text had been used to debug the TEX typesetting system in 1978, and it also provided the style pages in the design of [4]; thus it represents a wide variety of different things that happen in a 700-page book, compressed into five pages, two of which are less than half full. It involves the typesetting of 5211 characters, of which 576 are distinct when size variations are taken into account.

In terms of the abstract model of Section 2, the task of driving the authors' typesetting equipment can be described as the problem of implementing a sequence of commands having the following general form:

Lock all characters used on line k;

If time permits, issue G commands to bring in future characters;

Unlock all characters used on line $k-1$.

We do this for $k = 1, 2, \ldots,$ except that the pattern changes in special cases. The term *line* means a sequence of characters that are to be typeset at the same baseline; thus, a complex mathematical formula might actually occupy many lines.

There usually is time to preload future characters into the cache, because the time needed to transmit the information about what to set on line k is usually less than the time for the actual typesetting of that line by the CRS.

Notice that two consecutive lines are generally bound in the cache at once, since line $k-1$ is not unlocked until after line k has been locked, owing to buffering inside the typesetting machine. Typesetting is done by an asynchronous process that goes on in parallel with the transmission of data for future lines. If the typesetter is not finished with line $k-1$ when the characters on that line are unlocked, the controlling process stops before advancing k to $k+1$; this makes the unlocking operations legitimate. In emergency situations, when the ordinary policy would overload the cache, line $k-1$ is unlocked sooner and the controlling process pauses to make sure that the typesetter has caught up; the cache is also repacked at such times in order to make all of the available memory appear in consecutive locations.

CHAPTER THREE

RANDOM NUMBERS

Anyone who considers arithmetical methods of producing random digits is, of course, in a state of sin.
— JOHN VON NEUMANN (1951)

There wanted not some beams of light to guide men in the exercise of their Stocastick faculty.
— JOHN OWEN (1662)

3.1. INTRODUCTION

NUMBERS that are "chosen at random" are useful in many different kinds of applications. For example:

a) *Simulation.* When a computer is being used to simulate natural phenomena, random numbers are required to make things realistic. Simulation covers many fields, from the study of nuclear physics (where particles are subject to random collisions) to operations research (where people come into, say, an airport at random intervals).

b) *Sampling.* It is often impractical to examine all possible cases, but a random sample will provide insight into what constitutes "typical" behavior.

It is not easy to invent a foolproof random-number generator. This fact was convincingly impressed upon the author several years ago, when he attempted to create a fantastically good generator using the following peculiar approach:

Algorithm K (*"Super-random" number generator*). Given a 10-digit decimal number X, this algorithm may be used to change X to the number that should come next in a supposedly random sequence.

K1. [Choose number of iterations.] Set $Y \leftarrow \lfloor X/10^9 \rfloor$, the most significant digit of X. (We will execute steps K2 through K13, which cause X to be transformed in various weird and wonderful ways, exactly $Y+1$ times; that is, we will apply randomizing transformations a *random* number of times.)

K13. [Repeat?] If $Y > 0$, decrease Y by 1 and return to step K2. If $Y = 0$, the algorithm terminates with X as the desired "random" value. ▮

The moral of this story is that *random numbers should not be generated with a method chosen at random.* Some theory should be used.

1

FIGURE 1.1. The first page of test data.

2 RANDOM NUMBERS 3.1

Table 1

A COLOSSAL COINCIDENCE: THE NUMBER 6065038420 IS TRANSFORMED INTO ITSELF BY ALGORITHM K.

Step	X (after)		Step	X (after)	
K1	6065038420		K10	1620063735	
K12	1905867781	$Y = 5$	K11	1620063735	
			K12	6065038420	$Y = 0$

EXERCISES

▸ **1.** [*20*] Suppose that you wish to obtain a decimal digit at random, not using a computer. Shifting to exercise 16, let $f(x, y)$ be a function such that $0 \le x, y < m$ implies $0 \le f(x, y) < m$. The sequence is constructed by selecting X_0 and X_1 arbitrarily, and then letting

$$X_{n+1} = f(X_n, X_{n-1}) \qquad \text{for } n > 0.$$

What is the maximum period conceivably attainable in this case?

17. [*10*] Generalize the situation in the previous exercise so that X_{n+1} depends on the preceding k values of the sequence.

FIGURE 1.2. The second page of test data.

3.2.1 THE LINEAR CONGRUENTIAL METHOD 3

3.2. GENERATING UNIFORM RANDOM NUMBERS

IN THIS SECTION we shall consider methods for generating a sequence of random fractions, i.e., random *real numbers* U_n, *uniformly distributed between zero and one.* Since a computer can represent a real number with only finite accuracy, we shall actually be generating integers X_n between zero and some number m; the fraction

$$U_n = X_n/m$$

will then lie between zero and one.

3.2.1. The Linear Congruential Method

By far the most popular random-number generators in use today are special cases of the following scheme, introduced by D. H. Lehmer in 1949. [See *Proc. 2nd Symp. on Large-Scale Digital Calculating Machinery* (Cambridge: Harvard University Press, 1951), 141–146.] We choose four "magic numbers":

$$\begin{array}{rll} m, & \text{the modulus;} & m > 0. \\ a, & \text{the multiplier;} & 0 \le a < m. \\ c, & \text{the increment;} & 0 \le c < m. \\ X_0, & \text{the starting value;} & 0 \le X_0 < m. \end{array} \tag{1}$$

The desired sequence of random numbers $\langle X_n \rangle$ is then obtained by setting

$$X_{n+1} = (aX_n + c) \bmod m, \qquad n \ge 0. \tag{2}$$

This is called a *linear congruential sequence.*

Let w be the computer's word size. The following program computes the quantity $(aX + c) \bmod w$ (or something similar):

```
LDAN X          rA ← −X.
MUL  A          rAX ← (rA) · a.
SUB  TEMP       rA ← rA − rX.                          (2′)
JANN *+3        Exit if rA ≥ 9.
ADD  =w − 1=    rA ← rA + w − 1. (Cf. exercise 3.) ▌
```

Proof. We have $x = 1 + qp^e$ for some integer q that is not a multiple of p. By the binomial formula

$$\begin{aligned} x^p &= 1 + \binom{p}{1} qp^e + \cdots + \binom{p}{p-1} q^{p-1} p^{(p-1)e} + q^p p^{pe} \\ &= 1 + qp^{e+1}\left(1 + \frac{1}{p}\binom{p}{2} qp^e + \frac{1}{p}\binom{p}{3} q^2 p^{2e} + \cdots + \frac{1}{p}\binom{p}{p} q^{p-1} p^{(p-1)e}\right). \end{aligned}$$

By repeated application of Lemma P, we find that

$$\begin{aligned} &(a^{p^g} - 1)/(a - 1) \equiv 0 \text{ (modulo } p^g), \\ &(a^{p^g} - 1)/(a - 1) \not\equiv 0 \text{ (modulo } p^{g+1}). \end{aligned} \tag{6}$$

FIGURE 1.3. The third page of test data.

If $1 < k < p$, the binomial coefficient $\binom{p}{k}$ is divisible by p. (*Note:* A generalization of this result appears in exercise 3.2.2–11(a).) By Euler's theorem (exercise 1.2.4–28), $a^{\varphi(p^{e-f})} \equiv 1$ (modulo p^{e-f}); hence λ is a divisor of

$$\lambda(p_1^{e_1} \ldots p_t^{e_t}) = \operatorname{lcm}(\lambda(p_1^{e_1}), \ldots, \lambda(p_t^{e_t}))\,. \tag{9}$$

This algorithm in MIX is simply the following:

```
LDA  Y,6   A1. Add.
ADD  Y,5   Y_k + Y_j (overflow possible)
DEC5 1     A2. Advance. j ← j − 1.
ENT6 55    If k = 0, set k ← 55. ▮
```

That was on page 26. If we skip to page 49, $Y_1 + \cdots + Y_k$ will equal n with probability

$$\sum_{\substack{y_1+\cdots+y_k=n\\ y_1,\ldots,y_k\ge 0}} \prod_{1\le s\le k} \frac{e^{-np_s}(np_s)^{y_s}}{y_s!} = \frac{e^{-n}n^n}{n!}.$$

This is not hard to express in terms of n-dimensional integrals,

$$\frac{\int_{\alpha_n}^{n} dy_n \int_{\alpha_{n-1}}^{y_n} dy_{n-1} \;\ldots \int_{\alpha_1}^{y_2} dy_1}{\int_0^{n} dy_n \int_0^{y_n} dy_{n-1} \;\ldots \int_0^{y_2} dy_1}, \qquad \text{where} \qquad \alpha_j = \max(j-t, 0). \tag{24}$$

This together with (25) implies that, for all $s \ge 0$, we have

$$\lim_{n\to\infty} \frac{s}{\sqrt{n}} \sum_{\sqrt{n}s<k\le n} \binom{n}{k} \left(\frac{k}{n} - \frac{s}{\sqrt{n}}\right)^{k} \left(\frac{s}{\sqrt{n}} + 1 - \frac{k}{n}\right)^{n-k-1} = e^{-2s^2}. \tag{27}$$

EXERCISES—Special Set

17. [*HM26*] Let t be a fixed real number. For $0 \le k \le n$, let

$$P_{nk}(x) = \int_{n-t}^{x} dx_n \int_{n-1-t}^{x_n} dx_{n-1} \;\ldots \int_{k+1-t}^{x_{k+2}} dx_{k+1} \int_0^{x_{k+1}} dx_k \;\ldots \int_0^{x_2} dx_1;$$

another way to write those integrals, although I don't use it in my book, is

$$P_{nk}(x) = \int\limits_{n-t}^{x} dx_n \int\limits_{n-1-t}^{x_n} dx_{n-1} \;\ldots \int\limits_{k+1-t}^{x_{k+2}} dx_{k+1} \int\limits_{0}^{x_{k+1}} dx_k \;\ldots \int\limits_{0}^{x_2} dx_1;$$

Eq. (24)* is equal to

$$\sum_{0\le k<n} U_k' V_k' \bigg/ \sqrt{\sum_{0\le k<n} U_k'^{\,2}} \sqrt{\sum_{0\le k<n} V_k'^{\,2}}.$$

* Page 68.

FIGURE 1.4. The fourth page of test data.

3.3.3.3. This subsection doesn't exist. Finally, look at the new section 3.3.4, where there are some matrices. One way to get them leads to a display that's rather big:

$$U = \begin{pmatrix} -1479 & 616 & -2777 \\ -3022 & 104 & 918 \\ -227 & -983 & -130 \end{pmatrix}, \quad V = \begin{pmatrix} -888874 & 601246 & -2994234 \\ -2809871 & 438109 & 1593689 \\ -854296 & -9749816 & -1707736 \end{pmatrix}. \tag{34}$$

I'll try nine point type inside the matrices to make them smaller:

$$U = \begin{pmatrix} -1479 & 616 & -2777 \\ -3022 & 104 & 918 \\ -227 & -983 & -130 \end{pmatrix}, \quad V = \begin{pmatrix} -888874 & 601246 & -2994234 \\ -2809871 & 438109 & 1593689 \\ -854296 & -9749816 & -1707736 \end{pmatrix}. \tag{34}$$

So far none of my examples have involved em-fractions like $\frac{98765}{43210}$. That ends the test data, fortunately TEX is working fine.

FIGURE 1.5. The fifth page of test data.

Special actions occur at the beginning of a page: If the film inside the CRS has to move comparatively far in order to be in the proper position to start a new page, there is extra time to preload font information, hence the controlling process issues additional G commands. In particular, the characters for the first lines of the first page will generally have been brought into the cache by the time the typesetter is positioned at the top baseline.

Several dozen experiments were performed on Figure 1 in order to get some idea as to how the algorithm performs under various conditions. The cache size was varied so that it would be able to hold approximately 50, 75, 100, 125, or 150 characters; we refer to these sizes as C50, C75, ..., C150, respectively. The speed at which font information could be transmitted was varied so that there was free time to send either an average of six new characters per line (i.e., about six G instructions after each line), or about 4.5 new characters per line, or no such characters; in the latter case, no G commands were given, so the algorithm had to minimize the total number of characters transmitted. We refer to these transmission speeds as G6, G4.5, and G0. The algorithm was also run in four modes: (i) with full lookahead; (ii) with internal memory cut back so that only about 12 lines of data could be accommodated at once; (iii) with internal memory cut back to only about 6 lines; and (iv) with full lookahead, but with the priority queue decisions reversed so that the *worst* possible cache replacements were made whenever the algorithm had to take something from the cache. These four lookahead modes are called L∞, L12, L6, and L0, respectively.

Five cache sizes, three speeds, and four lookahead modes make for sixty combinations, and so sixty experiments were performed; the resulting numbers of faults are shown in Table 1.

TABLE 1. Faults that occur when typesetting Figure 1

	G0				G4.5				G6				
	L0	L6	L12	L∞	L0	L6	L12	L∞	L0	L6	L12	L∞	
C50	1881	1060	1040	1037	572	268	268	254	378	198*	204	197	C50
C75	1863	960	856	834	389	91	76	69	146	35	31*	32	C75
C100	1854	954	789	752	353*	79*	30	27	110	20*	0*	3	C100
C125	1821	941	786	699	381	83	26	12	22	22	0	0	C125
C150	1819	917	779	614	356	66	26	9	0*	22	0	0	C150

* Asterisks denote "anomalous" values that are surprisingly low.

These results are quite encouraging. Consider first the G0 case, when no "freeloading" is done: At least 576 faults must occur, since each distinct character must be brought in at least once; the table shows that a caching strategy with lookahead is able to make sure that only a few characters need to be brought in twice. The number of faults under G4.5 is substantially less, even for the unusually complicated text of Figure 1; and with G6 and a moderately large cache the faults disappear entirely.

The starred entries in Table 1 show interesting anomalies, where a lucky combination of circumstances led to fewer faults than would be expected. Consider, for example, the cases with G4.5 and L6 or L0, where the cache size C100 turned out to be slightly better than C125; the reason was that these inherently nonoptimal strategies made better guesses in the C100 case. Another interesting example is the case G6 and C150, where the supposedly pessimal strategy L0 actually did better than L6. The reason here is that L0 only pessimizes the choice of cache replacements. The other part of our algorithm, which looks ahead to find the next candidate for G to bring in, remains optimum; and when there are enough Gs, this part of the algorithm is strong enough to make the replacement strategy immaterial. On the other hand, the L6 restriction curtails the effectiveness of the G lookahead, as well as the replacement lookahead, so L6 can come out worse. The 22 faults occurred at the beginning of Figure 1.3, where a conversion from nine-point to ten-point type takes place; L6 wasn't prepared for so many changes all at once.

The most interesting anomaly arose in the case C100 and G6, when the suboptimal strategy L12 actually turned out to be better than the supposedly optimal L∞! A careful examination of what happened shows that this was a case of good luck for L12 and bad luck for L∞. It all started when the typesetting was going along routinely, about ten lines from the bottom of page 1; both L∞ and L12 were doing approximately the same thing, but with minor variations, so that their dynamic storage allocation patterns in the font cache were quite different. Both strategies had succeeded in looking rather far ahead, and were beginning to bring in the nine-point uppercase letters needed for the caption near the top of page 2. But when the "optimal" L∞ strategy had successfully brought in the nine-point 'O' and 'L', its cache had no free blocks big enough to bring in the 'S'. The restricted L12 strategy, on the other hand, had a fortuitous memory configuration that allowed it to bring in not only the 'S' but also the 'I' and 'N'. This put L12 three characters ahead of L∞, and it retained a three-character advantage all the way through page 2 and the beginning of page 3, where comparatively rapid font changes caused the lookahead to evaporate. Finally L12's lead manifested itself on the line before (1) on page 3; three faults occurred when L∞ had to bring in 'W' and the two pairs of quotation marks.

Notice that L12 was almost never a great deal worse than L∞ in any of the cases, so it appears that a restricted lookahead still makes a satisfactory approximation to optimal behavior. In the authors' application it turned out that there was enough core memory to look about 2500 lines ahead. Experiments showed, however, that L50 was essentially equivalent to L∞; thus the storage requirements could be reduced greatly from what we originally thought would be necessary.

Figure 2 on the next page shows a detailed trace of what went on in the experiment for case (L∞, G4.5, C125). The horizontal axis separates the 834 characters that were brought in during the time Figure 1 was being typeset; all but 12 of these were brought in during G commands, while the remaining 12 were faults. The vertical axis represents the 314 lines of text in Figure 1. The graph shows two zigzag paths, where the upper one represents each character's being loaded and the lower one represents each character's first use. Thus, the upper path is far above the lower path when characters are being preloaded many lines ahead, while the two paths touch each other when a fault has occurred. The lower path has a somewhat erratic behavior: We occasionally find a horizontal segment on that path, representing a line that introduces many new characters. (The worst cases are the line following 'EXERCISES— Special Set' and the line beginning '3.3.3.3. This subsection doesn't exist', both of which required 31 new characters to be preloaded in order to avoid faults.) The upper path, on the other hand, is more regular, because there is roughly the same amount of time for preloading characters on each line. Variations in the upper path occur when the characters to be brought in are especially large or small, or when the line being typeset is short (as at the end of a paragraph), or when the baselines are far apart; but those changes are comparatively minor.

Sometimes the cache is full of useful characters, so that the lookahead procedure stops and the current G commands are "passed"; that is, the G commands aren't used to preload anything. This phenomenon is indicated in Figure 2 by horizontal bars that stick out at the right of the upper path. The first such incidents occur near the bottom of Figure 1.2, and a more significant stoppage occurs during the big displayed equations near the bottom of Figure 1.3.

Before developing the algorithms described above, the authors did a hand simulation on some sample text using the assumptions (L∞, G6, C100), since those parameters appeared to be appropriate for the typesetting equipment that Stanford planned to acquire. The success of caching with those parameters, in spite of the multiplicity of fonts needed to typeset difficult technical material, encouraged us to proceed

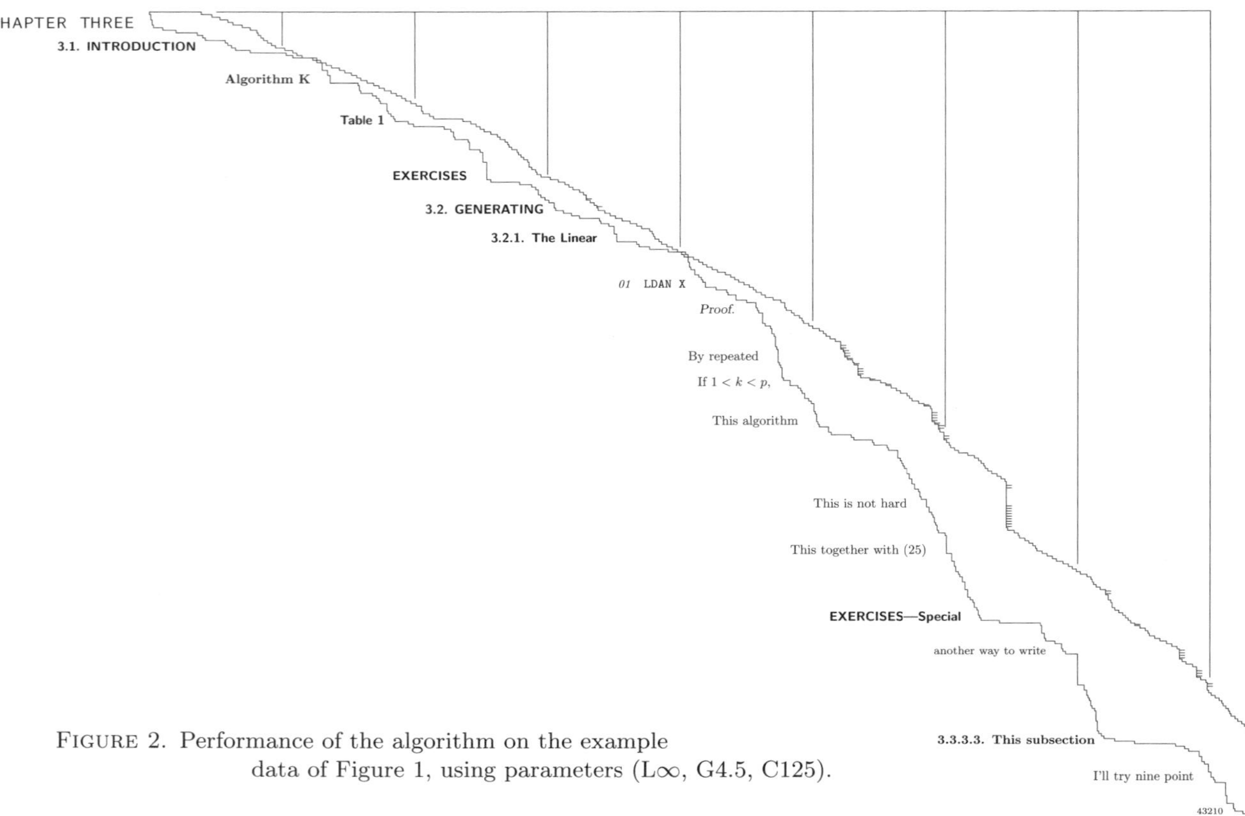

FIGURE 2. Performance of the algorithm on the example data of Figure 1, using parameters (L∞, G4.5, C125).

further. Two years later, after the hardware and software were put into production, we found that G4.5 was more appropriate than G6 because time-sharing interfered with transmissions to the typesetter; however, this effect was compensated by saving space in the typesetter software, so that C125 was more representative of the actual cache size. In fact, the new Alphatype model 400 arrived with additional cache memory, so that our current software corresponds to C155, and faults hardly ever occur.

Incidentally, the strategy of Section 3 does not minimize the number of times a character is brought into the cache, it only minimizes the number of faults. Consider, for example, a cache of size 2 and the job

$$L(1)\;\; L(2)\;\; U(1)\;\; G\;\; U(2)\;\; G\;\; L(3)\;\; L(1)\;\; U(1)\;\; U(3).$$

Strategy S will bring in 3 at the first G, then bring in 1 at the second; the alternative of passing on the first G and bringing in 3 on the second would be preferable if we were trying to minimize the number of brings.

Each time a character is brought into the cache and does not cause a fault, typesetting is not slowed down, but the amount of information that must be sent to the typesetter does increase, so it is desirable to try to minimize the number of characters brought into the cache. The algorithm of Sections 4 and 5 can be modified to "pass" a G if there are no dead characters in the cache (that is, if Q_0 is empty), provided that the lookahead pointer q is sufficiently far from p that it is reasonably safe to assume we will be able to avoid faults by acting on future Gs.

For example, suppose lookahead stops whenever it would require the replacement of a nondead character, provided that the algorithm has looked ahead so far that the next character to be brought in is 16 or more lines away from the current line being typeset. Let us call this variant U16. Then the algorithm may well be able to avoid rashly replacing characters that are not dead by holding back until a character becomes dead, without seriously risking future faults. Figure 2 shows 834 characters brought in when the parameters are (L∞, G4.5, C125); but if the distance between the upper path and lower path were constrained to be no more than about 16 or so, it is plausible to believe that we would end up bringing in characters fewer times, and might even be able to approach the optimum of 699 achieved in the case G0. The following data show what happens for L∞, G4.5, and C125:

	U∞	U24	U16	U8	U0
faults	12	24	24	26	105
brings	834	831	795	761	725

U16 brings 39 fewer characters into the cache, at a cost of 12 faults.

But Figure 1 is not a typical example. Therefore, further tests were made on "real" data. The text of Section 3.5 of [4] is representative of the difficulties of a normal mathematical paper, so it serves as a good indication of what we can usually expect. This second test case, which amounts to 28 typeset pages, involves the setting of 57,912 characters, 660 of which are distinct. When the algorithm was applied with parameters (L∞, G4.5, C125, U∞) there were only 17 faults, and all occurred near the very beginning. The total number of characters brought in to do the whole job was 2745; and with the U16 heuristic this dropped to 2131, while the number of faults remained at 17.

Several other experiments were made in the Section 3.5 file, holding all but one of the settings (L∞, G4.5, C125) fixed. When L∞ was changed to L12, there still were only 17 faults; restricting further to L6 increased them slightly to 44. And when L∞ was changed to the "pessimizing" L0, the result was 17 again! Thus, the lookahead process appears to be powerful enough to achieve optimality without the refinement of the priority queue, when we consider typical data, provided that the G speed and the cache size are suitably large.

As expected, the 17 faults vanished at speed G6. Reducing the speed to G3 increased the number of faults to 65; these occurred only at the beginning and at the switch to nine-point type for the exercises. With speed G1.5 there were 248 faults, and with speed G0.1 there were 1144; speed G0 gave 1569. This compares with

G6	G4.5	G3	G1.5	G0.1	G0
0	12	112	333	613	699

in the case of Figure 1.

Increasing the cache size to C150 did not reduce the number of faults below 17. With a setting of size C100 there were 26 faults, while C75 gave 201. Size C50 was not quite large enough to hold all of the characters bound in one of the nine-point lines; C52 gave 1406 faults.

We can summarize these experiments by saying that typical technical text can be typeset with negligibly few faults, provided that the algorithm of this paper is used in connection with the following resources:

i) a cache in the typesetter capable of holding about 125 character shape descriptions;

ii) time to preload about 4 characters per line without slowing down the typesetting process;

iii) enough memory in the host computer to look ahead about 12 lines (about 750 characters) in the text to be typeset.

This research was supported in part by NSF grants IST-7921977 and MCS-77-23728, by Office of Naval Research grant N00014-81-K-0269, and by the IBM Corporation.

References

[1] L. A. Belady, "A study of replacement algorithms for a virtual-storage computer," *IBM Systems Journal* **5** (1966), 78–101.

[2] L. A. Belady and F. P. Palermo, "On-line measurement of paging behavior by the multivalued MIN algorithm," *IBM Journal of Research and Development* **18** (1974), 2–19.

[3] M. R. Garey and D. S. Johnson, *Computers and Intractability* (San Francisco: W. H. Freeman, 1979).

[4] Donald E. Knuth, *Seminumerical Algorithms*, Volume 2 of *The Art of Computer Programming*, second edition (Reading, Massachusetts: Addison–Wesley, 1981).

[5] R. L. Mattson, J. Gecsei, D. R. Slutz, and I. L. Traiger, "Evaluation techniques for storage hierarchies," *IBM Systems Journal* **9** (1970), 78–117.

Addendum

The emphasis on applications to now-obsolete typesetting machines in this paper has probably caused many people to overlook the generality of the prefetching strategies that are discussed. I believe that these algorithms will permit considerable further exploitation, as discussed in the closing paragraph of Section 1.

Chapter 15

A Generalization of Dijkstra's Algorithm

[Originally published in Information Processing Letters **6** *(1977), 1–5.]*

E. W. Dijkstra [3] has introduced an important algorithm for finding shortest paths in a graph, when the distances on each arc of the graph are nonnegative. The purpose of this note is to present an algorithm that generalizes Dijkstra's, by analogy with the way in which tree structures generalize linear lists, or in which context-free languages generalize regular languages.

1. The Problem in General

Let R_+ be the nonnegative real numbers extended with the value $+\infty$. We shall say that a function $g(x_1, \ldots, x_k)$ from R_+^k into R_+ is a *superior function* if it is monotone nondecreasing in each variable and if

$$g(x_1, \ldots, x_k) \geq \max(x_1, \ldots, x_k) \quad \text{for all } x_1, \ldots, x_k.$$

When $k = 0$ the function g is simply a constant element of R_+, and it is trivially considered to be a superior function. When $k = 2$, several familiar functions $g(x, y)$ such as $\max(x, y)$ and $x + y$ and $\max(1, x) \cdot \max(1, y)$ are superior. Any functional composition of superior functions is a superior function of the variables occurring in the composition.

A *superior context-free grammar* is a context-free grammar in which all productions are of the general form

$$Y \to g(X_1, \ldots, X_k),$$

where Y, X_1, …, X_k are nonterminal symbols, g is a terminal symbol corresponding to a superior function (possibly a different superior function for each production), and the parentheses and commas are also terminal symbols. If $k > 0$, there are $k - 1$ commas; if $k = 0$ the right-hand side of the production is simply written 'g', a terminal symbol that corresponds to a nonnegative real constant.

For example, the following productions define a superior context-free grammar on the nonterminal symbols $A\ B\ C$ and the terminal symbols $(\ )\ ,\ a\ b\ c\ d\ e\ f$:

$$\begin{array}{lll} A \to a\,; & B \to c(A)\,; & C \to e\,; \\ A \to b(B,C)\,; & B \to d(A,C,A)\,; & C \to f(B,A)\,. \end{array}$$

Here a, b, c, d, e, and f are supposed to correspond to superior functions, and we may for example define

$$\begin{array}{rr} a = 4, & d(x,y,z) = x + \max(y,z), \\ b(x,y) = \max(x,y), & e = 9, \\ c(x) = x + 1, & f(x,y) = \tfrac{1}{2}(x + y + \max(x,y)). \end{array}$$

For every nonterminal symbol Y of a superior context-free grammar over the terminal alphabet T we let

$$L(Y) = \{\alpha \mid \alpha \in T^* \text{ and } Y \to^* \alpha\}$$

be the set of terminal strings derivable from Y. (As usual, '$\to^*$' denotes the relation "ultimately can produce" in a context-free derivation.) Every string α in $L(Y)$ is a composition of superior functions, so it corresponds to a uniquely defined nonnegative real number that we shall call $\mathrm{val}(\alpha)$. Thus, in the example above the language $L(A)$ contains the strings

$$\{a,\ b(c(a),e),\ b(c(a),f(c(a),a)),\ b(c(b(c(a),e)),e),\ \ldots\}$$

and the corresponding numerical values are

$$\{4,\ 9,\ 7,\ 10,\ \ldots\}.$$

The problem we shall solve is to compute the *smallest* values corresponding to these languages, namely

$$m(Y) = \min\{\mathrm{val}(\alpha) \mid \alpha \in L(Y)\},$$

for each nonterminal symbol Y.

2. Applications of the General Problem

(A) Consider $n+1$ cities $\{c_0, c_1, \ldots, c_n\}$ connected by a network of roads. For each road from c_i to c_j, introduce the production

$$C_i \to g_{ij}(C_j)$$

and the corresponding superior function

$$g_{ij}(x) = d_{ij} + x,$$

where $d_{ij} \geq 0$ is the length of the road. Then add one more production

$$C_0 \to 0,$$

where 0 corresponds to the constant function zero. In this superior context-free grammar, the elements of $L(C_i)$ correspond to the paths from c_i to c_0, and the corresponding values will be the lengths of those paths. For example, if there is a path from c_3 to c_4 to c_2 to c_0, one of the elements of $L(C_3)$ will be $g_{34}(g_{42}(g_{20}(0)))$, and its corresponding value is $d_{34} + d_{42} + d_{20} + 0$. Therefore $m(C_i)$ is the length of the *shortest path* from c_i to c_0, for all i.

The algorithm we shall develop for the general problem reduces to Dijkstra's algorithm in this special case.

(B) Given a context-free grammar, replace each production $Y \to \theta$ in which the nonterminal symbols of string θ are $X_1, \ldots, X_k$ from left to right (including repetitions) by

$$Y \to g_\theta(X_1, \ldots, X_k),$$

where

$$g_\theta(x_1, \ldots, x_k) = x_1 + \cdots + x_k + (\text{the number of terminal symbols in } \theta).$$

Then $m(Y)$ is the length of the shortest string derivable from Y in the given grammar. Alternatively if we let

$$g_\theta(x_1, \ldots, x_k) = \max(x_1, \ldots, x_k) + 1,$$

then $m(Y)$ is the minimum height of a parse tree for a string derivable from Y in the given grammar.

Finally, if we let

$$g_\theta(x_1, \ldots, x_k) = \max(x_1, \ldots, x_k),$$

we have $m(Y) = 0$ or ∞ according as $L(Y)$ is nonempty or empty. The algorithm we shall develop for the general problem reduces to the classical emptiness-testing algorithms of Bar-Hillel, Perles, and Shamir [2] or Greibach [5] in this special case.

(C) Consider the AND/OR graphs that arise in artificial intelligence applications as in Nilsson [9, Sections 4 and 5]. In this case we use one nonterminal symbol for each "problem" to be solved. If some problem-reduction operator shows that problem Y could be solved if we could solve all of problems $X_1, \ldots, X_k$, then we introduce the production

$$Y \to g(X_1, \ldots, X_k)$$

where $g(x_1, \ldots, x_k) = x_1 + \cdots + x_k +$ (the cost of solving Y, given solutions to $X_1, \ldots, X_k$). Then $m(Y)$ is the minimum cost of a solution to Y, provided that the cost of solving common subproblems is replicated. (All problem solutions are considered to be independent.) Our algorithm will therefore find a smallest AND/OR graph in this sense; but it is of limited utility for A.I. applications because it deals with the set of *all* "easy" subproblems, and that set is usually too large.

A special case of the algorithm to be described here has been published by Martelli and Montanari [7]. They deal only with AND/OR graphs whose functions $g(x_1, \ldots, x_k)$ have the linear form $x_1 + \cdots + x_k + c$ with $c > 0$; furthermore they restrict their discussion to AND/OR graphs that are acyclic. The algorithm below works also in the presence of positive cycles, and in this sense it is more general than the usual "dynamic programming" approach.

(D) Given $2n + 1$ probabilities $p_1, \ldots, p_n, q_0, \ldots, q_n$ that sum to 1, construct a context-free grammar with nonterminal symbols $C_{i,j}$ for $0 \le i \le j \le n$, where the productions are

$$\begin{aligned} &C_{i,i} \to 0 && \text{for } 0 \le i \le n, \\ &C_{i,j} \to g_{ij}(C_{i,k-1}, C_{k,j}) && \text{for } 0 \le i < k \le j \le n; \end{aligned}$$

and let

$$g_{ij}(x, y) = x + y + p_{i+1} + \cdots + p_j + q_i + \cdots + q_j.$$

Then $m(C_{0,n}) + p_1 + \cdots + p_n$ is the expected number of comparisons in an *optimum binary search tree* defined by the given probabilities, in the sense of [6, Section 6.2.2]. The general algorithm we shall describe here is not competitive with the special one in [6]; but it appears to be useful in connection with similar problems, such as that of constructing optimum programs for decision tables.

(E) The author has successfully used this approach to generalize the optimum code-generation algorithm of Aho and Johnson [1], treating the case of machines with asymmetric registers.

3. The General Algorithm

Given a superior context-free grammar, the following algorithm determines $m(Y)$ for all nonterminal Y. The algorithm operates on elements $\mu[Y]$ and $\nu[Y]$ for each Y. Initially all these elements are undefined; but when the algorithm terminates, $\mu[Y]$ will equal $m(Y)$ for all Y. The set D represents those Y for which $\mu[Y]$ has been defined.

Step 1. Set D to the empty set.

Step 2. If all nonterminals are in D, stop.

Step 3. For each nonterminal $Y \notin D$, compute

$$\nu[Y] = \min\{\, g(\mu[X_1], \ldots, \mu[X_k]) \mid Y \to g(X_1, \ldots, X_k) \text{ is a production and } \{X_1, \ldots, X_k\} \subseteq D\,\}.$$

(If this set is empty, $\nu[Y]$ is infinite.)

Step 4. Choose $Y \notin D$ with minimum $\nu[Y]$, and set $\mu[Y] \leftarrow \nu[Y]$.

Step 5. Include Y in D, and return to Step 2.

For example, consider the grammar given in Section 1. The computation proceeds as in Table 1, viewed after Step 3. Finally $\mu[C] \leftarrow 7$.

TABLE 1. An example calculation

D	$\mu[A]$	$\mu[B]$	$\mu[C]$	$\nu[A]$	$\nu[B]$	$\nu[C]$
$\emptyset$	—	—	—	4	∞	9
$\{A\}$	4	—	—	—	5	9
$\{A, B\}$	4	5	—	—	—	7

Actually the values $\nu[Y]$ do not need to be computed explicitly; they have been introduced here only for convenience in stating and proving the algorithm. By maintaining a priority queue of the current values of $\mu[Y]$ with $Y \notin D$, the running time of this algorithm is bounded by a constant times $m \log n + t$, where there are m productions and n nonterminals, and the total length of all productions is t. Further refinements are also possible, since all productions with Y on the left-hand side can be deleted from memory as soon as $\mu[Y]$ has been defined.

4. Proof of the Algorithm

We shall show that $\mu[Y] = m(Y)$ whenever Step 5 is performed; it will follow that the relation

$$X \in D \quad \text{implies} \quad \mu[X] = m(X) \tag{$*$}$$

is invariant throughout the algorithm.

It is clear that $\nu[Y] \geq m(Y)$ in Step 3; for if $\nu[Y] < \infty$, $(*)$ implies that $\nu[Y] = \mathrm{val}(g(\alpha_1, \ldots, \alpha_k))$ for some g and some $\alpha_1, \ldots, \alpha_k$, where

$$Y \to g(X_1, \ldots, X_k) \to^* g(\alpha_1, \ldots, \alpha_k)$$

and the terminal strings $\alpha_1, \ldots, \alpha_k$ satisfy $\mathrm{val}(\alpha_i) \geq m(X_i)$ for all i. Therefore the algorithm will be valid unless $\mu[Y] > m(Y)$ at some occurrence of Step 5. In that case there will exist a terminal string α such that $Y \to^* \alpha$ and $\mathrm{val}(\alpha) < \mu[Y]$.

Assume that α is a *shortest* terminal string such that $Z \to^* \alpha$ for some nonterminal $Z \notin D$ and such that $\mathrm{val}(\alpha) < \mu[Y]$, at some occurrence of Step 5. Then $\alpha = g(\alpha_1, \ldots, \alpha_k)$ for some $\alpha_1, \ldots, \alpha_k$, where the grammar contains the production $Z \to g(X_1, \ldots, X_k)$ and where $X_i \to^* \alpha_i$ for all i. If $\{X_1, \ldots, X_k\} \subseteq D$ we have $\mathrm{val}(\alpha_i) \geq m(X_i) = \mu[X_i]$ for all i, by $(*)$ and the definition of $m(X_i)$; hence $\mathrm{val}(\alpha) \geq g(\mu[X_1], \ldots, \mu[X_k])$, by the monotonicity of g in each variable. But $g(\mu[X_1], \ldots, \mu[X_k]) \geq \nu[Z]$ by Step 3, and $\nu[Z] \geq \nu[Y] = \mu[Y]$ by Step 4, contradicting $\mathrm{val}(\alpha) < \mu[Y]$. Therefore $X_i \notin D$ for some i. But now $\mathrm{val}(\alpha_i) \leq \mathrm{val}(\alpha)$ by the superiority of g; hence α_i is a shorter string having the stated properties of α. This contradiction completes the proof.

It is easy to prove that the algorithm defines the $\mu[Y]$ in nondecreasing order of their values, since the new element Y introduced into D by Steps 4 and 5 cannot make any of the quantities $\nu[Z]$ less than $\mu[Y]$ in the next occurrence of Step 3. The usual proofs of Dijkstra's algorithm rely on this monotonicity property; but the proof above shows that it isn't really a crucial fact, even when the algorithm has been substantially generalized.

5. Another Application

If the functions g of a superior context-free grammar satisfy the additional condition of strict inequality,

$$g(x_1, \ldots, x_k) > \max(x_1, \ldots, x_k),$$

we shall prove that there is a *unique* solution to the set of simultaneous equations

$$f(Y) = \min\{\, g(f(X_1), \ldots, f(X_k)) \mid Y \to g(X_1, \ldots, X_k) \text{ is a production} \,\}$$

for all nonterminal Y, namely $f(Y) = m(Y)$ for all Y.

In the first place, $m(Y)$ is a solution: For if $Y \to g(X_1, \ldots, X_k)$ is any production, then $g(m(X_1), \ldots, m(X_k))$ is infinite or equal to $\mathrm{val}(\alpha)$ for some terminal string α, where

$$Y \to g(X_1, \ldots, X_k) \to^* g(\alpha_1, \ldots, \alpha_k) = \alpha$$

for some $\alpha_1, \ldots, \alpha_k$. Thus $g(m(X_1), \ldots, m(X_k)) \geq m(Y)$ by the definition of $m(Y)$. Conversely if $m(Y) = \mathrm{val}(\alpha)$ then α has the form $g(\alpha_1, \ldots, \alpha_k)$ where there is a production $Y \to g(X_1, \ldots, X_k)$; and

$$\begin{aligned} g(m(X_1), \ldots, m(X_k)) &\leq g(\mathrm{val}(\alpha_1), \ldots, \mathrm{val}(\alpha_k)) \\ &= \mathrm{val}(\alpha) = m(Y). \end{aligned}$$

In the second place, the solution is unique: Suppose f_1 and f_2 are distinct solutions to the simultaneous minimization equations above. Let Y be a nonterminal such that $f_1(Y) \neq f_2(Y)$ and where $\min(f_1(Y), f_2(Y))$ is as small as possible. Without loss of generality assume that $f_1(Y) < f_2(Y)$; then $f_1(X) = f_2(X)$ for all X such that $f_1(X) < f_1(Y)$. There must exist a production $Y \to g(X_1, \ldots, X_k)$ with

$$g(f_1(X_1), \ldots, f_1(X_k)) = f_1(Y) < f_2(Y) \leq g(f_2(X_1), \ldots, f_2(X_k)).$$

But the strict inequality condition above implies that $f_1(X_i) < f_1(Y)$ for $1 \leq i \leq k$; hence $f_1(X_i) = f_2(X_i)$ for all i by our construction, and this is impossible.

Note that something like the strict inequality condition is necessary to guarantee uniqueness, because of the following example grammar:

$$\begin{array}{ll} A \to a & \qquad a = 5; \\ A \to b(B) & \qquad b(x) = x; \\ B \to c(A) & \qquad c(x) = x. \end{array}$$

The simultaneous minimization equations are

$$f(A) = \min(5, f(B)), \qquad f(B) = f(A);$$

hence the nonnegative solutions are

$$f(A) = f(B) = x \quad \text{for } 0 \leq x \leq 5.$$

The application discussed here arises, for example, in the study of "levels" in a flowchart, as defined by Floyd [4] and Mont-Reynaud [8].

Acknowledgments

I wish to thank Edsger W. Dijkstra and Nils J. Nilsson for their helpful comments on the first draft of this paper.

This research was supported in part by National Science Foundation grant MCS 72-03752 A03, by Office of Naval Research contract N00014-76-C-0330, and by the IBM Corporation.

References

[1] A. V. Aho and S. C. Johnson, "Optimal code generation for expression trees," *Journal of the Association for Computing Machinery* **23** (1976), 488–501.

[2] Y. Bar-Hillel, M. Perles, and E. Shamir, "On formal properties of simple phrase structure grammars," *Zeitschrift für Phonetik, Sprachwissenschaft und Kommunikationsforschung* **14** (1961), 143–172. Reprinted in Yehoshua Bar-Hillel, *Language and Information* (Reading, Massachusetts: Addison–Wesley, 1964), 116–150.

[3] E. W. Dijkstra, "A note on two problems in connexion with graphs," *Numerische Mathematik* **1** (1959), 269–271.

[4] R. W. Floyd, "Flowchart levels (preliminary draft)," unpublished manuscript (July 1965).

[5] Sheila Greibach, "Inverses of phrase structure generators," Mathematical Linguistics and Automatic Translation Report NSF-11 (Cambridge, Massachusetts: Harvard University, June 1963).

[6] Donald E. Knuth, *Sorting and Searching*, Volume 3 of *The Art of Computer Programming* (Reading, Massachusetts: Addison–Wesley, 1973).

[7] A. Martelli and U. Montanari, "Additive AND/OR graphs," Advance Papers of the Third International Joint Conference on Artificial Intelligence (1973), 1–11.

[8] Bernard Marcel Mont-Reynaud, *Hierarchical Properties of Flows, and the Determination of Inner Loops* (Ph.D. thesis, Stanford University, 1977).

[9] Nils J. Nilsson, *Problem-Solving Methods in Artificial Intelligence* (New York: McGraw–Hill, 1971).

Addendum

Special cases of the algorithm in this chapter are frequently rediscovered, because context-free grammars encompass such a wide variety of applications. There is, for instance, a strong connection with algorithms that deal with Horn clauses in logic (see Algorithm 7.1.1C in *The Art of Computer Programming*, Volume 4A).

An extension to dynamically changing grammars has been presented by G. Ramalingam and Thomas Reps, "An incremental algorithm for a generalization of the shortest-path problem," *Journal of Algorithms* **21** (1996), 267–305. Ramalingam and Reps also discuss further applications and generalizations of the non-dynamic case.

Yet another kind of extension applies to the solution of general systems of minimization equations such as those studied in Section 5 above. The example in that section is satisfied by $f(A) = f(B) = 0$, in spite of the fact that the algorithm of Section 3 yields $m(A) = m(B) = 5$ as a minimum solution with respect to terminal strings of finite length. Dijkstra-like algorithms that essentially find minimum solutions with respect to possibly infinite expansions have been given by George M. Adelson-Velsky and Eugene Levner, "Project scheduling in AND-OR graphs: A generalization of Dijkstra's algorithm," *Mathematics of Operations Research* **27** (2002), 504–517; Rolf H. Möhring, Martin Skutella, and Frederik Stork, "Scheduling with AND/OR precedence constraints," *SIAM Journal on Computing* **33** (2004), 393–415.

Chapter 16

Two-Way Rounding

[Originally published in SIAM Journal on Discrete Mathematics **8** *(1995), 281–290.]*

Given n real numbers $0 \le x_1 < 1, \ldots, 0 \le x_n < 1$, *and a permutation* σ *of* $\{1,\ldots,n\}$, *we can always find integers* $\hat{x}_1 \in \{0,1\}, \ldots, \hat{x}_n \in \{0,1\}$ *so that the partial sums* $\hat{x}_1 + \cdots + \hat{x}_k$ *and* $\hat{x}_{1\sigma} + \cdots + \hat{x}_{k\sigma}$ *differ from the unrounded values* $x_1 + \cdots + x_k$ *and* $x_{1\sigma} + \cdots + x_{k\sigma}$ *by at most* $n/(n+1)$, *for* $1 \le k \le n$. *The latter bound is best possible. The proof uses an elementary argument about flows in a certain network, and leads to a simple algorithm that finds an optimum way to round.*

Introduction

Many combinatorial optimization problems in integers can be solved or approximately solved by first obtaining a real-valued solution and then rounding to integer values. Spencer [12] proved that it is always possible to do the rounding so that partial sums in two independent orderings are properly rounded. His proof was indirect — a corollary of more general results [8] about discrepancies of set systems — and it guaranteed only that the rounded partial sums would differ by at most $1 - 2^{-2^n}$ from the unrounded values. The purpose of this note is to give a more direct proof, which leads to a sharper result.

Let $x_1, \ldots, x_n$ be real numbers and let σ be a permutation of $\{1,\ldots,n\}$. We will write

$$S_k = x_1 + \cdots + x_k\,, \qquad \Sigma_k = x_{1\sigma} + \cdots + x_{k\sigma}\,, \qquad 0 \le k \le n\,,$$

for the partial sums in two independent orderings. Our goal is to find integers $\hat{x}_1, \ldots, \hat{x}_n$ such that

$$\lfloor x_k \rfloor \le \hat{x}_k \le \lceil x_k \rceil\,,$$

and such that the rounded partial sums

$$\widehat{S}_k = \hat{x}_1 + \cdots + \hat{x}_k\,, \qquad \widehat{\Sigma}_k = \hat{x}_{1\sigma} + \cdots + \hat{x}_{k\sigma}$$

also satisfy

$$\lfloor S_k \rfloor \le \widehat{S}_k \le \lceil S_k \rceil\,, \qquad \lfloor \Sigma_k \rfloor \le \widehat{\Sigma}_k \le \lceil \Sigma_k \rceil\,, \tag{$*$}$$

for $0 \le k \le n$. Such $\hat{x}_1, \ldots, \hat{x}_n$ will be called a *two-way rounding* of $x_1, \ldots, x_n$ with respect to σ.

Lemma. *Two-way rounding is always possible.*

Proof. We can assume without loss of generality that $S_n = m$ is an integer, by adding an additional term and increasing n if necessary. We can also assume that $0 < x_k < 1$ for all k. Construct a network with nodes $\{s, a_1, \ldots, a_m, u_1, \ldots, u_n, v_1, \ldots, v_n, b_1, \ldots, b_m, t\}$ and the following arcs for $1 \le j \le m$ and $1 \le k \le n$:[1]

$$s \longrightarrow a_j \qquad \text{and} \qquad b_j \longrightarrow t\,;$$

$$u_k \longrightarrow v_k\,;$$

$$a_j \longrightarrow u_k \qquad \text{if} \qquad [j-1\,..\,j) \cap [S_{k-1}\,..\,S_k) \ne \emptyset\,;$$

$$v_{k\sigma} \longrightarrow b_j \qquad \text{if} \qquad [j-1\,..\,j) \cap [\Sigma_{k-1}\,..\,\Sigma_k) \ne \emptyset\,.$$

Each arc has capacity 1. This network supports a natural flow of m units, if we send 1 unit through each arc $s \longrightarrow a_j$ and $b_j \longrightarrow t$, and x_k units through $u_k \longrightarrow v_k$; the flow in $a_j \longrightarrow u_k$ is the measure of the interval $[j-1\,..\,j) \cap [S_{k-1}\,..\,S_k)$, and the flow in $v_{k\sigma} \longrightarrow b_j$ is similar. Deleting the arcs $s \longrightarrow a_j$ defines a cut of capacity m, so this must be a minimum cut.

Since the arc capacities are integers, the max-flow/min-cut theorem [3] implies that this network supports an integer flow of m units. Let $\hat{x}_k$, for $1 \le k \le n$, be the amount that flows through $u_k \longrightarrow v_k$ in one such flow. Then $\hat{x}_k \in \{0,1\}$. If $j = \lceil S_k \rceil$ we have $\widehat{S}_k = \hat{x}_1 + \cdots + \hat{x}_k =$ flow into $\{u_1, \ldots, u_k\} \le$ flow out of $\{a_1, \ldots, a_j\} = j$, because all arcs $a_i \longrightarrow u_l$ for $l \le k$ have $i \le j$. And if $j = \lfloor S_k \rfloor$ we have $\widehat{S}_k =$ flow into $\{u_1, \ldots, u_k\} \ge$ flow out of $\{a_1, \ldots, a_j\} = j$, because all arcs $a_i \longrightarrow u_l$ for

---

[1] Here and in the sequel $[a\,..\,b)$ denotes the half-open interval $\{\, x \mid a \le x < b \,\}$. This notation, due independently to C. A. R. Hoare and L. H. Ramshaw, is recommended in the book [6].

$i \leq j$ have $l \leq k$. A similar argument proves that $\lfloor \Sigma_k \rfloor \leq \widehat{\Sigma}_k \leq \lceil \Sigma_k \rceil$, hence $(*)$ holds. □

Corollary. *Given any fixed k, two-way rounding is possible with $\hat{x}_k = \lceil x_k \rceil$, as well as with $\hat{x}_k = \lfloor x_k \rfloor$.*

Proof. We may assume as before that $0 < x_k < 1$. The construction in the lemma establishes a feasible flow of x_k units in the arc $u_k \longrightarrow v_k$. It is well known that the polytope of all feasible flows has vertices whose coordinates are integers (see, for example, Application 19.2 in Schrijver [11]). Therefore the arc $u_k \longrightarrow v_k$ is saturated in at least one maximum flow, and it carries no flow at all in at least one other. □

Incidentally, it is important to impose a capacity of 1 on the arcs $u_k \longrightarrow v_k$ in the construction of this proof. Otherwise we might get solutions in which $\hat{x}_k = 2$. Condition $(*)$ does not by itself imply that $\hat{x}_k \leq \lceil x_k \rceil$ or that $\hat{x}_k \geq \lfloor x_k \rfloor$.

1. An Application

Sometimes it is desirable to round "spreadsheet" data to larger units while preserving row and column sums and the grand total.

Theorem 1. *Given mn real numbers x_{ij} for $1 \leq i \leq m$, $1 \leq j \leq n$, we can round them to integers $\hat{x}_{ij}$ in such a way that*

$$\lfloor x_{ij} \rfloor \leq \hat{x}_{ij} \leq \lceil x_{ij} \rceil, \qquad \text{for all } i \text{ and } j;$$

$$\left\lfloor \sum_{j=1}^{n} x_{ij} \right\rfloor \leq \sum_{j=1}^{n} \hat{x}_{ij} \leq \left\lceil \sum_{j=1}^{n} x_{ij} \right\rceil, \qquad \text{for all } i;$$

$$\left\lfloor \sum_{i=1}^{m} x_{ij} \right\rfloor \leq \sum_{i=1}^{m} \hat{x}_{ij} \leq \left\lceil \sum_{i=1}^{m} x_{ij} \right\rceil, \qquad \text{for all } j;$$

$$\left\lfloor \sum_{i=1}^{m} \sum_{j=1}^{n} x_{ij} \right\rfloor \leq \sum_{i=1}^{m} \sum_{j=1}^{n} \hat{x}_{ij} \leq \left\lceil \sum_{i=1}^{m} \sum_{j=1}^{n} x_{ij} \right\rceil .$$

Proof. Let $a_i = \sum_{j=1}^{n} x_{ij}$, $b_j = \sum_{i=1}^{m} x_{ij}$, and $s = \sum_{i=1}^{m} a_i = \sum_{j=1}^{n} b_j$, and consider the $(m+1) \times (n+1)$ array

$$\begin{array}{ccccc} x_{11} & x_{12} & \ldots & x_{1n} & \alpha_1 \\ x_{21} & x_{22} & \ldots & x_{2n} & \alpha_2 \\ \vdots & \vdots & \ddots & \vdots & \vdots \\ x_{m1} & x_{m2} & \ldots & x_{mn} & \alpha_m \\ \beta_1 & \beta_2 & \ldots & \beta_n & s \end{array}$$

where $\alpha_i = -a_i$ and $\beta_j = -b_j$. Apply two-way rounding to these numbers, when ordered by rows and by columns. The resulting integers $\hat{x}_{ij}$ and $\hat{\alpha}_i$ satisfy the condition $\sum_{i=1}^{k}\bigl(\sum_{j=1}^{n}\hat{x}_{ij}+\hat{\alpha}_i\bigr)=0$ for all k, by $(*)$, hence $\sum_{j=1}^{n}\hat{x}_{ij}+\hat{\alpha}_i=0$ for all i; it follows that

$$\lfloor a_i\rfloor = -\lceil -a_i\rceil \le -\hat{\alpha}_i = \sum_{j=1}^{n}\hat{x}_{ij} \le -\lfloor -a_i\rfloor = \lceil a_i\rceil .$$

Similarly $\lfloor b_j\rfloor \le \sum_{i=1}^{m}\hat{x}_{ij} \le \lceil b_j\rceil$ for all j. We also have $\sum_{i=1}^{m}\alpha_i+s=0$; hence $\sum_{i=1}^{m}\hat{\alpha}_i+\hat{s}=0$. The sum $\sum_{i=1}^{m}\sum_{j=1}^{n}\hat{x}_{ij}$ therefore equals $\hat{s}$, which is either $\lfloor s\rfloor$ or $\lceil s\rceil$. □

2. A Sharper Bound

Notice that $(*)$ is equivalent to the conditions

$$|S_k-\widehat{S}_k|<1 \qquad\text{and}\qquad |\Sigma_k-\widehat{\Sigma}_k|<1\,, \qquad\text{for}\quad 0\le k\le n\,,$$

since $\widehat{S}_k$ and $\widehat{\Sigma}_k$ are integers. Let us say that two-way rounding has discrepancy bounded by δ if $|S_k-\widehat{S}_k|\le\delta$ and $|\Sigma_k-\widehat{\Sigma}_k|\le\delta$ for all k. A slight extension of the construction in the lemma makes it possible to prove a stronger result:

Theorem 2. *If $0<x_k<1$ for $1\le k\le n$ and if $x_1+\cdots+x_n=m$ is an integer, the sequence $(x_1,\ldots,x_n)$ can be two-way rounded with discrepancy bounded by $(2m+1)/(2m+2)$.*

Proof. We will prove that two-way rounding bounded by δ is possible for all $\delta>(2m+1)/(2m+2)$. Only finitely many roundings exist; therefore the stated result follows by taking the limit as δ decreases to $(2m+1)/(2m+2)$.

The proof uses a network like that of the lemma, but we omit certain arcs that would lead to discrepancies near 1. More precisely, if ϵ is any fixed positive number $<1/(2m+2)$, we have

$$\begin{aligned} a_j &\longrightarrow u_k &\quad &\text{if } j-1+\epsilon<S_k \text{ and } j-\epsilon>S_{k-1}\,;\\ v_{k\sigma} &\longrightarrow b_j &\quad &\text{if } j-1+\epsilon<\Sigma_k \text{ and } j-\epsilon>\Sigma_{k-1}\,. \end{aligned}$$

We also allow these arcs to have infinite capacity. But the capacities of the "source" arcs $s\longrightarrow a_j$, the "middle" arcs $u_k\longrightarrow v_k$, and the "sink" arcs $b_j\longrightarrow t$ all remain 1.

(Since $\epsilon < 1/2$ and $S_{k-1} < S_k$, the joint conditions $S_{k-1} < j - \epsilon$ and $S_k > j - 1 + \epsilon$ are equivalent to saying that

$$[j - 1 + \epsilon \mathinner{..} j - \epsilon) \cap [S_{k-1} \mathinner{..} S_k) \neq \emptyset.$$

But we will also be considering the ϵ-reduced network in other circumstances below, when values of $\epsilon \geq 1/2$ will be meaningful even though $[j - 1 + \epsilon \mathinner{..} j - \epsilon)$ is empty.)

The minimum cut in this reduced network has size m. For if any $m - 1$ of the unit-capacity arcs are cut, we will prove that we can still connect s to t. Suppose we remove p source arcs, q middle arcs, and r sink arcs, where $p + q + r = m - 1$. We send $1 - 2\epsilon$ units of flow from s through each of the $m - p$ remaining source arcs. From every a_j reached in this way, we send as many units of flow from $a_j \longrightarrow u_k$ as the size of the interval $[j - 1 + \epsilon \mathinner{..} j - \epsilon) \cap [S_{k-1} \mathinner{..} S_k)$. Some of the flow now gets stuck, if u_k is one of the q vertices for which the arc $u_k \longrightarrow v_k$ was removed. But at most $1 - 2\epsilon$ units flow into each u_k, so we still have at least $(m - p - q)(1 - 2\epsilon) = (r + 1)(1 - 2\epsilon)$ units of flow arriving at $\{v_1, \ldots, v_n\}$. Now consider an "antiflow" of $1 - 2\epsilon$ units from t back through each of the $m - r$ remaining sink arcs $b_j \longrightarrow t$. From every such b_j we send the antiflow back through $v_{k\sigma} \longrightarrow b_j$ according to the size of $[j-1+\epsilon \mathinner{..} j-\epsilon) \cap [\Sigma_{k-1} \mathinner{..} \Sigma_k)$. In this way $(m-r)(1-2\epsilon)$ units of antiflow come from t to $\{v_1, \ldots, v_n\}$. Each vertex v_k contains at most x_k units of flow and at most x_k units of antiflow. We know that the total flow plus antiflow at $\{v_1, \ldots, v_n\}$ is at least $(r + 1)(1 - 2\epsilon) + (m - r)(1 - 2\epsilon) = m + 1 - (2m + 2)\epsilon > m = x_1 + \cdots + x_n$. Therefore some vertex v_k must contain both flow and antiflow. And this establishes the desired link between s and t.

Since m is the size of a minimum cut and all capacities are integers, the network supports an integer flow of value m. Let $\hat{x}_k$ be the flow from u_k to v_k; we will prove that $(\hat{x}_1, \ldots, \hat{x}_n)$ is a two-way rounding with discrepancy $< \delta = 1 - \epsilon$. Notice that

$$|\widehat{S}_k - S_k| < 1 - \epsilon \quad \Longleftrightarrow \quad \lfloor S_k + \epsilon \rfloor \leq \widehat{S}_k \leq \lceil S_k - \epsilon \rceil \,.$$

If $j = \lceil S_k - \epsilon \rceil$ we have $\widehat{S}_k = \hat{x}_1 + \cdots + \hat{x}_k =$ flow into $\{u_1, \ldots, u_k\} \leq$ flow out of $\{a_1, \ldots, a_j\} = j$, because all arcs $a_i \longrightarrow u_l$ for $l \leq k$ have $[i-1+\epsilon \mathinner{..} i-\epsilon) \cap [S_{l-1} \mathinner{..} S_l) \neq \emptyset$, hence $i-1+\epsilon < S_l$ and $i \leq \lceil S_l - \epsilon \rceil \leq j$. Similarly, if $j = \lfloor S_k + \epsilon \rfloor$ we have $\widehat{S}_k \geq$ flow out of $\{a_1, \ldots, a_j\} = j$, because all arcs $a_i \longrightarrow u_l$ for $i \leq j$ have $l \leq k$. (If $l > k$ we would have $S_{l-1} \geq S_k \geq j - \epsilon \geq i - \epsilon$, contradicting $S_{l-1} < i - \epsilon$.) A similar proof shows that $\lfloor \Sigma_k + \epsilon \rfloor \leq \widehat{\Sigma}_k \leq \lceil \Sigma_k - \epsilon \rceil$. $\square$

3. A Lower Bound

The bound of Theorem 2 is, in fact, best possible, in the sense that no better bound can be guaranteed as a function of m.

Theorem 3. *For all positive integers m there exists a sequence of real numbers $(x_1, \ldots, x_n)$ with sum m and a permutation σ of $\{1, \ldots, n\}$ that cannot be two-way rounded with discrepancy $< (2m+1)/(2m+2)$.*

Proof. Let $n = 2m + 2$ and $\epsilon = 1/n$. Define

$$x_1 = x_2 = x_3 = \epsilon\,; \qquad x_{m+3} = (2m-1)\epsilon\,;$$

$$x_{k+3} = 2\epsilon\,,\; x_{k+m+3} = 2m\epsilon\,, \quad \text{for} \quad 1 \le k < m\,;$$

$$1\sigma = 2\,,\; 2\sigma = 1\,,\; 3\sigma = m+3\,,\; (2m+2)\sigma = 3\,;$$

$$(2k+2)\sigma = k+3\,,\; (2k+3)\sigma = k+m+3\,, \quad \text{for} \quad 1 \le k < m\,.$$

For example, when $m = 4$ we have $(x_1, \ldots, x_{10}) = (.1, .1, .1, .2, .2, .2, .7, .8, .8, .8)$ and $(1\sigma, \ldots, 10\sigma) = (2, 1, 7, 4, 8, 5, 9, 6, 10, 3)$. Hence

$$(S_1, \ldots, S_{10}) = (.1,\ .2,\ .3,\ .5,\ .7,\ .9,\ 1.6,\ 2.4,\ 3.2,\ 4.0)\,;$$

$$(\Sigma_1, \ldots, \Sigma_{10}) = (.1,\ .2,\ .9,\ 1.1,\ 1.9,\ 2.1,\ 2.9,\ 3.1,\ 3.9,\ 4.0)\,.$$

We will prove that this sequence and permutation cannot be two-way rounded with discrepancy less than $(2m+1)/(2m+2) = 0.9$; the same proof technique will work for any $m \ge 1$.

The main point is that whenever S_k or Σ_k has the form $l \pm 0.1$ where l is an integer, it must be rounded to l in order to keep the discrepancy small. This forces $\widehat{S}_1 = \widehat{\Sigma}_1 = 0$, $\widehat{\Sigma}_3 = \widehat{\Sigma}_4 = 1$, $\widehat{\Sigma}_5 = \widehat{\Sigma}_6 = 2$, $\widehat{\Sigma}_7 = \widehat{\Sigma}_8 = 3$, $\widehat{\Sigma}_9 = 4$, hence $\hat{x}_1 = \hat{x}_2 = \hat{x}_3 = \hat{x}_4 = \hat{x}_5 = \hat{x}_6 = 0$. But then $\widehat{S}_6 = \hat{x}_1 + \cdots + \hat{x}_6 = 0$ differs by 0.9 from S_6. $\square$

4. A Uniform Bound

Although Theorem 3 proves that Theorem 2 is "optimal," we can do still better if m is greater than $\frac{1}{2}n$, because we can replace each x_k by $1 - x_k$. This replaces m by $n - m$, and the bound on discrepancy decreases to $(2n - 2m + 1)/(2n - 2m + 2)$. Then we can restore the original x_k and change $\hat{x}_k$ to $1 - \hat{x}_k$. This computation preserves $|S_k - \widehat{S}_k|$ and $|\Sigma_k - \widehat{\Sigma}_k|$, so it preserves the discrepancy.

Further improvement is also possible when $m = \lfloor n/2 \rfloor$, if we look at the construction closely. The following theorem gives a uniform bound in terms of n, without any assumption about the value of $x_1 + \cdots + x_n$.

Theorem 4. *Any sequence* $(x_1, \ldots, x_n)$ *and permutation* $(1\sigma, \ldots, n\sigma)$ *can be two-way rounded with discrepancy bounded by* $n/(n+1)$.

Proof. We will show in fact that the discrepancy can always be bounded by $(n-1)/n$, when $x_1 + \cdots + x_n = m$ is an integer. This special case implies the general case if we set $x_{n+1} = \lceil S_n \rceil - S_n$ and increase n by 1.

If $2m+2 \le n$ or $2n-2m+2 \le n$, the result follows from Theorem 2 and possible complementation. Therefore we need only show that a discrepancy of at most $(n-1)/n$ is achievable when $m = \lfloor n/2 \rfloor$.

Consider first the case $n = 2m+1$. We use the network in the proof of Theorem 2, but now we allow ϵ to be any number $< 1/n$. Suppose, as in the former proof, that we can disconnect s from t by deleting p source arcs, q middle arcs, and r sink arcs, where $p+q+r = m-1$. Let q be minimum over all such ways to disconnect the network. We construct flows and antiflows as before, and we say that x_k is *green* if v_k contains positive flow, *red* if v_k contains positive antiflow. No x_k is both green and red, since there is no path from s to t. The previous proof showed that there are at least $(r+1)(1-2\epsilon)$ units of green flow and $(m-r)(1-2\epsilon)$ units of red flow, hence there are at least $m+1-(2m+2)\epsilon$ units of flow altogether. If we can raise this lower bound by ϵ, we will have a contradiction, because $m+1-(2m+1)\epsilon > m$.

Suppose $q > 0$, and let $u_k \longrightarrow v_k$ be a middle arc that was deleted. At most two arcs emanate from v_k in the network. Since q is minimum, there must in fact be two; otherwise we could restore $u_k \longrightarrow v_k$ and delete a non-middle arc. The two arcs from v_k must be consecutive, from $v_k \longrightarrow b_j$ and $v_k \longrightarrow b_{j+1}$, say. Furthermore the arcs $b_j \longrightarrow t$ and $b_{j+1} \longrightarrow t$ have not been cut. If $k = l\sigma$ we have $\Sigma_{l-1} < j - \epsilon$ and $\Sigma_l > j + \epsilon$. Our lower bound on antiflow can now be raised by 2ϵ, because it was based on the weak assumption that no antiflow runs back from $[j-\epsilon \,..\, j+\epsilon)$. This improved lower bound leads to a contradiction; hence $q = 0$.

Divide the interval $[0 \,..\, m)$ into $3m$ regions, namely "tiny left" regions of the form $[j-1 \,..\, j-1+\epsilon)$, "inner" regions of the form $[j-1+\epsilon \,..\, j-\epsilon)$, and "tiny right" regions of the form $[j-\epsilon \,..\, j)$, for $1 \le j \le m$. If we color the points of $[S_{k-1} \,..\, S_k)$ with the color of x_k, our lower bound $(r+1)(1-2\epsilon)$ for green flow was essentially obtained by noting that $m-p = r+1$ of the inner regions are purely green. Similarly, if we color the points of $[\Sigma_{k-1} \,..\, \Sigma_k)$ with the color of $x_{k\sigma}$, our lower bound for red flow was obtained by noting that $m-r = p+1$ inner regions in this second coloring are purely red. Notice that there is complete symmetry between red and green, because we can invert the network and replace σ by σ^{-1}.

Call an element x_k *large* if it exceeds $1-\epsilon$. If any x_k is large, the interval $[S_{k-1} \mathinner{..} S_k)$ occupies more than ϵ units outside of an inner region; this allows us to raise the lower bound by ϵ and obtain a contradiction. Therefore no element is large. It follows that no element x_k can intersect more than 2 tiny regions, when x_k is placed in correspondence with $[S_{k-1} \mathinner{..} S_k)$ or with $[\Sigma_{k-1} \mathinner{..} \Sigma_k)$.

Let's look now at the $2m$ tiny regions. Each of them must contain at least some red in the first coloring; otherwise we would have at least $(p+1)(1-2\epsilon)$ red units packed into at most $2m-1$ tiny regions and p inner regions, hence $(p+1)(1-2\epsilon) \le (2m-1)\epsilon + p(1-2\epsilon)$, contradicting $\epsilon < 1/n$. This means there must be at least $m+1$ red elements x_k, since no red element is large and since m non-large red intervals can intersect all the tiny regions only if they also cover all the inner regions (at least one of which is green). Similarly, there must be at least $m+1$ green elements. But this is impossible, since there are only $2m+1$ elements altogether. Therefore the network has minimum cut size m, and the rest of the proof of Theorem 2 goes through as before.

Now suppose $n = 2m$. Then we can carry out a similar argument, but we need to raise the lower bound by 2ϵ. Again we can assume that $q = 0$. We can also show without difficulty that there cannot be *two* large elements. When $n = 2m$ the argument given above shows that at least $2m-1$ of the tiny regions must contain some red, in the first coloring.

Suppose there are only $m-1$ red elements. Then, in the first coloring, $m-2$ of them intersect 2 tiny intervals and the other is large and intersects 3; we have raised the red lower bound by ϵ. But $(p+1)(1-2\epsilon)+\epsilon$ red units cannot be packed into $2m-1$ tiny regions and p inner regions, because $(p+1)(1-2\epsilon)+\epsilon > (n-1)\epsilon + p(1-2\epsilon)$.

A symmetrical argument shows that there cannot be only $m-1$ green elements. Therefore exactly m elements are red and exactly m are green. Suppose no element is large. Then we have at least one purely green tiny interval in the first coloring and at least one purely red tiny interval in the second — another contradiction. Thus, we may assume that there is one large red element, and that the $2m$ tiny intervals in the first coloring contain a total of less than ϵ units of green. In particular, each of them contains some red. Either the first interval $[0 \mathinner{..} \epsilon)$ or the last interval $[m-\epsilon \mathinner{..} m)$ is intersected by a non-large red element, which intersects at most ϵ units of space in tiny intervals. The other $m-1$ red elements intersect at most 2ϵ units of tiny space each, so at most $(2m-1)\epsilon$ such units are red. This final contradiction completes the proof. □

The result of Theorem 4 is best possible, because we can easily prove (as in Theorem 3) that the values

$$x_1 = \frac{1}{n+1}, \qquad x_k = \begin{cases} (n-1)/(n+1), & k \text{ even, } 2 \le k \le n \\ 2/(n+1), & k \text{ odd, } 3 \le k \le n \end{cases}$$

and a "shuffle" permutation that begins

$$k\sigma = \begin{cases} 2k-1 & \text{for } 1 \le 2k-1 \le n, \quad n \text{ odd} \\ 2k & \text{for } 1 \le 2k \le n, \quad n \text{ even} \end{cases}$$

cannot be two-way rounded with discrepancy less than $n/(n+1)$.

5. An Algorithm

So far we have discussed only worst-case bounds. But a particular two-way rounding problem, defined by given values $(x_1, \ldots, x_n)$ and a given permutation $(1\sigma, \ldots, n\sigma)$, will usually be solvable with smaller discrepancy than guaranteed by Theorems 2 and 4. A closer look at the construction of Theorem 2 leads to an efficient algorithm that finds the best possible discrepancy in any given case.

Theorem 5. *Let ϵ be any positive number. There exists a solution with discrepancy less than $1-\epsilon$ to a given two-way rounding problem if and only if the network constructed in the proof of Theorem 2 supports an integer flow of value m.*

Proof. The final paragraph in the proof of Theorem 2 demonstrates the "if" half. Conversely, suppose $\hat{x}_1, \ldots, \hat{x}_n$ is a solution with discrepancy $< 1-\epsilon$. If $\hat{x}_k = 1$, let $j = \widehat{S}_k$. Then $j - 1 = \widehat{S}_{k-1}$, so the condition $|\widehat{S}_{k-1} - S_{k-1}| < 1-\epsilon$ implies $S_{k-1} < j-\epsilon$. Also $|\widehat{S}_k - S_k| < 1-\epsilon$ implies $S_k > j-1+\epsilon$. Therefore there is an arc $a_j \longrightarrow u_k$. Similarly, there is an arc $v_{k\sigma} \longrightarrow b_j$ when $\hat{x}_{k\sigma} = 1$ and $j = \widehat{\Sigma}_k$. So the network supports an integer flow of value m. □

In other words, the optimum discrepancy $\delta = 1-\epsilon$ is obtained when ϵ is just large enough to reduce the network to the point where no m-unit flow can be sustained. We can in fact find an optimum rounding as follows: Let

$$f(j,k) = \min(j - S_{k-1}, S_k - j + 1)$$

be the "desirability" of the arc $a_j \longrightarrow u_k$, and let

$$g(j,k\sigma) = \min(j - \Sigma_{k-1}, \Sigma_k - j + 1)$$

be the desirability of $v_{k\sigma} \longrightarrow b_j$. (Thus the arcs $a_j \longrightarrow u_k$, $v_{k\sigma} \longrightarrow b_j$ are included in the network of Theorem 2 if and only if their desirability is greater than ϵ.) Sort these arcs by desirability, and add them one by one to the initial arcs $\{s \longrightarrow a_j, u_k \longrightarrow v_k, b_j \longrightarrow t\}$ until an integer flow of m units is possible. Then let $\hat{x}_k$ be the flow in $u_k \longrightarrow v_k$, for all k; this flow has discrepancy equal to 1 minus the desirability of the last arc added, and no smaller discrepancy is possible.

Notice that the arc $a_j \longrightarrow u_k$ has desirability $> \frac{1}{2}$ if and only if $S_{k-1} < j - \frac{1}{2} < S_k$, so at most m such arcs are present. Since all x_k lie between 0 and 1, at most $m + n - 1$ arcs of the form $a_j \longrightarrow u_k$ will have positive desirability, because $a_j \longrightarrow u_k$ and $a_{j+1} \longrightarrow u_k$ will both be desirable if and only if $S_{k-1} < j < S_k$.

The following simple algorithm turns out to be quite efficient, assuming that m is at most $n/2$: Begin with the network consisting of arcs $\{s \longrightarrow a_j, u_k \longrightarrow v_k, b_j \longrightarrow t\}$ for $1 \le j \le m$ and $1 \le k \le n$, plus any additional arcs of desirability $> 1/2$. Call an arc $a_j \longrightarrow u_k$ or $v_{k\sigma} \longrightarrow b_j$ "special" if its desirability lies between $1/\min(2m + 2, n)$ and $1/2$, inclusive; fewer than $2m + 2n$ arcs are special. Then, for $j = 1, \ldots, m$, send one unit of flow from a_j to t along an "augmenting path," using the well-known algorithm of Ford and Fulkerson [3, pages 17–19] but specialized for unit-capacity arcs. In other words, construct a breadth-first search tree from a_j until encountering t; then choose a path from a_j to t and reverse the orientation of all arcs on that path. If t is not reachable from a_j, add special arcs to the network, in order of decreasing desirability, until t is reachable.

6. Computational Experience

The running time of this algorithm is bounded by $O(mn)$ steps, but in practice it runs much faster on random data. For example, Tables 1 and 2 show the results of various tests when the input permutation σ is random and when the values $(x_1, \ldots, x_n)$ are selected as follows: Let $y_1, \ldots, y_n$ be independent uniform integers in the range $1 \le y_k \le N$, where N is a large integer (chosen so that arithmetic computations will not exceed 31 bits). Increase one or more of the y's by 1, if necessary, until $y_1 + \cdots + y_n$ is a multiple of m; then set $x_k = y_k/d$, where $d = (y_1 + \cdots + y_n)/m$. Reject $(x_1, \ldots, x_n)$ and start over, if some $x_k \ge 1$. (In practice, rejection occurs about half the time when $m = n/2$, but almost never when $m \ll n/2$.)

Table 1 shows the optimum discrepancies found, and Table 2 shows the running time in memory references or "mems" [7, pages 464–465] divided by n. All entries in these tables are given in the form $\mu \pm \sigma$,

TABLE 1. Empirical optimum discrepancies

	$m = 1$	$m = 2$	$m = \lfloor \lg n \rfloor$	$m = \lfloor \sqrt{n} \rfloor$	$m = n/2$
$n = 10$	$.566 \pm .06$	$.619 \pm .07$	$.627 \pm .07$	$.627 \pm .07$	$.622 \pm .08$
$n = 100$	$.537 \pm .02$	$.575 \pm .03$	$.664 \pm .03$	$.710 \pm .03$	$.759 \pm .02$
$n = 1000$	$.513 \pm .007$	$.527 \pm .01$	$.582 \pm .01$	$.662 \pm .02$	$.794 \pm .02$
$n = 10000$	$.504 \pm .002$	$.509 \pm .003$	$.535 \pm .005$	$.612 \pm .01$	$.818 \pm .01$
$n = 100000$	$.502 \pm .001$	$.503 \pm .001$	$.513 \pm .002$	$.570 \pm .005$	$.838 \pm .007$

TABLE 2. Empirical running time, in mems/n

	$m = 1$	$m = 2$	$m = \lfloor \lg n \rfloor$	$m = \lfloor \sqrt{n} \rfloor$	$m = n/2$
$n = 10$	10 ± 4	19 ± 6	27 ± 8	27 ± 8	37 ± 11
$n = 100$	2.9 ± 1.3	6 ± 2	18 ± 5	29 ± 7	76 ± 15
$n = 1000$	0.9 ± 0.5	1.9 ± 0.7	8.5 ± 2.2	25 ± 6	152 ± 32
$n = 10000$	0.3 ± 0.2	0.6 ± 0.2	3.6 ± 0.8	22 ± 7	289 ± 49
$n = 100000$	0.1 ± 0.1	0.2 ± 0.1	1.4 ± 0.4	17 ± 4	540 ± 72

where μ is the sample mean and σ is an estimate of the standard deviation; more precisely, σ is the square root of an unbiased estimate of the variance. The number of test runs $t(n)$ for each experiment was $10^6/n$; thus, 10^5 runs were made for each m when $n = 10$, but only 10 runs were made for each m when $n = 10^5$. The actual confidence interval for the tabulated μ values is therefore approximately $2\sigma/\sqrt{t(n)} = .002\sigma\sqrt{n}$.

Notice that when $m \ll n$, the optimum discrepancy is nearly $\frac{1}{2}$. Indeed, this is obvious on intuitive grounds: When n is large, approximately ϵn values of k will have S_k within $\frac{1}{2}\epsilon$ of $\{\frac{1}{2}, \frac{3}{2}, \ldots, m - \frac{1}{2}\}$, and approximately $\epsilon^2 n$ will also have equally good values Σ_j where the permutation σ has $j\sigma = k$. So we are essentially looking for a perfect matching in a bipartite graph with m vertices in each part and $\epsilon^2 n$ edges. For fixed m as $n \longrightarrow \infty$, the matching will exist when $\epsilon^2 n$ is sufficiently large, hence the mean optimum discrepancy is $\frac{1}{2} + O(n^{-\frac{1}{2}})$.

However, the behavior of the mean optimum discrepancy when $m = n/2$ is not clear. It appears to approach 1, but quite slowly, perhaps as $1 - c/\log n$.

When n is fixed and m varies, the mean optimum discrepancy is not maximized when $m = n/2$. For example, when $n = 10$, Table 1 shows that it is .622 when $m = 5$ but .627 when $m = 3$.

The running times shown in Table 2 do not include the work of constructing the network or sorting the special arcs by desirability. Those operations are easily analyzed, and in practice they take $am + bn$ steps for some constants a and b, because a straightforward bucket sort is satisfactory for this application. Therefore only the running time of the subsequent flow calculations is of interest.

The average running time to compute the flows appears to be $o(n)$ when $m \le \sqrt{n}$, and approximately proportional to $n^{1.3}$ (or perhaps $n \log n$) when $m = n/2$. So it is much less than the obvious upper bound mn of the Ford–Fulkerson scheme. The author tried to obtain still faster results by using more sophisticated max-flow algorithms, but these "improved" algorithms actually turned out to run more than an order of magnitude slower.

For example, the algorithm of Dinitz, as improved by Karzanov and others, seems at first to be especially well suited to this application because the network of Theorem 2 is "simple" in the sense discussed by Papadimitriou and Steiglitz [10, pages 212–214]: Every internal vertex has in-degree 1 or out-degree 1, hence edge-disjoint paths are vertex-disjoint; the running time with M unit-capacity arcs and N vertices is $O(MN^{1/2}) = O(n^{3/2})$. Using binary search to find the optimum number of special arcs gives us a guaranteed worst-case performance of $O\bigl(\min(m, n^{1/2})n \log n\bigr)$. Unfortunately, in practice the performance of that algorithm actually matches this worst-case estimate, even on random data. For example, when $m = n/2$ the observed running time in mems/n was 15284 ± 2455 when $n = 10000$, compared to 289 ± 49 by the simple algorithm. Each flow calculation consumed more than $1000n$ mems, and binary search required 14 flow calculations to be carried out.

When modern preflow push/relabel algorithms are specialized to unit-capacity networks of the type considered here, they behave essentially like the Dinitz algorithm and are no easier to implement (see Goldberg, Plotkin, and Vaidya [5]). Such algorithms do allow networks to change dynamically by adding arcs from s and/or deleting arcs to t (see Gallo, Grigoriadis, and Tarjan [4]); but our application requires adding or deleting special arcs in the *middle* of the network, so the techniques of [4] do not apply. Thus the simple Ford–Fulkerson algorithm seems to be a clear winner for this application, in spite of a lack of performance guarantees.

7. Comments and Open Problems

How complex can the networks of Theorem 2 be? If we have any bipartite graph with m vertices in each part and with n edges, and if every edge can be extended to a perfect matching, then we can find real numbers $(x_1, \ldots, x_n)$ in the range $0 < x_k \le 1$ and a permutation $(1\sigma, \ldots, n\sigma)$ such that $x_1 + \cdots + x_n = m$ and the two-way roundings are in one-to-one correspondence with the perfect matchings of the given graph. For we can take $(x_1, \ldots, x_n) = t_1\alpha_1 + \cdots + t_n\alpha_n$ where $t_1 + \cdots + t_n = 1$ and α_k is the characteristic vector of a perfect matching that uses edge k. The sum of x_k over all the edges touching any vertex is 1. Represent an edge from u to v by the ordered pair (u, v), and label the edges $1, \ldots, n$ in lexicographic order of these pairs; then define the permutation $1\sigma, \ldots, n\sigma$ by lexicographic order of the dual pairs (v, u). It follows that if k is the final edge for vertex j in the first part, we have $S_k = j$; and if $k\sigma$ is the final edge for vertex j in the second part, we have $\Sigma_k = j$. The correspondence between matchings and roundings is now evident.

This construction shows that the networks arising in Theorem 2 are general enough to mimic the networks that arise in bipartite matching problems, but only when the bipartite graphs contain no unmatchable edges; and the corollary preceding Theorem 1 shows that the latter restriction cannot be removed. This restriction on network complexity might account for the excellent performance we obtain with the simple Ford–Fulkerson algorithm.

If the capacity constraint on $u_k \longrightarrow v_k$ is removed, our network becomes equivalent to a network for bipartite matching, in which we want to match $\{a_1, \ldots, a_m\}$ to $\{b_1, \ldots, b_m\}$ through edges $a_j \text{---} b_{j'}$ whenever $a_j \longrightarrow u_k$ and $v_k \longrightarrow b_{j'}$. The problem of finding the *best* such match, when the edge $a_j \text{---} b_{j'}$ is ranked by the minimum of the desirabilities $f(j, k)$ and $g(j', k)$, is then a *bottleneck assignment problem* [2, 3].

The problem of optimum two-way rounding is, however, more general than the bottleneck assignment problem, because the unit capacity constraint on $u_k \longrightarrow v_k$ is significant. Consider, for example, the case $n = 7$, $m = 3$, $(x_1, \ldots, x_7) = (8, 8, 24, 11, 11, 11, 11)/28$, $(1\sigma, \ldots, 7\sigma) = (2, 1, 3, 5, 4, 7, 6)$. Then $(S_1, \ldots, S_7) = (\Sigma_1, \ldots, \Sigma_7) = (8, 16, 40, 51, 62, 73, 84)/28$, and the arcs $\{a_j \longrightarrow u_k, v_k \longrightarrow b_j\}$ ranked by desirability are:

$a_3 \longrightarrow u_6,\ v_7 \longrightarrow b_3$	desirability $= \min(22/28, 17/28) = 17/28$
$a_1 \longrightarrow u_2,\ v_1 \longrightarrow b_1,\ a_2 \longrightarrow u_4,\ v_5 \longrightarrow b_2$	desirability 16/28
$a_1 \longrightarrow u_3,\ v_3 \longrightarrow b_1,\ a_2 \longrightarrow u_3,\ v_3 \longrightarrow b_2$	desirability 12/28
$a_3 \longrightarrow u_7,\ v_6 \longrightarrow b_3$	desirability 11/28
$a_1 \longrightarrow u_1,\ v_2 \longrightarrow b_1$	desirability 8/28
$a_3 \longrightarrow u_5,\ v_4 \longrightarrow b_3$	desirability 6/28
$a_2 \longrightarrow u_5,\ v_4 \longrightarrow b_2$	desirability 5/28

Thus the edges $a_j \text{---} b_{j'}$ ranked by desirability are:

$a_1 \text{---} b_1,\ a_1 \text{---} b_2,\ a_2 \text{---} b_1,\ a_2 \text{---} b_2$	(12/28 via u_3, v_3)
$a_3 \text{---} b_3$	(11/28 via u_6, v_6 and via u_7, v_7)
$a_1 \text{---} b_1$	(8/28 via u_1, v_1 and via u_2, v_2)
$a_2 \text{---} b_3,\ a_3 \text{---} b_2$	(6/28 via u_4, v_4 or via u_5, v_5)
$a_2 \text{---} b_2$	(5/28 via u_4, v_4 and via u_5, v_5)

The bottleneck assignment problem is solved by matching a_1 — b_1, a_2 — b_2, and a_3 — b_3 with desirability $\min(12/28, 12/28, 11/28) = 11/28$. But this matching does not correspond to a valid two-way rounding because it uses the intermediate arc $u_3 \longrightarrow v_3$ twice; it rounds x_3 to 2 and x_6 (or x_7) to 1. The optimum two-way rounding uses another route from a_1 to b_1 and has desirability $\min(8/28, 12/28, 11/28) = 8/28$, discrepancy $1 - 8/28 = 20/28$; it rounds x_1 (or x_2), x_3, and x_6 (or x_7) to 1, the other x's to 0.

In closing, we note that a conjecture of József Beck [8, 12] remains a fascinating open problem: Is there a constant K such that three-way rounding is always possible with discrepancy at most K? (In three-way rounding the partial sums are supposed to be well approximated with respect to a third permutation $(1\tau, \ldots, n\tau)$, in addition to $(1, \ldots, n)$ and $(1\sigma, \ldots, n\sigma)$.) It suffices [8, 12] to prove this when $x_k = \frac{1}{2}$ for all k. An upper bound of $O(\log n)$ is known; in fact, G. Bohus [1] has proved that the discrepancy of r-way rounding is always $O(r \log n)$.

Can any of the methods of this paper be extended to find better bounds on the discrepancy of arbitrary set systems (or at least of set systems more general than those for two-way rounding), in the sense of [12]?

Acknowledgments

I wish to thank Joel Spencer for proposing the problem and for showing me a simple construction that forces discrepancy $n/(n+1)$. Thanks also to Noga Alon, Svante Janson, and Serge Plotkin for several stimulating discussions as I was working out the solution described above. Shortly after I had proved Theorems 2–4, a somewhat similar construction was found independently by Jacek Ossowski, who described it in terms of common systems of distinct representatives instead of network flows; see §9.2 in [9].

References

[1] Géza Bohus, "On the discrepancy of 3 permutations," *Random Structures & Algorithms* **1** (1990), 215–220.

[2] Jack Edmonds and D. R. Fulkerson, "Bottleneck extrema," *Journal of Combinatorial Theory* **8** (1970), 299–306.

[3] L. R. Ford, Jr., and D. R. Fulkerson, *Flows in Networks* (Princeton, New Jersey: Princeton University Press, 1962).

[4] Giorgio Gallo, Michael D. Grigoriadis, and Robert E. Tarjan, "A fast parametric maximum flow algorithm and applications," *SIAM Journal on Computing* **18** (1989), 30–55.

[5] Andrew V. Goldberg, Serge A. Plotkin, and Pravin M. Vaidya, "Sublinear-time parallel algorithms for matching and related problems," *Journal of Algorithms* **14** (1993), 180–213.

[6] Ronald L. Graham, Donald E. Knuth, and Oren Patashnik, *Concrete Mathematics* (Reading, Massachusetts: Addison–Wesley, 1989).

[7] Donald E. Knuth, *The Stanford GraphBase* (New York: ACM Press, 1994).

[8] L. Lovász, J. Spencer, and K. Vesztergombi, "Discrepancy of set-systems and matrices," *European Journal of Combinatorics* **7** (1986), 151–160.

[9] L. Mirsky, *Transversal Theory* (New York: Academic Press, 1971).

[10] Christos H. Papadimitriou and Kenneth Steiglitz, *Combinatorial Optimization* (Englewood Cliffs, New Jersey: Prentice–Hall, 1982).

[11] Alexander Schrijver, *Theory of Linear and Integer Programming* (London: Wiley, 1986).

[12] Joel Spencer, "Ten Lectures on the Probabilistic Method," *CBMS-NSF Regional Conference Series in Applied Mathematics* **52** (Philadelphia, Pennsylvania: SIAM, 1987), Lecture 5.

Addendum

An important generalization of Theorem 4 was found by Benjamin Doerr ["Linear discrepancy of totally unimodular matrices," *Combinatorica* **24** (2004), 117–125] and independently by Tom Bohman and Ron Holzman ["Linear versus hereditary discrepancy," *Combinatorica* **25** (2005), 39–47]: *If A is any totally unimodular matrix with $m \geq 2$ rows and n columns, and if $x = (x_1, \ldots, x_n)^T$ is any vector of real numbers, there is an integer vector $\hat{x} = (\hat{x}_1, \ldots, \hat{x}_n)^T$ such that all entries of $A(x - \hat{x})$ are at most $1 - 1/\min(m, n+1)$ in absolute value.* (A matrix is said to be *totally unimodular* if all determinants obtainable by choosing k rows and k columns, for any k, are either -1, 0, or $+1$. In particular, all entries of a totally unimodular matrix must themselves be either 0 or ±1. Given any permutation σ, the $2n \times n$ matrix A for which $Ax = (S_1, \ldots, S_n, \Sigma_1, \ldots, \Sigma_n)^T$ can be shown to be totally unimodular by induction on n as follows: A nonzero subdeterminant of A that omits column j is a subdeterminant of the $2(n-1) \times (n-1)$ matrix A'_j that arises when column j and two redundant rows are erased. Therefore we need consider only subdeterminants with $k = n$. If row 1 is chosen for such a subdeterminant, we can subtract it from all rows $> n$ that contain 1 in column 1, thus obtaining a subdeterminant of A'_1. If rows $j-1$ and j are both chosen, for some j with $1 < j \le n$, we can zero out column j in all rows $> n$ and obtain a subdeterminant of A'_j. A similar reduction occurs if we choose row $n+1$, or both of rows $n+j-1$ and $n+j$ for some j. The only remaining case arises when n is even and the subdeterminant comes from rows 2, 4, ..., $2n$; but that subdeterminant is zero, because $S_n = \Sigma_n$.)

An interesting counterexample to József Beck's conjecture about 3-way rounding was found by Alantha Newman, Ofer Neiman, and Aleksandar Nikolov, "Beck's three permutations conjecture: A counterexample and some consequences," *IEEE Symposium on Foundations of Computer Science* **53** (2012), 253–262. They proved that the three permutations on $\{0, 1, \ldots, 3^k - 1\}$ defined by

$$(a_1 \ldots a_k)_3 \;\mapsto\; \bigl(((a_1 + j) \bmod 3) \ldots ((a_k + j) \bmod 3)\bigr)_3$$

for $j \in \{0, 1, 2\}$ have discrepancy at least $k/3 + 1$.

Chapter 17

Matroid Partitioning

[The following notes, written when the author was visiting the University of Oslo in 1973, were circulated as Stanford Computer Science Report STAN-CS-73-342 (March 1973).]

This report discusses a modified version of Edmonds's algorithm for the partitioning of a set into subsets that are independent in various given matroids. If $\mathcal{M}_1, \ldots, \mathcal{M}_k$ are matroids defined on a finite set E, the algorithm yields a simple necessary and sufficient condition for whether or not the elements of E can be colored with k colors such that (i) all elements E_j of color j are independent in $\mathcal{M}_j$, and (ii) the number of elements of color j lies between given limits, $n_j \le |E_j| \le n'_j$. The algorithm either finds such a coloration or it finds a proof that none exists, after making at most $n^3 + n^2k$ tests of independence in the given matroids, where n is the number of elements in E.

Let $\mathcal{M}_1, \ldots, \mathcal{M}_k$ be matroids (otherwise known as "pregeometries," in the terminology of Crapo and Rota [1]), over the n-element set E. Edmonds [2] has given an efficient algorithm to determine whether or not the elements of E can be partitioned into k disjoint subsets, $E = E_1 \cup \cdots \cup E_k$, in such a way that E_j is independent in $\mathcal{M}_j$ for all j. The purpose of this note is to present his algorithm in a somewhat different manner, which indicates how he might have discovered it in the first place. We shall also extend the algorithm slightly so that bounds are placed on the number of elements in the subsets E_j.

In order to make this report somewhat colorful, we shall imagine that the elements of E are being painted with k colors, so that E_j contains the elements of color j. We assume that the basic definitions of matroid theory need not be stated, since by now there are dozens of papers in which those definitions occupy the first two pages. Edmonds's paper [2] indicates the wide variety of applications for matroid partitioning.

Derivation of an Algorithm

The natural way to get the elements colored is to start with them all blank and successively to paint them. Many combinatorial algorithms have the following general form: "Starting with a certain configuration, try to find a better configuration by some reasonably straightforward method. If this attempt succeeds, replace the initial configuration by the improved one, and start again. If it fails, prove that no better configuration exists." Of course we are not always able to carry out the latter proof; but in many important cases, such a proof is possible, hence a rather simple algorithm emerges. Matroid partitioning is such a case.

Suppose we have painted certain elements of E, and let E_j be the set of elements that have color j. We assume that E_j is independent in $\mathcal{M}_j$. Let $E_0 = E \setminus (E_1 \cup \cdots \cup E_k)$ be the unpainted elements. If x is some element not of color j, we could paint it with that color if $x \cup E_j$ were independent in $\mathcal{M}_j$. (We write '$x \cup E_j$' as shorthand for '$\{x\} \cup E_j$'.) On the other hand, if $x \cup E_j$ is dependent, there is a unique circuit $P \subseteq x \cup E_j$, and we can paint x with color j if the color of any element y of $P \cap E_j$ is scraped off. Then perhaps we can paint y with some other color.

A sequence of such repaintings might be denoted by, say,

$$x \to y \to z \to 0_3$$

meaning "paint x with the current color of y, then repaint y with the current color of z, then repaint z with color 3." In general we may write

$$x \to y \iff x \cup E_j \setminus y \text{ is independent in } \mathcal{M}_j$$

when $y \in E_j$ and $x \notin E_j$; and

$$x \to 0_j \iff x \cup E_j \text{ is independent in } \mathcal{M}_j$$

where x is an element of $E \setminus E_j$ and 0_j is a special symbol distinct from the elements of E. (We may think of 0_j as a "standard" element of color j, whose color never needs to be washed off.) Notice that if $x \to 0_j$ then $x \to y$ for all $y \in E_j$.

In effect, this arrow notation defines a directed graph on the $n + k$ vertices $E \cup \{0_1, \ldots, 0_k\}$, and $x \to y \to z \to 0_3$ is an oriented path from x to 0_3. We shall denote oriented paths in the usual way by writing

$$x \to^{+} y \iff \text{there is a path } x = x_0 \to x_1 \to \cdots \to x_m = y,\ m \geq 1.$$

If x is uncolored and there is a path $x \to^+ 0_t$, this path specifies a repainting that results in a net increase of one more element painted with the tth color. This path would give us a way to decrease the number of unpainted elements. However, we have overlooked an important consideration: All the "$\to$" relationships have been calculated with respect to a particular choice of the E_j, and some repaintings may invalidate future ones. In fact there do exist examples of paths $x \to^+ 0_t$ that correspond to no correct repainting.

Fortunately this problem does not arise when we consider *shortest* paths instead of arbitrary paths.

Lemma. *In terms of the notation above, let*

$$x = x_0 \to x_1 \to \cdots \to x_m = 0_t,$$

where $x_i \not\to x_j$ for $j > i+1$. Then if x_i is painted the color of x_{i+1}, for $0 \le i < m$, the resulting elements of color j are independent in $\mathcal{M}_j$ for $1 \le j \le k$.

Proof. The result is trivial when $m = 1$. If $m > 1$, consider what happens after making just the mth step of the repainting: Let x_{m-1} have color s, and let

$$E'_j = \begin{cases} E_j \cup x_{m-1}, & \text{if } j = t; \\ E_j \setminus x_{m-1}, & \text{if } j = s; \\ E_j, & \text{otherwise.} \end{cases}$$

Let $\to'$ denote relations in the directed graph that corresponds to these new color classes E'_j; and let $x'_i = x_i$ for $0 \le i < m-1$, $x'_{m-1} = 0_s$. The lemma will follow by induction, if we prove that $x'_0 \to' x'_1 \to' \cdots \to' x'_{m-1}$ and that $x'_i \not\to' x'_j$ for $j > i+1$.

To prove that $x'_i \to' x'_{i+1}$, the only nontrivial case occurs when x'_{i+1} has color t. In this case, $i+1 < m-1$ and we must show that the set $I = x_i \cup E'_t \setminus x_{i+1}$ is independent in $\mathcal{M}_t$. If it were dependent, it would contain a unique circuit P; and P must contain both x_i and x_{m-1}, since $I \setminus x_i$ and $I \setminus x_{m-1}$ are independent. Furthermore since $x_i \not\to 0_t$ there is a circuit $Q \subseteq x_i \cup E_t$ in $\mathcal{M}_t$. But then P and Q are distinct circuits contained in $x_i \cup E'_t$, contradicting the independence of E'_t in $\mathcal{M}_t$.

On the other hand if $x'_i \to' x'_j$ for $j > i+1$, we reach an immediate contradiction unless x'_j has color s and $j < m-1$. In the latter case we find that $x_i \cup E'_s \setminus x_j$ is independent but $x_i \cup E_s \setminus x_j$ is dependent; thus there is a unique circuit $P \subseteq x_i \cup E_s \setminus x_j$, and x_{m-1} and x_i are both in this circuit, so $x_i \to x_{m-1}$. This contradiction completes the proof. ☐

This lemma tells us that the number of unpainted elements in an existing coloration can be reduced by one whenever we are able to find a path from an uncolored element to 0_t, for some t. Thus we would have a coloring algorithm if we could show conversely that a better coloration exists only when there is such a path.

Indeed, it isn't hard to convince oneself that this converse is true: Consider any painting E_0, E_1, ..., E_k where each E_j has $|E_j| = n_j$ elements, and suppose there is another one E'_0, E'_1, ..., E'_k where E'_j has $n_j + \delta_{jt} - \delta_{j0}$ elements. (Thus, the second coloration has one more element of color t.) Then there is some element x in E'_t that is independent of E_t, because E_t has rank n_t in $\mathcal{M}_t$ and it could not span all of the $n_t + 1$ elements in E'_t. We can repaint x with color t; then if x was painted color s, we can find some y in E'_s that is independent of $E_s \setminus x$, etc. Each repainting brings the E_j closer to the E'_j, so the process eventually terminates by finding an uncolored element to paint.

Derivation of Good Characterizations

So we know that the path method above will indeed lead to a good algorithm for matroid partitioning. However, experience with other algorithms, for which matroid partitioning provides a generalization, encourages us to look for more: We would like to find a simple *reason* that a painting cannot be extended, so that skeptics who don't necessarily believe that our computer program is correct can see for themselves that our best painting is optimum. Such an explanation is far more satisfactory than if we merely say, "The computer has made an exhaustive search and found nothing better." A short demonstration that no improvement exists is what Edmonds has called a *good characterization.* In this way the programmer can present the user with a convincing answer, whether the algorithm succeeds or not.

Therefore let us try to find a good characterization. Let t be some fixed value, $1 \le t \le k$, and suppose there is no oriented path $x \to^+ 0_t$ satisfying the conditions of the lemma, for any uncolored $x \in E_0$. Let

$$B_j = \{x \mid x \in E_j \text{ and } x \to^+ 0_t\},$$
$$A_j = E_j \setminus B_j,$$

for $0 \le j \le k$. Then B_0 is empty, for if $x \in B_0$ the *shortest* path $x \to^+ 0_t$ would satisfy the conditions of the lemma. Let

$$A = A_0 \cup A_1 \cup \cdots \cup A_k,$$
$$B = B_1 \cup \cdots \cup B_k,$$

so that we have partitioned E into two disjoint sets A and B. Experience with other algorithms suggests that we might be able to use these sets A and B to obtain a "good characterization."

If x is independent of A_j in $\mathcal{M}_j$, then either x is independent of E_j in $\mathcal{M}_j$, or $x \in B_j$, or $x \to y$ for some $y \in B_j$. These three cases imply that either $x \in B$ or B_j is empty. In other words, the following statement holds for $1 \le j \le k$:

if $x \in A$, and either $B_j \neq \emptyset$ or $j = t$, then x depends on A_j in $\mathcal{M}_j$.

A little fiddling around with this condition, and simplifying, leads to the good characterization that is desired:

Theorem 1. *Let $\mathcal{M}_1, \ldots, \mathcal{M}_k$ be matroids on a set E. It is possible to find disjoint subsets $E_1, \ldots, E_k$ of E, such that E_j is independent in $\mathcal{M}_j$ and $|E_j| = n_j$, if and only if*

$$|A| \le |E| - \sum_{j=1}^{k} \max(n_j - r_j(A), 0)$$

for all $A \subseteq E$, where r_j is the rank function in $\mathcal{M}_j$.

Proof. The condition is necessary, for if $E_1, \ldots, E_k$ is such a collection of subsets and $A \subseteq E$ then $|E_j \cap A| \le r_j(A)$, hence

$$|E_j \cap (E \setminus A)| \ge n_j - r_j(A).$$

Also clearly $|E_j \cap (E \setminus A)| \ge 0$. Summing over j gives

$$|E \setminus A| \ge \sum_{j=1}^{k} \max(n_j - r_j(A), 0),$$

which is the condition of the theorem.

Conversely, if we have disjoint subsets $E_1, \ldots, E_k$ with E_j independent in $\mathcal{M}_j$ and $|E_j| \le n_j$ and $|E_t| < n_t$, the algorithm sketched above will be able to increase $|E_t|$ without changing the number of elements in the other sets E_j. This must be so, for if the algorithm fails, the set A constructed above satisfies the condition $r_j(A) = |A_j|$ or ($A_j = E_j$ and $j \neq t$), for all j. Therefore $|B_j| = |E_j| - |A_j| \le \max(n_j - r_j(A), 0)$; and $|B_t| < n_t - r_t(A)$. Hence

$$|B| = |E| - |A| < \sum_{j=1}^{k} \max(n_j - r_j(A), 0)$$

contradicts the condition of the theorem. □

The special case of this theorem in which all $\mathcal{M}_j$ are identical and all $n_j = r_j(E)$ was proved by Edmonds [3].

A similar characterization applies when we ask whether or not all elements can be painted.

Theorem 2. *Let $\mathcal{M}_1, \ldots, \mathcal{M}_k$ be matroids on a set E. It is possible to find disjoint subsets $E_1, \ldots, E_k$ of E, such that E_j is independent in $\mathcal{M}_j$ and $|E_j| \le n'_j$ and $E = E_1 \cup \cdots \cup E_k$, if and only if*

$$|A| = \sum_{j=1}^{k} \min(r_j(A), n'_j)$$

for all $A \subseteq E$, where r_j is the rank function in $\mathcal{M}_j$.

Proof. The condition is necessary, since we have $|A| = \sum_{j=1}^{k} |E_j \cap A| \le \sum_{j=1}^{k} \min(r_j(A), n'_j)$ in any such partitioning.

Conversely, the condition is sufficient. Consider an algorithm that looks for paths $x \to^+ 0_t$ where $x \in E_0$ and $|E_t| < n'_t$, and paints such elements x, until this is no longer possible. A construction like that preceding Theorem 1 can be used, but with

$$B_j = \{x \mid x \in E_j \text{ and } x \to^+ 0_t \text{ for some } t \text{ with } |E_t| < n'_t\}.$$

Then we find $|A_j| = r_j(A)$ or $|A_j| = n'_j$ for $1 \le j \le k$. Hence either all elements are painted, or A_0 is nonempty and $|A| = |A_0| + \cdots + |A_k| > \sum_{j=1}^{k} \min(r_j(A), n'_j)$. □

Theorem 2 is implicit in the paper of Edmonds [2], who proved it when all the n'_j are infinite. To get the general case, simply "truncate" $\mathcal{M}_j$ by saying that a set is dependent in $\mathcal{M}_j$ whenever it contains more than n'_j elements. Furthermore we can derive Theorem 1 from Theorem 2, by setting $n'_j = n_j$ and introducing a new matroid $\mathcal{M}_0$ with all sets independent and $n'_0 = |E| - (n_1 + \cdots + n_k)$.

However, the following theorem seems to be a mild generalization of Edmonds's theorem, not so readily deducible from it:

Theorem 3. *Let $\mathcal{M}_1, \ldots, \mathcal{M}_k$ be matroids on a set E, and let (n_j, n'_j) be pairs of numbers with $n_j \le n'_j$ for $1 \le j \le k$. It is possible to find disjoint subsets $E_1, \ldots, E_k$ of E, such that E_j is independent in $\mathcal{M}_j$ and $n_j \le |E_j| \le n'_j$ and $E = E_1 \cup \cdots \cup E_k$, if and only if both the conditions of Theorems 1 and 2 hold for all $A \subseteq E$.*

Proof. Consider an algorithm that first looks for a painting satisfying Theorem 1; if it fails, it finds a set A that violates the first condition. If it succeeds, it continues to extend the painting as in Theorem 2. If this extension fails, it finds a set A that violates the second condition. □

The Algorithm

The proof of Theorem 3 leads essentially to the following algorithm, which either finds a partition $E_1, \ldots, E_k$ as specified in the theorem, or finds a set A that proves that no such partition is possible. For ease in description, the algorithm is presented here in an ad hoc ALGOL-like language, without optimization.

```
   begin E_0 := E;
   for j := 1 until k do E_j := ∅;
   for x ∈ E do color(x) := 0;
   for j := 1 until k do color(0_j) := j;
   for j := 1 until k do for i := 1 until n_j do augment(j);
   while E_0 ≠ ∅ do augment(0);
   for j := 1 until k do output E_j;
 exit: end.

 procedure augment (integer value t);
   begin for x ∈ E do succ(x) := none;
   A := E;  B := if t > 0 then {0_t} else {0_j | |E_j| < n'_j};
   comment Later succ(x) will be set to y if we find a
         shortest path x → y →* 0_j for some 0_j now in B.
         Also A = {x | succ(x) = none};
   while B ≠ ∅ do
      begin C := ∅;
      for y ∈ B do for x ∈ A do
         begin j := color(y);
         if x ∪ E_j \ y is independent in M_j then
            begin succ(x) := y;  A := A \ x;  C := C ∪ x;
            if color(x) = 0 then go to repaint;
            end;
         end;
      B := C;
      end;
   output A;
   output "This set A violates the condition of Theorem ";
   output if t > 0 then "1" else "2";  go to exit;
 repaint: while x ∈ E do
      begin y := succ(x);  j := color(x);  E_j := E_j \ x;
      j := color(y);  E_j := E_j ∪ x;  color(x) := j;  x := y;
      end;
   end.
```

The innermost loop of this algorithm is the test whether $x \cup E_j \setminus y$ is independent in $\mathcal{M}_j$, and it is performed at most $O(n^2 + nk)$ times per call of *augment*, where $n = |E|$. Hence it is performed at most $O(n^3 + n^2k)$ times in all. However in practice this estimate is probably much too high, since the loop will terminate quickly. (The loop '**for** $x \in A$' should consider the elements $x \in E_0$ before the x's that have nonzero color.) It is an open question whether this $O(n^3)$ upper bound can be reduced.

Discussion

Consider a very special case of matroid partitioning, known as the "bipartite matching" or "distinct representatives" problem: Given an $n \times k$ matrix of 0s and 1s, we wish to encircle exactly one 1 in every row and at most one 1 in every column. Here $E = \{1, \ldots, n\}$, element x corresponds to row x, matroid $\mathcal{M}_j$ corresponds to column j, and $(n_j, n'_j) = (0, 1)$ for all j. A set E_j of rows is independent in $\mathcal{M}_j$ if and only if $E_j = \emptyset$ or $E_j = \{x\}$, where row x contains a 1 in column j. In this case the test for independence is, of course, extremely simple, and the algorithm runs in $O(n^3 + n^2k)$ units of time. Hopcroft and Karp [7] have shown how to reduce the running time to $O(n^{2.5})$ when $n = k$.

If this example is slightly generalized so that a set E_j is independent in $\mathcal{M}_j$ if and only if row x contains a 1 in column j for all $x \in E_j$, and if we allow arbitrary n_j and n'_j, we have the problem of encircling exactly one 1 in each row, and between n_j and n'_j of them in column j. The algorithm works in $O(n^3 + n^2k)$ units of time for this case also. Ford and Fulkerson [5, 6] call this special case the "system of restricted representatives" (SRR) problem, for which they proved the analog of Theorem 3. The condition of Theorem 1 can be simplified to

$$|A| \;\le\; |E| - \sum_{j=1}^{k} \{n_j \mid r_j(A) = 0\}$$

in the SRR problem, by altering the set A when $0 < r_j(A) < n_j$. Similarly, an alteration of A when $0 < r_j(A) < n'_j$ will simplify the SRR condition of Theorem 2 to

$$|A| \;\le\; \sum_{j=1}^{k} \{n'_j \mid r_j(A) \ne 0\}.$$

Another important case occurs when $k = 2$ and when $\mathcal{M}_2$ is taken as the orthogonal complement (or dual) to some matroid $\mathcal{M}$. Then the algorithm can be used to find maximum-cardinality *intersections* of $\mathcal{M}_1$ and $\mathcal{M}$. (See Edmonds [4, page 82].)

The algorithm can probably be generalized to allow the E_j's to overlap, with each x appearing at least n_x and at most n'_x times, and where we require each set $S_x = \{j \mid x \in E_j\}$ to be independent in some given matroid $\mathcal{M}_x$ on the elements $\{1, \ldots, k\}$. Edmonds [4, page 83] shows essentially that matroid intersection would give such an algorithm if all the lower bounds were zero.

This research was supported by the National Science Foundation under grant number GJ-992, and the Office of Naval Research under grant number N-00014-67-A-0112-0057 NR 044-402. The research was also supported in part by the John Simon Guggenheim foundation and Norges Almenvitenskapelige Forskningsråd.

References

[1] Henry H. Crapo and Gian-Carlo Rota, *On the Foundations of Combinatorial Theory: Combinatorial Geometries*, preliminary edition (Cambridge, Massachusetts: MIT Press, 1970).

[2] Jack Edmonds, "Matroid partition," in *Mathematics of the Decision Sciences*, part 1, edited by G. B. Dantzig and A. F. Veinott, Jr. (Providence, Rhode Island: American Mathematical Society, 1968), 335–345.

[3] Jack Edmonds, "Lehman's switching game and a theorem of Tutte and Nash-Williams," *Journal of Research of the National Bureau of Standards* **69B** (1965), 73–77.

[4] Jack Edmonds, "Submodular functions, matroids, and certain polyhedra," in *Combinatorial Structures and their Applications*, Proceedings of a conference held at the University of Calgary in 1969, edited by Richard Guy, Haim Hanani, Norbert Sauer, and Johanan Schönheim (New York: Gordon and Breach, 1970), 69–87.

[5] L. R. Ford, Jr., and D. R. Fulkerson, "Network flow and systems of representatives," *Canadian Journal of Mathematics* **10** (1958), 78–84.

[6] L. R. Ford, Jr., and D. R. Fulkerson, *Flows in Networks* (Princeton, New Jersey: Princeton University Press, 1962), §2.10.

[7] John E. Hopcroft and Richard M. Karp, "An $n^{5/2}$ algorithm for maximum matchings in bipartite graphs," *SIAM Journal on Computing* **2** (1973), 225–231.

Chapter 18

Irredundant Intervals

[Originally published in ACM Journal of Experimental Algorithmics **1** (1996), article 1, 19 pages.]

This expository note presents simplifications of a theorem due to Győri and an algorithm due to Franzblau and Kleitman: Given a family F of m intervals on a linearly ordered set of n elements, we can construct in $O(m+n)^2$ steps an irredundant subfamily having maximum cardinality, as well as a generating family having minimum cardinality. The algorithm is of special interest because it solves a problem analogous to finding a maximum independent set, but on a class of objects that is more general than a matroid. This note is also a complete, runnable computer program, which can be used for experiments in conjunction with the public-domain software of The Stanford GraphBase. *An electronic source file for this program is downloadable from the Internet site*

`http://www-cs-faculty.stanford.edu/~knuth/preprints.html`

under the name 'P151 Irredundant intervals'.

1. Introduction. Let's say that a family of sets is *irredundant* if its members can be arranged in a sequence with the following property: Each set contains a point that isn't in any of the preceding sets.

If F is a family of sets, we write $F^{\cup}$ for the family of all nonempty unions of elements of F. When F and G are families with $F \subseteq G^{\cup}$, we say that G *generates* F. If F is irredundant and G generates F, we obviously have $|F| \le |G|$, because each set in the sequence requires a new generator.

In the special case that the members of F are intervals of the real line, András Frank conjectured that the largest irredundant subfamily of F has the same cardinality as F's smallest generating family. This conjecture was proved by Ervin Győri [4], who noted that such a result was a minimax theorem of a new type, apparently unrelated to any of the other well-known minimax theorems of graph theory and combinatorics. A constructive proof was found shortly afterwards by Franzblau and Kleitman [3], who sketched an algorithm to find a generating family and an irredundant subfamily of equal cardinality. (Győri, Franzblau, and Kleitman were led to these results while studying the more general problem of finding a minimum number of subrectangles that cover a given polygon. Further information about polygon covers appears in references [3] and [1] below.)

The purpose of this note is to describe the beautiful algorithm of Franzblau and Kleitman in full detail. Indeed, the `CWEB` source file that generated this document is a computer program that can be used in connection with the Stanford GraphBase [7] to find maximum irredundant subfamilies and minimum generating families of any given collection of intervals. Perhaps this new exposition will shed new light on the class of optimization problems for which an efficient algorithm exists.

According to the conventions of `CWEB` [9], the sections of this document are sequentially numbered 1, 2, 3, etc. In this respect we are returning to a style of exposition used by Euler and Gauss and their contemporaries. A `CWEB` program is also essentially a hypertext; therefore this document may also be regarded as experimental in another sense, as an attempt to find new forms of exposition appropriate to modern technology.

Note: Győri used the term "U-increasing" for an irredundant family; Franzblau and Kleitman called such intervals "independent." Since a family of sets is a hypergraph, it seems unwise to deviate from the standard meaning of independent edges, yet "U-increasing" is not an especially appealing alternative. We will see momentarily that the term "irredundant" is quite natural in theory and practice.

2. A far-reaching generalization of Győri's theorem was proved recently by Frank and Jordán [2], who introduced a large new family of minimax theorems related to linking systems. In particular, Frank and Jordán extended Győri's results to intervals on a circle instead of a line. But no combinatorial algorithm is known as yet for the circular case.

Can any or all of the Franzblau–Kleitman methods be "lifted" to such more general problems? We will return to this tantalizing question after becoming familiar with Franzblau and Kleitman's remarkable algorithm.

3. Theory. It is wise to study the theory underlying the Franzblau–Kleitman algorithm before getting into the program itself.

4. A family of sets is called *redundant* if it is not irredundant. Any family that contains a redundant subfamily is redundant, since any family contained in an irredundant family is irredundant.

5. If F is a family of sets and s is an arbitrary set, let $F|s$ denote the sets of F that are contained in s. This operation is left-associative by convention: $F|s|t = (F|s)|t = F|(s \cap t)$.

We also write $\bigcup F$ for $\bigcup\{f \mid f \in F\}$; thus $F|\bigcup F = F$.

(An index to all the main notations and definitions that we will use appears at the end of this note.)

6. Lemma. *A finite family F is redundant if and only if there is a nonempty set s such that every point of s belongs to at least two members of $F|s$.* (The set s need not belong to F.)

Proof. If F is irredundant there is no such s, because $F|s$ is irredundant and its last set in the assumed sequence contains a point that isn't in any of the others. But if F is redundant, it contains a minimal redundant subfamily F_0; then we have

$$f \subseteq \bigcup(F_0 \setminus \{f\}) \qquad \text{for all } f \in F_0\,,$$

since $F_0 \setminus \{f\}$ is irredundant. It follows that every point of $s = \bigcup F_0$ is contained in at least two members of F_0, hence in at least two members of $F|s$ (since $F_0 \subseteq F|s$). □

7. Corollary. *A finite family F of intervals on a line is redundant if and only if there is an interval s such that every point of s belongs to at least two intervals of $F|s$.* (The set s need not belong to F.)

Proof. Intervals are nonempty. By the proof of the preceding lemma, it suffices to consider sets s that can be written $\bigcup F_0$ for some minimal redundant subfamily F_0. In the special case of intervals, $\bigcup F_0$ must be a single interval; otherwise F_0 would not be minimal. □

8. Henceforth we will restrict consideration to finite families F of intervals on a linearly ordered set. It suffices, in fact, to deal with integer elements; we will consider subintervals of the n-element set $[0\,..\,n)$. (The notation $[a\,..\,b)$ stands here for the set of all integers x such that $a \le x < b$.)

If F is a family of sets and x is a point, we will write $N_x\,F$ for the number of sets that contain x. The corollary just proved can therefore be stated as follows: "F is irredundant if and only if every interval $s \subseteq \bigcup F$ contains a point x with $N_x\,F|s \le 1$." This characterization provides a polynomial-time test for irredundancy.

9. Irredundant intervals have an interesting connection to the familiar computer-science concept of *binary search trees* (see, for example, [6, §6.2.2]): A family of intervals is irredundant if and only if we can associate its intervals with a binary tree whose nodes are each labeled with an integer x and an interval containing x. All nodes in the left subtree of such a node correspond to intervals that are strictly less than x, in the sense that all elements of those intervals are $< x$; all nodes in the right subtree correspond to intervals that are strictly greater than x. The root of the binary tree corresponds to the interval that is last in the assumed irredundant ordering. Its distinguished integer x is an element that appears in no other interval.

Given such a tree, we obtain a suitable irredundant ordering by traversing it recursively from the leaves to the root, in postorder [5, §2.3.1]. Conversely, given an irredundant ordering, we can construct a binary tree recursively, proceeding from the root to the leaves.

10. An example might be helpful at this point. Suppose $n = 9$ and

$$f_1 = [0\,..\,8)\,,\quad f_2 = [0\,..\,7)\,,\quad f_3 = [1\,..\,6)\,,$$
$$f_4 = [1\,..\,5)\,,\quad f_5 = [3\,..\,9)\,,\quad f_6 = [2\,..\,9)\,.$$

Then $\{f_1, f_2, f_3, f_4, f_5\}$ and $\{f_1, f_3, f_5, f_6\}$ are irredundant. (Indeed, a family of intervals is irredundant whenever its members have no repeated left endpoints or no repeated right endpoints.) These subfamilies are in fact *maximally* irredundant—they become redundant when any other interval of the family is added. Therefore maximal irredundant subfamilies need not have the same cardinality; irredundant subfamilies do not form the independent sets of a matroid.

On the other hand, irredundant sets of intervals do have matroid-like properties. For example, if F is irredundant and $F \cup \{g\}$ is redundant, there is an $f \in F$ such that $F \cup \{g\} \setminus \{f\}$ is irredundant. (The proof is by induction on $|F|$: There is $x \in f \in F$ such that $F = F_l \cup \{f\} \cup F_r$, where F_l and F_r correspond to the left and right subtrees of the root in the binary tree representation. If $x \in g$, the family $F_l \cup \{g\} \cup F_r$ is irredundant; if $x < g$, there is $f' \in F_r$ with $F_l \cup \{f\} \cup (F_r \cup \{g\} \setminus \{f'\})$ irredundant, by induction; if $x > g$ there is $f' \in F_l$ with $(F_l \cup \{g\} \setminus \{f'\}) \cup \{f\} \cup F_r$ irredundant.) Such near-matroid behavior makes families of intervals especially instructive.

11. Let's say that an interval s is *good* for F if $N_x\, F|s \leq 1$ for some $x \in s$; otherwise s is *bad*. Franzblau and Kleitman introduced a basic reduction procedure for any family F of intervals that possesses a bad interval s. Their procedure is analogous to modification along an augmenting path in other combinatorial algorithms.

Let $[a_1 \mathinner{..} b_1)$, $\ldots$, $[a_k \mathinner{..} b_k)$ be the maximal intervals in $F|s$, ordered so that $a_1 < \cdots < a_k$ and $b_1 < \cdots < b_k$. In our applications below we will have $s = [a_1 \mathinner{..} b_k)$. For example, we might have the following picture:

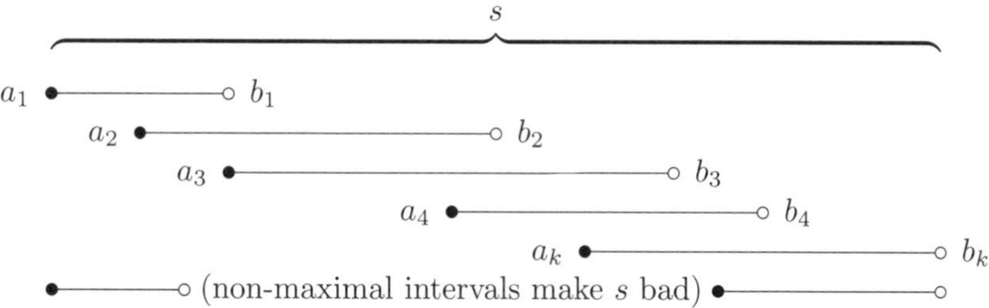

If $a_{j+1} < b_j$ for $1 \leq j < k$, the family F *reduced in* s is defined to be

$$F{\downarrow}s \;=\; F \setminus \{[a_1 \mathinner{..} b_1), \ldots, [a_k \mathinner{..} b_k)\} \,\cup\, \{[a_2 \mathinner{..} b_1), \ldots, [a_k \mathinner{..} b_{k-1})\}\,.$$

In the simplest case we have $k = 1$ and the reduced family is simply $F \setminus \{[a_1 \mathinner{..} b_1)\}$.

If s is a *minimal* bad interval for F, we can prove that $a_{j+1} < b_j$ for $1 \leq j < k$, hence $F{\downarrow}s$ is well-defined. Indeed, point a_{j+1} must be in some interval $[c \mathinner{..} d)$ other than $[a_{j+1} \mathinner{..} b_{j+1})$, since s is bad. We can assume that $c < a_{j+1}$; otherwise all intervals of $F|s$ would be contained in $[a_1 \mathinner{..} a_{j+1})$ or $[a_{j+1} \mathinner{..} b_k)$, and both of these subintervals would be bad, contradicting the minimality of s. But $c < a_{j+1}$ implies that $[c \mathinner{..} d)$ is contained in some maximal $[a_i \mathinner{..} b_i)$ with $i \leq j$. Hence $a_{j+1} < b_i \leq b_j$.

The notation $F{\downarrow}s$ is defined to be left-associative, like $F|s$; that is, $F{\downarrow}s{\downarrow}t = (F{\downarrow}s){\downarrow}t$ and $F{\downarrow}s\,|t = (F{\downarrow}s)|t$.

12. Lemma. *If s is a minimal bad interval for F, we have $F \subseteq (F{\downarrow}s)^{\cup}$.*

Proof. We must show that $[a_j \mathinner{..} b_j) \in (F{\downarrow}s)^{\cup}$ for $1 \le j \le k$. Let $G = F{\downarrow}s \,|\, [a_j \mathinner{..} b_j)$; we will prove that $[a_j \mathinner{..} b_j) = \bigcup G$. If $x \in [a_j \mathinner{..} b_j)$, the badness of s implies that $x \in t$ for some $t \in F|s$ with $t \ne [a_j \mathinner{..} b_j)$; let t be contained in the maximal interval $[a_i \mathinner{..} b_i)$. If $i = j$, we have $t \in G$; if $i < j$, we have $x \in [a_j \mathinner{..} b_i) \subseteq [a_j \mathinner{..} b_{j-1}) \in G$; and if $i > j$, we have $x \in [a_i \mathinner{..} b_j) \subseteq [a_{j+1} \mathinner{..} b_j) \in G$. □

13. Lemma. *Suppose s is a minimal bad interval for F, while t is a good interval. Then t is good also for $F{\downarrow}s$.*

Proof. Let $s = [a \mathinner{..} b)$, $t = [c \mathinner{..} d)$, $l = \min(a, c)$, $r = \max(b, d)$. Suppose $N_x\, F|[l \mathinner{..} d) \ge 2$ and $N_x\, F|[c \mathinner{..} r) \ge 2$ for all $x \in t$. Then $l = a < c < d < b = r$, because t is good for F. By the minimality of s, there is some $x \in [a \mathinner{..} d)$ with $N_x\, F|[a \mathinner{..} d) \le 1$. Since $N_y\, F|[a \mathinner{..} d) \ge 2$ for all $y \in t$, we have $x < c$. Furthermore, x is in some interval of $(F|s) \setminus (F|[a \mathinner{..} d))$, because s is bad; so x is in some maximal $[a_j \mathinner{..} b_j)$ with $a \le a_j \le x < c < d < b_j \le b$. It follows that none of the intervals $[a_1 \mathinner{..} b_1)$, $\ldots$, $[a_k \mathinner{..} b_k)$, $[a_2 \mathinner{..} b_1)$, $\ldots$, $[a_k \mathinner{..} b_{k-1})$ are contained in t, hence $F{\downarrow}s \,|\, t = F|t$.

On the other hand, suppose $N_x\, F|[l \mathinner{..} d) \le 1$ for some $x \in t$. We will show that $N_x\, F{\downarrow}s \,|\, t \le 1$. This assertion can fail only if x lies in some interval $[a_{j+1} \mathinner{..} b_j) \subseteq t$ newly added to $F{\downarrow}s$. Then $x \in [a_j \mathinner{..} b_j) \subseteq [l \mathinner{..} d)$, and $x \in [a_{j+1} \mathinner{..} b_{j+1}) \not\subseteq [l \mathinner{..} d)$, hence $b_j \le d < b_{j+1}$; it follows that j is uniquely determined, and the only interval containing x in $F{\downarrow}s \,|\, t$ is $[a_{j+1} \mathinner{..} b_j)$. A similar argument applies if $N_x\, F|[c \mathinner{..} r) \le 1$. □

14. Corollary. *If s is a minimal bad interval for F, we have $N_x\, F{\downarrow}s = N_x\, F - N_x\, s$.*

Proof. The proof of the preceding lemma shows in particular that none of the intervals $[a_{j+1} \mathinner{..} b_j)$ are already present in F before the reduction. (If such an interval were present, it would be a good interval t with $N_x\, F{\downarrow}[l \mathinner{..} d) \ge 2$ and $N_x\, F{\downarrow}[c \mathinner{..} r) \ge 2$ for all $x \in t$.) And if $x \in s$, suppose x lies in $[a_i \mathinner{..} b_i)$, $[a_{i+1} \mathinner{..} b_{i+1})$, $\ldots$, $[a_j \mathinner{..} b_j)$; then it lies in $[a_{i+1} \mathinner{..} b_i)$, $\ldots$, $[a_j \mathinner{..} b_{j-1})$ after reduction, a net change of -1. □

15. The Franzblau–Kleitman algorithm has a very simple outline: We let $G_0 = F$ and repeatedly set $G_{k+1} = G_k{\downarrow}s_k$, where s_k is the leftmost minimal bad interval for G_k, until we finally reach a family G_r in which no bad intervals remain. This must happen sooner or later, because $|G_k| = |F| - k$. The final irredundant family $G = G_r$ generates F, because $F \subseteq G_k^{\cup}$ for all k by the lemma of §12. Franzblau and

Kleitman proved the nontrivial fact that $|G|$ is the size of the maximum irredundant subfamily of F; hence G is a minimum generating family.

16. It is tempting to try to prove the optimality of G by a simpler, inductive approach in which we "grow" F one interval at a time, updating its maximum irredundant set and minimum generating set appropriately. But experiments show that the maximum irredundant set can change drastically when F receives a single new interval, so this direct primal-dual approach seems doomed to failure. The indirect approach is more difficult to prove, but no more difficult to program. So we will proceed to develop further properties of Franzblau and Kleitman's reduction procedure [3]. The key fact is a remarkable theorem that we turn to next.

17. Theorem. *The same final family $G = G_r$ is obtained when s_k is chosen to be an arbitrary (not necessarily leftmost) minimal bad interval of G_k in the reduction algorithm. Moreover, the same multiset $\{s_0, \ldots, s_{r-1}\}$ of minimal bad intervals arises, in some order, regardless of the choices made at each step.*

Proof. We use induction on r, the maximum number of steps to convergence among all reduction procedures that begin with a family F. If $r = 0$, the result is trivial, and if F has only one minimal bad interval the result is immediate by induction. Suppose therefore that s and t are distinct minimal bad intervals of F. We will prove later that t is a minimal bad interval for $F{\downarrow}s$, and that $F{\downarrow}s{\downarrow}t = F{\downarrow}t{\downarrow}s$. Let r' be the maximum distance to convergence from $F{\downarrow}s$, and r'' the maximum from $F{\downarrow}t$; then r' and r'' are less than r, and induction proves that the final result from $F{\downarrow}s$ is the final result from $F{\downarrow}s{\downarrow}t = F{\downarrow}t{\downarrow}s$, which is the final result from $F{\downarrow}t$. (Readers familiar with other reduction algorithms, like that of [8], will recognize this as a familiar "diamond lemma" argument. We construct a diamond-shaped diagram with four vertices: F, $F{\downarrow}s$, $F{\downarrow}t$, and a common outcome of $F{\downarrow}s$ and $F{\downarrow}t$.) This completes the proof, except for two lemmas that will be demonstrated below; their proofs have been deferred so that we could motivate them first. □

18. This theorem and the lemma of §13 have an important corollary: *Let $S = \{s_0, \ldots, s_{r-1}\}$ be the multiset of minimal bad intervals determined by the algorithm from F, and let t be any interval. Then $S|t$ is the multiset of minimal bad intervals determined by the algorithm from $F|t$.* This holds because an interval $s \subseteq t$ is bad for F if and only if it is bad for $F|t$. Minimal bad intervals within t never appear again once they are removed, and we can remove them first.

Reducing a minimal bad interval s when s is contained in a bad interval t may make t good, or leave it bad, or make it minimally bad. If s is minimally bad for F, it might also be minimally bad for $F{\downarrow}s$.

All elements of S are bad in the original family F, but they need not all be *minimally* bad.

19. Now we are ready for the *coup de grâce* and the *pièce de résistance.* After the reduction algorithm has computed the irredundant generating family $G = G_r$ and the multiset S of minimal bad intervals, we can construct an irredundant subfamily F' of F with $|F'| = |G|$ by constructing a binary search tree as described in §9. The procedure is recursive, starting with an initial interval $t = [a \mathinner{..} b)$ that contains F: The tree defined for $F|t$ is empty if $F|t$ is empty. Otherwise it has a root node labeled with x and with any interval of $F|t$ containing x, where x is an integer such that $N_x\, G^{(t)} = 1$; here $G^{(t)}$ is the final generating set that is obtained when the reduction procedure is applied to $F|t$. A suitable interval containing x exists, because every element of $G^{(t)}$ is an intersection of intervals in $F|t$. The left subtree of the root node is the binary search tree for $F|\bigl(t \cap [a \mathinner{..} x)\bigr)$; the right subtree is the binary search tree for $F|\bigl(t \cap [x+1 \mathinner{..} b)\bigr)$.

The number of nodes in this tree is $|G|$. For if x is the integer in the label of the root, G has one interval containing x, and its other intervals are $G|[a \mathinner{..} x)$ and $G|[x+1 \mathinner{..} b)$.

The family $G^{(t)}$ is not the same as $G|t$; but we do have $|G^{(t)}| = \bigl|G|t\bigr|$ when t has the special form $[a \mathinner{..} x)$ or $[x+1 \mathinner{..} b)$, because in such cases $F{\downarrow}s\,|t$ has the same cardinality as $F|t$ when s is a minimal bad interval and $s \not\subseteq t$. For example, if $t = [a \mathinner{..} x)$ and $b_j \le x < b_{j+1}$, we have $F{\downarrow}s\,|t = F|t \setminus \{[a_1 \mathinner{..} b_1), \ldots, [a_j \mathinner{..} b_j)\} \cup \{[a_2 \mathinner{..} b_1), \ldots, [a_{j+1} \mathinner{..} b_j)\}$.

20. It is not necessary to compute each $G^{(t)}$ from scratch by starting with $F|t$ and applying the reduction algorithm until it converges, because the binary tree construction algorithm requires only a knowledge of the incidence function $N_x\, G^{(t)}$. This function is easy to compute, because $N_x\, F{\downarrow}s = N_x\, F - N_x\, s$ by §14; therefore

$$N_x\, G^{(t)} = N_x\, F|t - N_x\, S|t\,.$$

21. All the basic ideas of Franzblau and Kleitman's algorithm have now been explained. But we must still carry out a careful analysis of some fine points of reduction that were claimed in the proof of the main theorem. If s and t are distinct minimal bad intervals, the lemma of §13 implies that no bad subintervals of t appear in $F{\downarrow}s$; we also need to verify that t itself remains bad.

Lemma. *If s is a minimal bad interval for F and t is a bad interval such that $s \not\subseteq t$, then t is bad for $F{\downarrow}s$.*

Proof. Let $s = [a\,..\,b)$ and $t = [c\,..\,d)$. We can assume by left-right symmetry that $a < c$. Then $b < d$, by minimality of s. Assume that t isn't bad for $F{\downarrow}s$. The subfamily $F|t$ must contain at least one of the maximal intervals $[a_j\,..\,b_j)$ of $F|s$ that are deleted during the reduction; hence $c \le a_j < b_{j-1} \le b$.

Let j be minimal with $c \le a_j$. Then

$$F{\downarrow}s\,|t = F|t \setminus \{[a_j\,..\,b_j), \ldots, [a_k\,..\,b_k)\} \cup \{[a_j\,..\,b_{j-1}), \ldots, [a_k\,..\,b_{k-1})\}\,;$$

so the elements of t that are covered once less often are the elements of $[b_{j-1}\,..\,b_k)$. Suppose $y \in [b_{i-1}\,..\,b_i)$ for some $i \ge j$. Then $y \in [a_i\,..\,b_i) \subseteq s \cap t$, and y is in some other interval $f \subseteq s$. Every maximal interval containing f must be $[a_l\,..\,b_l)$ for some $l \ge i$, hence $f \subseteq s \cap t$. Thus $N_y\, F|(s \cap t) \ge 2$ for all $y \in [b_{j-1}\,..\,b_k)$. But $s \cap t$ is good, so there must be a point $x \in [c\,..\,b_{j-1})$ with $N_x\, F|(s{\cap}t) \le 1$. We also have $N_x\, F|t \ge 2$, since t is bad, so there's an interval in $F|t \setminus F|(s \cap t)$ that contains x. This interval contains $[x\,..\,b) = [x\,..\,b_k)$. Consequently $N_y\, F|t \ge 3$ for all $y \in [b_{j-1}\,..\,b_k)$. $\square$

22. Lemma. *If s and t are minimal bad intervals for F and $s \neq t$, we have $F{\downarrow}s{\downarrow}t = F{\downarrow}t{\downarrow}s$.*

Proof. Let the maximal intervals in $F|s$ and $F|t$ be $[a_1 \mathinner{..} b_1)$, ..., $[a_k \mathinner{..} b_k)$ and $[c_1 \mathinner{..} d_1)$, ..., $[c_l \mathinner{..} d_l)$, respectively, where $a_1 < c_1$. The lemma is obvious unless $F|(s \cap t)$ is nonempty, so we assume that $c_1 < b_k < d_l$. Let $x \in s \cap t$ have $N_x\, F|(s \cap t) = 1$, and let f be the interval of $F|(s \cap t)$ that contains x. Let p be maximal with $a_p < c_1$, and let q be minimal with $d_q > b_k$. Since $N_x\, F|s > 1$, there is an interval $[a_j \mathinner{..} b_j)$ containing x with $j \le p$; thus $x \in [a_p \mathinner{..} b_p)$. Similarly $x \in [c_q \mathinner{..} d_q)$. Furthermore, if $p < k$ we have $[a_{p+1} \mathinner{..} b_{p+1}) \subseteq s \cap t$; hence either $[a_{p+1} \mathinner{..} b_{p+1}) = f$ or $a_{p+1} > x$. If $q > 1$ we have either $[c_{q-1} \mathinner{..} d_{q-1}) = f$ or $d_{q-1} \le x$.

If $p = k$ or $f \neq [a_{p+1} \mathinner{..} b_{p+1})$, any newly added intervals $[a_{j+1} \mathinner{..} b_j)$ for $p \le j < k$ in $F{\downarrow}s$ are properly contained in $[c_q \mathinner{..} d_q)$, so they remain in $F{\downarrow}s{\downarrow}t$. Thus we can easily describe the compound operation $F{\downarrow}s{\downarrow}t$ in detail:

$$\begin{aligned}&\text{Delete } [a_1 \mathinner{..} b_1), \ldots, [a_k \mathinner{..} b_k);\quad \text{insert } [a_2 \mathinner{..} b_1), \ldots, [a_k \mathinner{..} b_{k-1});\\ &\qquad \text{delete } [c_1 \mathinner{..} d_1), \ldots, [c_l \mathinner{..} d_l);\quad \text{insert } [c_2 \mathinner{..} d_1), \ldots, [c_l \mathinner{..} d_{l-1}).\end{aligned}$$

No two of these intervals are identical, so $F{\downarrow}t{\downarrow}s$ gives the same result. (If $f = [c_{q-1} \mathinner{..} d_{q-1})$, the family $F{\downarrow}t$ has f replaced by $[c_q \mathinner{..} d_{q-1}) \subseteq f \subseteq [a_i \mathinner{..} b_i)$ for some $i \le p$, so $[c_q \mathinner{..} d_{q-1})$ is not maximal in $F{\downarrow}t\,|s$.)

The remaining case $f = [a_{p+1} \mathinner{..} b_{p+1}) = [c_{q-1} \mathinner{..} d_{q-1})$ needs to be considered specially, since we can't delete this interval twice. The following picture might help clarify the situation:

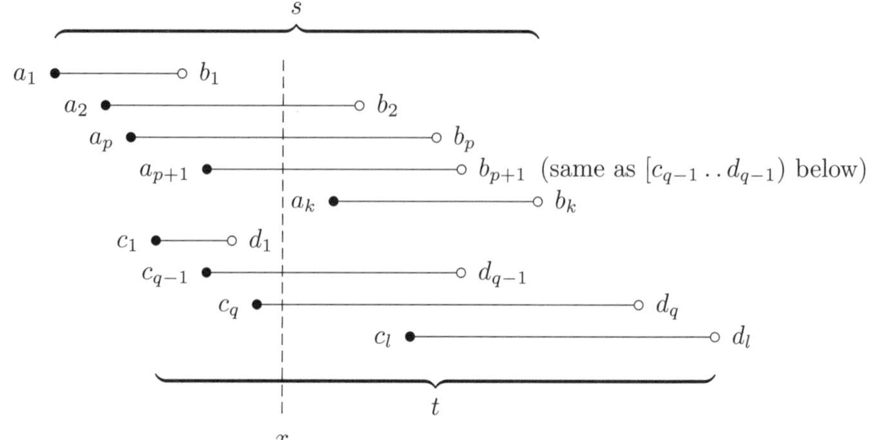

Suppose g is an interval of $F|t$ that is contained in $[c_j \mathinner{..} d_j)$ if and only if $j = q - 1$. Then g contains a point $< c_q$. If $x \in g$ we have $g = f$, since $g \subseteq s \cap t$. Otherwise $g \subseteq [c_{q-1} \mathinner{..} x) \subseteq [c_{q-1} \mathinner{..} b_p) = [a_{p+1} \mathinner{..} b_p)$. It follows that the maximal intervals of $F{\downarrow}s\,|t$ are

$$[c_1 \mathinner{..} d_1),\ \ldots,\ [c_{q-2} \mathinner{..} d_{q-2}),\ [c_{q-1} \mathinner{..} b_p),\ [c_q \mathinner{..} d_q),\ \ldots,\ [c_l \mathinner{..} d_l).$$

These intervals are replaced in $F{\downarrow}s{\downarrow}t$ by

$$[c_2 \mathinner{..} d_1),\ \ldots,\ [c_{q-1} \mathinner{..} d_{q-2}),\ [c_q \mathinner{..} b_p),\ [c_{q+1} \mathinner{..} d_q),\ \ldots,\ [c_l \mathinner{..} d_{l-1}).$$

Thus $F{\downarrow}s{\downarrow}t$ is formed almost as in the previous case, but with $[a_{p+1} \mathinner{..} b_p)$ and $[c_q \mathinner{..} d_{q-1})$ replaced by $[c_q \mathinner{..} b_p)$. And we get precisely the same intervals in $F{\downarrow}t{\downarrow}s$. □

(Is there a simpler proof?)

23. Practice. The computer program in the remainder of this note operates on a family of intervals defined by a graph on $n+1$ vertices $\{0,1,\ldots,n\}$. We regard an edge between u and v as the half-open interval $[u\,..\,v)$, when $u<v$.

Graphs are represented as in the algorithms of the Stanford GraphBase [7], and the reader of this program is supposed to be familiar with the elementary conventions of that system.

The program reads two command-line parameters, m and n, and an optional third parameter representing a random-number seed. (The seed value is zero by default.) The Franzblau–Kleitman algorithm is then applied to the graph $random\_graph(n+1, m, 0, 0, 0, 0, 0, 0, 0, seed)$, a random graph with vertices $\{0,1,\ldots,n\}$ and m edges. Alternatively, the user can specify an arbitrary graph as input by typing the single command-line parameter `-g`⟨filename⟩; in the latter case the named file should describe the graph in *save_graph* format, as in the MILES_SPAN program of [7].

When the computation is finished, a minimal generating family and a maximal irredundant subfamily will be printed on the standard output file.

If a negative value is given for n, the random graph is reversed from left to right; each interval $[a\,..\,b)$ is essentially replaced by $[-b\,..\,-a)$ (but minus signs are suppressed in the output). This feature lends credibility to the correctness of our highly asymmetric algorithm and program, because we can verify the fact that the minimum generating family of the mirror image of F is indeed the mirror image of F's minimum generating family.

In practice, the algorithm tends to be interesting only when m and n are roughly equal. If n is large compared to m, we can remove any vertices of degree zero; such vertices aren't the endpoint of any interval. If m is large compared to n, we can almost always find n irredundant intervals by inspection. The running time in general is readily seen to be $O(mn+n^2)$.

```
#define panic(k)
        {
          fprintf(stderr, "Oops,␣we're␣out␣of␣memory!␣(Case␣%d)\n", k);
          return k;
        }
#include "gb_graph.h"     /* the GraphBase data structures */
#include "gb_rand.h"      /* the random_graph generator */
#include "gb_save.h"      /* the restore_graph generator */
  ⟨Preprocessor definitions⟩
```

```
Graph *F;     /* the graph that defines intervals */
Graph *G;     /* a graph of intervals that generate F */
⟨ Subroutines 33 ⟩
main(argc, argv)
    int argc;     /* the number of command-line arguments, plus 1 */
    char *argv[];     /* the command-line arguments */
{
  register Vertex *t, *u, *v, *w, *x;
    /* current vertices of interest */
  register Arc *a, *b, *c;     /* current arcs of interest */
  ⟨ Scan the command-line options and generate F 24 ⟩;
  ⟨ Compute G and S by the Franzblau–Kleitman algorithm 26 ⟩;
  if (gb_trouble_code) panic(1);
  ⟨ Construct an irredundant subfamily of F with the cardinality of G 32 ⟩;
  ⟨ Print the results 36 ⟩;
  return 0;     /* this is the normal exit */
}
```

fprintf: **int** ( ), `<stdio.h>`.
gb_trouble_code: **long**, GB_GRAPH §14.
random_graph: **Graph** *( ), GB_RAND §5.
restore_graph: **Graph** *( ), GB_SAVE §4.
save_graph: **long** ( ), GB_SAVE §20.
stderr: **FILE** *, `<stdio.h>`.

24. ⟨ Scan the command-line options and generate F 24 ⟩ ≡

```
{
  int m = 0, n = 0, seed = 0;
  if (argc ≥ 3 ∧ sscanf(argv[1], "%d", &m) ≡ 1 ∧ sscanf(argv[2], "%d",
        &n) ≡ 1) {
    if (argc > 3) sscanf(argv[3], "%d", &seed);
    if (m < 0) {
      m = −m;
        /* we assume the user meant to negate n instead of m */
      n = −n;
    }
    if (n ≥ 0) F = random_graph(n + 1, m, 0, 0, 0, 0, 0, 0, 0, seed);
    else {
      G = random_graph(−n + 1, m, 0, 0, 0, 0, 0, 0, 0, seed);
      ⟨ Set F to the mirror image of G 25 ⟩;
      gb_recycle(G);
    }
  }
  else if (argc ≡ 2 ∧ strncmp(argv[1], "-g", 2) ≡ 0)
    F = restore_graph(argv[1] + 2);
  else {
    fprintf(stderr, "Usage:␣%s␣m␣n␣[seed]␣␣|␣␣%s␣-gfoo.gb\n",
          argv[0], argv[0]);
    return 2;
  }
  if (¬F) {
    fprintf(stderr,
          "Sorry,␣can't␣create␣the␣graph!␣(error␣code␣%ld)\n",
          panic_code);
    return 3;
  }
}
printf("Applying␣Franzblau--Kleitman␣to␣%s:\n", F→id);
```

This code is used in section 23.

25. ⟨ Set F to the mirror image of G 25 ⟩ ≡

```
F = gb_new_graph(G->n);
if (¬F) panic(4);
make_compound_id(F, "reflect(", G, ")");
for (v = G->vertices, u = F->vertices + F->n − 1; v < G->vertices + G->n;
        u−−, v++) {
    v->clone = u;
    u->name = gb_save_string(v->name);
}
for (v = G->vertices; v < G->vertices + G->n; v++)
    for (a = v->arcs; a; a = a->next) gb_new_arc(v->clone, a->tip->clone, 1);
```

This code is used in section 24.

---

a: **register Arc** ∗, §23.
arcs: **struct arc_struct** ∗, GB_GRAPH §9.
argc: **int**, §23.
argv: **char** ∗[ ], §23.
clone = macro, §27.
F: **Graph** ∗, §23.
fprintf: **int** ( ), `<stdio.h>`.
G: **Graph** ∗, §23.
gb_new_arc: **void** ( ), GB_GRAPH §30.
gb_new_graph: **Graph** ∗( ), GB_GRAPH §23.
gb_recycle: **void** ( ), GB_GRAPH §40.
gb_save_string: **char** ∗( ), GB_GRAPH §35.
id: **char** [ ], GB_GRAPH §20.
make_compound_id: **void** ( ), GB_GRAPH §26.
n: **long**, GB_GRAPH §20.
name: **char** ∗, GB_GRAPH §9.
next: **struct arc_struct** ∗, GB_GRAPH §10.
panic = macro ( ), §23.
panic_code: **long**, GB_GRAPH §5.
printf: **int** ( ), `<stdio.h>`.
random_graph: **Graph** ∗( ), GB_RAND §5.
restore_graph: **Graph** ∗( ), GB_SAVE §4.
sscanf: **int** ( ), `<stdio.h>`.
stderr: **FILE** ∗, `<stdio.h>`.
strncmp: **int** ( ), `<string.h>`.
tip: **struct vertex_struct** ∗, GB_GRAPH §10.
u: **register Vertex** ∗, §23.
v: **register Vertex** ∗, §23.
vertices: **Vertex** ∗, GB_GRAPH §20.

26. The main algorithm. We follow the outline of §15.

```
⟨ Compute G and S by the Franzblau–Kleitman algorithm 26 ⟩ ≡
  ⟨ Make G a copy of F, retaining only leftward arcs 27 ⟩;
  for (v = G→vertices + 1; v < G→vertices + G→n; v++) ⟨ Reduce all minimal
        bad intervals with right endpoint v and record them in S 28 ⟩;
```

This code is used in section 23.

27. The algorithm doesn't need pointers from the left endpoint of an interval to the right endpoint; leftward pointers are sufficient. (This observation makes the reduction procedure faster.)

While copying F, we remove its rightward arcs, and we assign length 1 to all its leftward arcs. Later on, we will represent intervals of S by recording them in F as leftward arcs of length -1.

We also clear two utility fields of F's vertices, *count* and *link*, since they will be used by the algorithm later.

```
#define clone u.V
        /* the vertex in F that matches a vertex in G, or vice versa */

⟨ Make G a copy of F, retaining only leftward arcs 27 ⟩ ≡
  switch_to_graph(Λ);     /* prepare to return to graph F later */
  G = gb_new_graph(F→n);
    /* a graph with nameless vertices and no arcs */
  if (¬G) panic(5);
  for (u = F→vertices, v = G→vertices; u < F→vertices + F→n; u++, v++) {
    u→clone = v; v→clone = u;
    u→count = 0; u→link = Λ;
    v→name = gb_save_string(u→name);
    for (a = u→arcs, b = Λ; a; a = a→next)
      if (a→tip ≥ u) {      /* we will remove the non-leftward arc a */
        if (b) b→next = a→next;
        else u→arcs = a→next;
      } else {      /* we will copy the leftward arc a */
        gb_new_arc(v, a→tip→clone, 0);      /* the length in G is 0 */
        a→len = 1;      /* but in F the length is 1 */
        b = a;      /* b points to the last non-removed arc */
      }
  }
  if (gb_trouble_code) panic(6);
  switch_to_graph(F);     /* now we can add arcs to F again */
```

This code is used in section 26.

---

a: **register Arc** ∗, §23.
arcs: **struct arc_struct** ∗, GB_GRAPH §9.
b: **register Arc** ∗, §23.
count = macro, §28.
F: **Graph** ∗, §23.
G: **Graph** ∗, §23.
gb_new_arc: **void** ( ), GB_GRAPH §30.
gb_new_graph: **Graph** ∗( ), GB_GRAPH §23.
gb_save_string: **char** ∗( ), GB_GRAPH §35.
gb_trouble_code: **long**, GB_GRAPH §14.
len: **long**, GB_GRAPH §10.
link = macro, §28.
n: **long**, GB_GRAPH §20.
name: **char** ∗, GB_GRAPH §9.
next: **struct arc_struct** ∗, GB_GRAPH §10.
panic = macro ( ), §23.
switch_to_graph: **void** ( ), GB_GRAPH §39.
tip: **struct vertex_struct** ∗, GB_GRAPH §10.
u: **register Vertex** ∗, §23.
u: **util**, GB_GRAPH §9.
v: **register Vertex** ∗, §23.
V: **struct vertex_struct** ∗, GB_GRAPH §8.
vertices: **Vertex** ∗, GB_GRAPH §20.

28. Here's the most interesting part of the program, algorithmwise. Given a vertex v, we want to find the largest u, if any, such that $[u\,..\,v)$ is bad for the intervals in G. So we sweep through the intervals $[u\,..\,v)$ from right to left, decreasing u until it reaches a limiting value t. Here t is the least upper bound on a left endpoint that could guarantee double coverage of all points in $[u\,..\,v)$.

The utility field $x\text{→}count$ records the number of intervals with left endpoint x and right endpoint w in the range $u < w \le v$. Another utility field $x\text{→}link$ is used to link vertices with nonzero counts together so that we can clear the counts to zero again afterwards.

It's easy to see that each iteration of the **while** loop in this section takes at most $O(m+n)$ steps. The actual computation time is, however, usually much faster.

This program is designed to work correctly when G contains more than one arc from u to v. Duplicate arcs are discarded as a special case of the reduction procedure.

```
#define count z.I
        /* coverage decreases by this much when we pass to the left */
#define link y.V      /* pointer to a vertex whose count field needs to be
          zeroed later */
```

⟨ Reduce all minimal bad intervals with right endpoint v and record them in S 28 ⟩ ≡

```
{
  while (1) {
    int coverage = 0;
      /* the number of intervals ⊆ [t..v) that contain u */
    int potential = 0;     /* sum of x→count for x < t */
    Vertex *cleanup = Λ;
      /* head of the list of vertices with nonzero count */

    for (u = v, t = v − 1; u > t; u−−) {
      coverage −= u→count;     /* we prepare to decrease u */
      ⟨ Update the counts for all intervals ending at u 29 ⟩;
      if (coverage + potential < 2) {
          /* there's no bad interval ending at v */
        ⟨ Clean up all count fields 30 ⟩;
        goto done_with_v;
      }
      while (coverage < 2) {
        t−−;
        coverage += t→count; potential −= t→count;
      }
    }
```

```
    gb_new_arc(v→clone, u→clone, −1);
      /* [u . . v) is minimally bad; we record it in S = −F */
    ⟨ Replace G by G↓[u . . v) 31 ⟩;
    ⟨ Clean up all count fields 30 ⟩;      /* now we'll try again */
  }
done_with_v: ;
}
```

This code is used in section 26.

29. ⟨ Update the counts for all intervals ending at u 29 ⟩ ≡

```
for (a = u→arcs; a; a = a→next) {
  w = a→tip;
  if (w→count ≡ 0) {
    w→link = cleanup;
    cleanup = w;
  }
  w→count++;
  if (w ≥ t) coverage++;
  else potential++;
}
```

This code is used in section 28.

30. ⟨ Clean up all count fields 30 ⟩ ≡

```
for (w = cleanup; w; w = w→link) w→count = 0;
```

This code is used in section 28.

---

a: **register Arc** ∗, §23.
arcs: **struct arc_struct** ∗, GB_GRAPH §9.
clone = macro, §27.
G: **Graph** ∗, §23.
gb_new_arc: **void** ( ), GB_GRAPH §30.
I: **long**, GB_GRAPH §8.
next: **struct arc_struct** ∗, GB_GRAPH §10.
t: **register Vertex** ∗, §23.
tip: **struct vertex_struct** ∗, GB_GRAPH §10.
u: **register Vertex** ∗, §23.
v: **register Vertex** ∗, §23.
V: **struct vertex_struct** ∗, GB_GRAPH §8.
w: **register Vertex** ∗, §23.
y: **util**, GB_GRAPH §9.
z: **util**, GB_GRAPH §9.

31. The reduction process is kind of cute too.

⟨ Replace G by $G{\downarrow}[u\,..\,v)$ 31 ⟩ ≡

```
for (a = v→arcs, c = Λ, w = v;  a;  c = a, a = a→next)
   if (a→tip ≥ u ∧ a→tip < w)  w = a→tip, b = c;      /* now [w..v) is the
         longest interval from v inside [u..v); we'll remove it */
if (b) b→next = b→next→next;
else v→arcs = v→arcs→next;
   /* the remaining job is to shorten the other maximal arcs in [u..v) */
for (t = v − 1;  w > u;  t−−) {
   for (a = t→arcs, x = w;  a;  a = a→next)
      if (a→tip ≥ u ∧ a→tip < x)  x = a→tip, b = a;
   if (x < w)  b→tip = w, w = x;
         /* [x..t) is the longest interval from t */
}
```

This code is used in section 28.

32. The dénouement. Now we build a binary tree in the original graph F, by filling in some of the utility fields of F's vertices. If a node in the tree is labeled with x and with the interval $[u\,..\,v)$, we represent it by $x\to left = u$ and $x\to right = v$; the subtrees of this node are $x\to llink$ and $x\to rlink$. The root of the whole tree is $F\to root$.

The *rlink* field happens to be the same as the *count* field, but this overloading is no problem because the *rlink* is never changed or examined until after the *count* has been reset to zero for the last time.

#define *left* $x.V$ /* left endpoint of interval labeling this node */
#define *right* $w.V$ /* right endpoint of interval labeling this node */
#define *llink* $v.V$ /* left subtree of this node */
#define *rlink* $z.V$ /* right subtree of this node */
#define *root* $uu.V$ /* root node of the binary tree for this graph */

⟨ Construct an irredundant subfamily of F with the cardinality of G 32 ⟩ ≡
$F\to root = make\_tree(F\to vertices, F\to vertices + F\to n - 1)$;

This code is used in section 23.

---

a: **register Arc** ∗, §23.
arcs: **struct arc_struct** ∗, GB_GRAPH §9.
b: **register Arc** ∗, §23.
c: **register Arc** ∗, §23.
count = macro, §28.
F: **Graph** ∗, §23.
G: **Graph** ∗, §23.
make_tree: **Vertex** ∗( ), §33.
n: **long**, GB_GRAPH §20.
next: **struct arc_struct** ∗, GB_GRAPH §10.
t: **register Vertex** ∗, §23.
tip: **struct vertex_struct** ∗, GB_GRAPH §10.
u: **register Vertex** ∗, §23.
uu: **util**, GB_GRAPH §20.
v: **register Vertex** ∗, §23.
v: **util**, GB_GRAPH §9.
V: **struct vertex_struct** ∗, GB_GRAPH §8.
vertices: **Vertex** ∗, GB_GRAPH §20.
w: **register Vertex** ∗, §23.
w: **util**, GB_GRAPH §9.
x: **register Vertex** ∗, §23.
x: **util**, GB_GRAPH §9.
z: **util**, GB_GRAPH §9.

33. With a little care we could maintain a stack within F itself, but it's easier to use recursion in C. Let's just hope the system programmers have given us a large enough runtime stack to work with.

This subroutine is based on the trick explained in §20.

```
⟨ Subroutines 33 ⟩ ≡
  Vertex *make_tree(t, w)
      Vertex *t, *w;
  {
    register Vertex *u, *v, *x;
    register Arc *a;
    ⟨ Find a vertex x with N_x F|[t..w) − N_x S|[t..w) = 1 34 ⟩;
    if (¬x) return Λ;      /* F|[t..w) is empty */
    ⟨ Find an interval [u..v) such that x ∈ [u..v) ⊆ [t..w) 35 ⟩;
    x→left = u;
    x→right = v;
    x→llink = make_tree(t, x);
    x→rlink = make_tree(x + 1, w);
    return x;
  }
```

See also section 37.

This code is used in section 23.

34. At this point all *count* fields of F are 0 and all *link* fields are Λ.

A subtle bug is avoided here when we realize that a vertex might already be in the cleanup list when its *count* is zero.

```
⟨ Find a vertex x with N_x F|[t..w) − N_x S|[t..w) = 1 34 ⟩ ≡
  {
    register int coverage = 0;     /* coverage in F minus S */
    Vertex *cleanup = w + 1;       /* w + 1 is a sentinel value */
    for (v = w, x = Λ; v > t; v−−) {
      coverage −= v→count;      /* now coverage refers to N_{v−1} */
      for (a = v→arcs; a; a = a→next) {
        u = a→tip;
        if (u ≥ t) {
          if (u→link ≡ Λ) u→link = cleanup, cleanup = u;
          u→count += a→len;     /* the length is +1 for F, −1 for S */
          coverage += a→len;
        }
      }
      if (coverage ≡ 1) {
        x = v − 1; break;
      }
    }
```

```
    if (¬x ∧ cleanup ≤ w) fprintf(stderr, "This␣can't␣happen!\n");
    while (cleanup ≤ w) {
      v = cleanup→link;
      cleanup→count = 0;
      cleanup→link = Λ;
      cleanup = v;
    }
  }
```

This code is used in section 33.

35. ⟨Find an interval $[u \mathinner{..} v)$ such that $x \in [u \mathinner{..} v) \subseteq [t \mathinner{..} w)$ 35⟩ ≡

```
  for (v = w; v > x; v−−) {
    for (a = v→arcs; a; a = a→next)
      if (a→len > 0) {
        u = a→tip;
        if (u ≤ x ∧ u ≥ t) goto done;
      }
  }
  done:
```

This code is used in section 33.

36. ⟨Print the results 36⟩ ≡

```
  printf("Minimum␣generating␣family:");
  for (v = G→vertices + 1; v < G→vertices + G→n; v++)
    for (a = v→arcs; a; a = a→next)
      printf("␣[%s..%s)", a→tip→name, v→name);
  printf("\nMaximum␣irredundant␣family:");
  postorder_print(F→root);
  printf("\n");
```

This code is used in section 23.

---

arcs: **struct arc_struct** ∗, GB_GRAPH §9.
count = macro, §28.
F: **Graph** ∗, §23.
fprintf: **int** ( ), <stdio.h>.
G: **Graph** ∗, §23.
left = macro, §32.
len: **long**, GB_GRAPH §10.
link = macro, §28.
llink = macro, §32.
n: **long**, GB_GRAPH §20.
name: **char** ∗, GB_GRAPH §9.
next: **struct arc_struct** ∗, GB_GRAPH §10.
postorder_print: **void** ( ), §37.
printf: **int** ( ), <stdio.h>.
right = macro, §32.
rlink = macro, §32.
root = macro, §32.
stderr: **FILE** ∗, <stdio.h>.
tip: **struct vertex_struct** ∗, GB_GRAPH §10.
vertices: **Vertex** ∗, GB_GRAPH §20.

37. There's just one subroutine to go. This is textbook stuff.

```
⟨Subroutines 33⟩ +≡
  void postorder_print(x)
      Vertex *x;
  {
    if (x) {
      postorder_print(x→llink);
      postorder_print(x→rlink);
      printf("␣%s[%s..%s)", x→name, x→left→name, x→right→name);
    }
  }
```

38. Comments and extensions. The program just presented incorporates several refinements to the implementation sketched by Franzblau and Kleitman in [3], and the author hopes that readers will enjoy finding them in the code. The Stanford GraphBase provides convenient data structures, by means of which it was possible to make the program short and sweet. However, most of the key ideas (except for the *make_tree* procedure) can be found in [3].

39. Lubiw [10] discovered that the algorithm of Franzblau and Kleitman can be generalized so that it finds optimum irredundant subfamilies and generating families in an appropriate sense when the points of the underlying line have been given arbitrary nonnegative *weights*. It should be interesting and instructive to extend the program above so that it handles this more general problem.

---

left = macro, §32.
llink = macro, §32.
make_tree: **Vertex** $*$(), §33.
name: **char** $*$, GB_GRAPH §9.
printf: **int** (), `<stdio.h>`.
right = macro, §32.
rlink = macro, §32.

40. The introductory remarks in §2 mention the recent breakthrough by Frank and Jordán [2], who showed (among many other things) that Győri's theorem can be extended to intervals on a circle as well as a line. Such a generalization was surprising because the size of a minimum generating set might be strictly larger than the size of a maximum irredundant subfamily of cyclic intervals. For example, the n intervals $[k \mathinner{.\,.} k+2)$ for $0 \le k < n$ on the ring of integers mod n are obviously redundant; if we leave out any one of them, the remaining $n-1$ intervals will cover all n points. However, these cyclic intervals cannot be generated by fewer than n subintervals: No $n-1$ subintervals of length 1 will do the job, and if $[k \mathinner{.\,.} k+2)$ is one of the generators the remaining $n-1$ intervals require $n-1$ further generators because they are irredundant.

Győri's minimax principle is restored, however, if we change the definition of irredundant families. We can say that F is irredundant if each $f \in F$ has a distinguished element $f_* \in f$, such that whenever f and f' are distinct sets of F we have either $f_* \notin f'$ or $f'_* \notin f$. If F is irredundant in this sense, and if G generates F, it is not difficult to prove that $|F| \le |G|$: There is a $g_f \in G$ for each $f \in F$, with the property that $f_* \in g_f \subseteq f$; our new definition guarantees that $g_f \ne g_{f'}$ when $f \ne f'$.

According to this new definition, the intervals $[k \mathinner{.\,.} k+d)$ for $0 \le k < n$, modulo n, are irredundant whenever $n > 2(d-1)$, because we can let $[k \mathinner{.\,.} k+d)_* = k$. Frank and Jordán showed that if F is any family of intervals modulo n with the property that each intersection $f \cap f'$ of two of its members is either empty or a single interval, then the size of F's smallest generating family is the size of its largest irredundant subfamily under the new definition.

41. For intervals on a line, Győri [4] had already observed that both definitions of irredundancy are equivalent. Suppose a system of representatives $f_* \in f$ is given for all f in some family F of intervals on a line, such that $f \ne f'$ implies $f_* \notin f'$ or $f'_* \notin f$. If we cannot arrange those intervals in a sequence $f^{(1)}, f^{(2)}, \ldots, f^{(n)}$ such that $f_*^{(j)} \notin f^{(1)} \cup \cdots \cup f^{(j-1)}$ for $1 < j \le n$, there must be some cycle of intervals such that $f_*^{(1)} \in f^{(2)}$, $f_*^{(2)} \in f^{(3)}$, $\ldots$, $f_*^{(m)} \in f^{(1)}$, where $m > 2$. Consider the shortest such cycle, and suppose

$$f_*^{(1)} = \min\bigl(f_*^{(1)}, f_*^{(2)}, \ldots, f_*^{(m)}\bigr).$$

We cannot have $f^{(k-1)} < f_*^{(k)}$ for $1 < k \le m$, because $f_*^{(m)} \in f^{(1)}$; let the index $k > 1$ be minimum such that $f^{(k-1)}$ is not strictly less than $f_*^{(k)}$. Then $f^{(k-1)}$ must be strictly greater than $f_*^{(k)}$, and we have

$$f_*^{(1)} < f_*^{(k)} < f_*^{(k-1)}.$$

There is some j with $1 < j < k$ and $f_*^{(j-1)} < f_*^{(k)} < f_*^{(j)}$; since $f_*^{(j-1)} \in f^{(j)}$ we have $f_*^{(k)} \in f^{(j)}$, a shorter cycle. This contradiction shows that no cycles exist.

42. Frank and Jordán gave another criterion for irredundancy that works also for general families of intervals on a circle when large intervals might wrap around so that their intersection $f \cap f'$ consists of two disjoint intervals. In such cases they allow $\{f_*, f'_*\} \subseteq f \cap f'$, but only if f_* and f'_* lie in different components of $f \cap f'$. For example, the intervals $[k \mathinner{..} k+d)$ for $0 \le k < n$, modulo n, are irredundant by this definition for all $n > d$. Once again the minimax theorem for generating families and irredundant subfamilies remains valid, in this extended sense.

43. The algorithms presented by Frank and Jordán [2] for such problems require linear programming as a subroutine. Therefore it would be extremely interesting to find a purely combinatorial procedure, analogous to the algorithm of Franzblau and Kleitman, either for the wrap-restricted situation of §40 or for the more general setup of §42.

44. References.

[1] Joseph C. Culberson and Robert A. Reckhow, "Covering polygons is hard," *Journal of Algorithms* **17** (1994), 2–44.

[2] András Frank and Tibor Jordán, "Minimal edge-coverings of pairs of sets," *Journal of Combinatorial Theory* **B65** (1995), 73–110.

[3] Deborah S. Franzblau and Daniel J. Kleitman, "An algorithm for constructing polygons with rectangles," *Information and Control* **63** (1984), 164–189.

[4] Ervin Győri, "A minimax theorem on intervals," *Journal of Combinatorial Theory* **B37** (1984), 1–9.

[5] Donald E. Knuth, *Fundamental Algorithms*, Volume 1 of *The Art of Computer Programming* (Reading, Massachusetts: Addison–Wesley, 1968).

[6] Donald E. Knuth, *Sorting and Searching*, Volume 3 of *The Art of Computer Programming* (Reading, Massachusetts: Addison–Wesley, 1973).

[7] Donald E. Knuth, *The Stanford GraphBase* (New York: ACM Press, 1994). Available from the server `ftp.cs.stanford.edu`, via anonymous ftp, in directory `pub/sgb`.

[8] Donald E. Knuth and P. B. Bendix, "Simple word problems in universal algebras," in *Computational Problems in Abstract Algebra*, Proceedings of a conference held at Oxford University in 1967, edited by John Leech (Oxford: Pergamon, 1970), 263–297. Reprinted in *Automation of Reasoning*, edited by Jörg H. Siekmann and Graham Wrightson, **2** (Springer, 1983), 342–376. [Reprinted with revisions as Chapter 19 of the present volume.]

[9] Donald E. Knuth and Silvio Levy, *The CWEB System of Structured Documentation* (Reading, Massachusetts: Addison–Wesley, 1994).

[10] Anna Lubiw, "A weighted min-max relation for intervals," *Journal of Combinatorial Theory* **B53** (1991), 151–172.

45. The author thanks the referees of this note for many valuable suggestions that greatly improved the presentation.

46. Index. The following list shows the section numbers where each identifier makes an appearance. Underlined numbers indicate a place of definition. Single-letter identifiers are indexed only when they are defined. Other indexable things, such as notational conventions and the names of people whose work is cited, also can be found here.

47. Names of the sections.

⟨ Clean up all *count* fields 30 ⟩ Used in section 28.
⟨ Compute G and S by the Franzblau–Kleitman algorithm 26 ⟩ Used in section 23.
⟨ Construct an irredundant subfamily of F with the cardinality of G 32 ⟩ Used in section 23.
⟨ Find a vertex x with $N_x\,F|[t\,..\,w) - N_x\,S|[t\,..\,w) = 1$ 34 ⟩ Used in section 33.
⟨ Find an interval $[u\,..\,v)$ such that $x \in [u\,..\,v) \subseteq [t\,..\,w)$ 35 ⟩ Used in section 33.
⟨ Make G a copy of F, retaining only leftward arcs 27 ⟩ Used in section 26.
⟨ Print the results 36 ⟩ Used in section 23.
⟨ Reduce all minimal bad intervals with right endpoint v and record them in S 28 ⟩ Used in section 26.
⟨ Replace G by $G{\downarrow}[u\,..\,v)$ 31 ⟩ Used in section 28.
⟨ Scan the command-line options and generate F 24 ⟩ Used in section 23.
⟨ Set F to the mirror image of G 25 ⟩ Used in section 24.
⟨ Subroutines 33, 37 ⟩ Used in section 23.
⟨ Update the counts for all intervals ending at u 29 ⟩ Used in section 28.

Chapter 19

Simple Word Problems in Universal Algebras

[Written with Peter B. Bendix. Originally published in Computational Problems in Abstract Algebra, the proceedings of a conference held at the Mathematical Institute of Oxford University between 29 August and 2 September 1967, edited by John Leech (Oxford: Pergamon Press, 1970), 263–297.]

An algorithm is described that is capable of solving certain word problems: It sometimes decides whether or not two words, composed of variables and operators, can be proved equal as a consequence of a given set of identities satisfied by the operators. Although the general word problem is well known to be unsolvable, this algorithm provides results in many interesting cases. For example, if we are given the binary operator $\cdot$, *the unary operator* $^-$, *the nullary operator* e, *and the three identities* $a \cdot (b \cdot c) = (a \cdot b) \cdot c$, $a \cdot a^- = e$, $a \cdot e = a$, *the algorithm is able to deduce the laws* $a^- \cdot a = e$, $e \cdot a = a$, $a^{--} = a$, *etc. Furthermore it can show that an equation such as* $a \cdot b = b \cdot a^-$ *is* not *a consequence of the given axioms.*

The method is based on a well-ordering of the set of all words, such that each identity can be construed as a "reduction" in the sense that the right-hand side of the identity represents a word smaller in the ordering than the left-hand side. A set of reduction identities is said to be "complete" when two words are equal as a consequence of the identities if and only if they reduce to the same word by a series of reductions. The method used in this algorithm is essentially to test whether a given set of identities is complete; if it is not complete, the algorithm often finds a new consequence of the identities, which can be added to the set. The process is repeated until either a complete set is achieved or until an anomalous situation occurs that cannot at present be handled.

Results of several computational experiments are discussed.

Introduction

The purpose of this paper is to examine a general technique for solving certain algebraic problems that are traditionally treated in an ad hoc, trial-and-error manner. The technique is precise enough that it can be done by computer, but it is also simple enough that it is useful for hand calculation, as an aid to working with unfamiliar types of algebraic axioms.

Given a set of operators and some identities satisfied by those operators, the general problem treated here is to examine the consequences of the given identities; that is, we want to determine when two formulas are equal because of the identities. The general approach suggested here may be described in very informal terms as follows: We regard an identity of the form $\alpha = \beta$ as a "reduction," where one side of the identity, say β, is considered to be "simpler" than the other side α, and we agree to simplify any formula having the form of α to the form of β. For example, the axiom $a^{-1}(ab) = b$ can be considered as a reduction rule in which we are to replace any formula of the form $a^{-1}(ab)$ by b. (The associative law for multiplication is not necessarily being assumed here.) We shall demonstrate below that the most fruitful way to obtain new consequences of reductions is to take pairs of reductions $\alpha_1 = \beta_1$, $\alpha_2 = \beta_2$ and to find a formula that has the form of α_1, in which one of the subformulas corresponding to an operator of α_1 also has the form of α_2. If the latter subformula is replaced by β_2 and the resulting formula is equated to β_1, a useful new identity often results. For example, let $\alpha_1 = \alpha_2 = a^{-1}(ab)$, and let $\beta_1 = \beta_2 = b$; then the formula $(x^{-1})^{-1}(x^{-1}(xy))$ has the form of α_1, while its subformula $(x^{-1}(xy))$ corresponding to the multiplication of a by b in α_1 has the form of α_2. Therefore we can equate $(x^{-1})^{-1}(x^{-1}(xy))$ both to xy and to $(x^{-1})^{-1}y$.

The general procedure that has been described so vaguely in the preceding paragraph is formalized rigorously in Sections 1–6 of this paper. Section 7 presents more than a dozen examples of how the method has given successful results for many different axiom systems of interest. The success of this technique seems to indicate that it might be desirable to teach its general principles to students in introductory algebra courses.

The formal development in Sections 1–6 below is primarily a precise statement of what hundreds of mathematicians have been doing for many decades. Thus, no great claims of originality are intended for most of the concepts or methods used. But the overall viewpoint of this paper appears to be novel, and so it seems desirable to present here a self-contained treatment of the underlying theory. The main new contribution of this paper is intended to be an extension of some methods used by Trevor Evans [4]: We allow operators to be of arbitrary degree, and we make use of a well-ordering of words that allows us to deal with axioms such as the associative law. Furthermore some of the techniques and results of the examples in Section 7 appear to be of independent interest.

1. Words

In the following sections we will deal with four fixed sequences of quantities:

a) An infinite sequence of *variables* v_1, v_2, v_3, ..., which are distinguishable symbols from an infinite alphabet.

b) A finite sequence of *operators* f_1, f_2, ..., f_N, which are distinguishable symbols from another alphabet, disjoint from the variables.

c) A finite sequence of *degrees* d_1, d_2, ..., d_N, which are nonnegative integers. We say d_j is the degree of operator f_j.

d) A finite sequence of *weights* w_1, w_2, ..., w_N, which are nonnegative integers. We say w_j is the weight of operator f_j.

An operator whose degree is 0, 1, 2, 3, ... will be called a nullary, unary, binary, ternary, ... operator, respectively. Nullary operators take the place in this discussion of what are traditionally called "constants" or "generators." We will assume that there is at least one nullary operator.

Two special conditions are placed on the sequences defined above:

1) *Each nullary operator has positive weight.* Thus if $d_j = 0$, $w_j > 0$.

2) *Each unary operator has positive weight, with the possible exception of* f_N. Thus if $d_j = 1$ and $j < N$ we must have $w_j > 0$.

The reason for these two restrictions will become clear in the proof of Theorem 1.

Certain sequences of variable and operator symbols are called *words* ("well-formed formulas"), which are defined inductively as follows: A variable v_j standing alone is a word; and

$$f_j \alpha_1 \ldots \alpha_d \tag{1.1}$$

is a word if α_1, ..., α_d are words and $d = d_j$. Notice that if f_j is a nullary operator, the symbol f_j standing alone is a word.

The *subwords* of a word α are defined to be (i) the entire word α itself, and (ii) the subwords of $\alpha_1, \ldots, \alpha_d$ if α has the form (1.1). Clearly the number of subwords of α is the number of symbols in α, and in fact each symbol of α is the initial symbol of a unique subword. Furthermore, assuming that α and β are words, β is a subword of α if and only if β is a substring of α, that is, $\alpha = \varphi\beta\psi$ for some strings of symbols φ and ψ.

Let us say a *nontrivial subword* is a subword that contains at least one operator symbol; thus it's a subword that doesn't simply have the trivial form "v_j" for some variable v_j. The number of nontrivial subwords of a word α is clearly the number of operator symbols in α.

This definition of words and subwords is, of course, just one of many ways to define what is essentially an "ordered tree structure," and we may make use of the well-known properties of tree structure.

Let us write $n(x, \alpha)$ for the number of occurrences of the symbol x in the word α. A *pure word* α is one containing no variables at all; thus α is pure if $n(v_j, \alpha) = 0$ for all j. The *weight* of a pure word is

$$w(\alpha) = \sum_j w_j n(f_j, \alpha), \tag{1.2}$$

the sum of the weights of its individual symbols. Pure words always exist, because we are assuming that there is at least one nullary operator. Since every nullary operator has positive weight, every pure word has positive weight.

The set of all pure words can be ordered by the following relation: $\alpha > \beta$ if and only if either

1) $w(\alpha) > w(\beta)$; or
2) $w(\alpha) = w(\beta)$ and $\alpha = f_j\alpha_1 \ldots \alpha_{d_j}$, $\beta = f_k\beta_1 \ldots \beta_{d_k}$, and either
   2a) $j > k$; or
   2b) $j = k$ and we have $\alpha_1 = \beta_1, \ldots, \alpha_{t-1} = \beta_{t-1}$, and $\alpha_t > \beta_t$, for some t with $1 \le t \le d_j$.

It is not difficult to design an algorithm that will decide whether or not $\alpha > \beta$, given two pure words α and β; details will be omitted here.

Theorem 1. *The relation "$>$" defines a well-ordering on the set of all pure words.*

Proof. First it is necessary to prove that $\alpha > \beta > \gamma$ implies $\alpha > \gamma$, and that for any pure words α and β we have exactly one of the three possibilities $\alpha > \beta$, $\alpha = \beta$, $\alpha < \beta$. These properties are readily verified by a somewhat tedious case analysis, so it is clear that "$>$" is at least a linear ordering.

We must also prove that there is no infinite sequence of pure words with

$$\alpha_1 > \alpha_2 > \alpha_3 > \cdots. \tag{1.3}$$

Since the words are ordered first on weight, we need only show that there is no infinite sequence (1.3) of pure words *having the same weight* w.

Suppose α is a pure word of weight w with n_j occurrences of symbols that have degree d_j. It is easy to prove inductively that the length of α is

$$n_0 + n_1 + n_2 + \cdots \;=\; 1 + 0 \cdot n_0 + 1 \cdot n_1 + 2 \cdot n_2 + \cdots;$$

thus $n_0 = 1 + n_2 + 2n_3 + \cdots$. Since each nullary operator has positive weight, we have $w \geq n_0$; so there are only a finite number of choices for n_0, n_2, n_3, $\ldots$, if we are to have a word of weight w. Furthermore if each unary operator has positive weight, we would have $w \geq n_1$ and there would be only finitely many pure words of weight w. Therefore (1.3) is impossible unless f_N is a unary operator of weight zero.

Suppose $w_N = 0$, $d_N = 1$, and define the function $h(\alpha)$ to be the word obtained from α by erasing all occurrences of f_N. Clearly $h(\alpha)$ has the same weight as α. And by the argument in the preceding paragraph, only finitely many words $h(\alpha)$ exist of weight w. To complete the proof of the theorem, we will show that there is no infinite sequence (1.3) with $h(\alpha_1) = h(\alpha_2) = h(\alpha_3) = \cdots$.

Let $h(\alpha) = s_1 s_2 \ldots s_n$; then α has the form $f_N^{r_1} s_1 f_N^{r_2} s_2 \ldots f_N^{r_n} s_n$, where r_1, $\ldots$, r_n are nonnegative integers. Define $r(\alpha) = (r_1, \ldots, r_n)$, an n-tuple of nonnegative integers. It is now easy to verify that, if $h(\alpha) = h(\beta)$, we have $\alpha > \beta$ if and only if the relation $r(\alpha) > r(\beta)$ holds in lexicographic order. Since it is well known that lexicographic order is a well-ordering, the proof of Theorem 1 is complete. □

Notice that if f_j were a unary operator of weight zero and $j < N$, we would not have a well-ordering, since there would be a sequence of pure words of the form $f_N\alpha > f_j f_N \alpha > f_j f_j f_N \alpha > \cdots$. And if we had nullary operators of weight zero, other counterexamples would arise; for example if f_1 is nullary and f_2 is binary, both of weight zero, then

$$f_2 f_2 f_1 f_1 f_1 > f_2 f_1 f_2 f_2 f_1 f_1 f_1 > f_2 f_1 f_2 f_1 f_2 f_2 f_1 f_1 f_1 > \cdots.$$

Such examples account for the restrictions that we have imposed on the degrees and the weights.

2. Substitutions

Most of §1 was concerned with pure words, and it is now time to consider the variables that can enter. If α is a string of symbols containing variables, we let $v(\alpha)$ be the largest subscript of any variable that occurs. If α involves no variables, we let $v(\alpha) = 0$.

Suppose θ_1, θ_2, ..., θ_n, and α are strings of symbols, with $n \geq v(\alpha)$. We will write

$$S(\theta_1, \theta_2, \ldots, \theta_n; \alpha) \tag{2.1}$$

for the string obtained from α by substituting θ_j for each occurrence of v_j, $1 \leq j \leq n$. For example if $v(\alpha) = 2$, $S(v_2, v_1; \alpha)$ is obtained from α by interchanging the variables v_1 and v_2.

We say a word β *has the form of* a word α if β can be obtained by substitution from α, that is, if there exist words θ_1, θ_2, ..., θ_n such that $\beta = S(\theta_1, \theta_2, \ldots, \theta_n; \alpha)$.

It is not difficult to prove that two substitutions can always be replaced by one, in the sense that

$$\begin{aligned} &S(\varphi_1, \ldots, \varphi_m; S(\theta_1, \ldots, \theta_n; \alpha)) = \\ &\qquad S(S(\varphi_1, \ldots, \varphi_m; \theta_1), \ldots, S(\varphi_1, \ldots, \varphi_m; \theta_n); \alpha). \end{aligned} \tag{2.2}$$

So if γ has the form of β and β has the form of α, the word γ also has the form of α.

It is comparatively easy to design an algorithm that decides whether or not β has the form of α, given two words β and α. Briefly, let $\alpha = y_1 y_2 \ldots y_m$, where each y_j is a variable or an operator. Then β must have the form $\beta = \beta_1 \beta_2 \ldots \beta_m$ where, if y_j is an operator, $y_j = \beta_j$; and if $y_j = y_k$ is a variable, then $\beta_j = \beta_k$ is a word, for $1 \leq j \leq k \leq m$.

Let w_0 be the minimum weight of a pure word; thus w_0 is the minimum weight of a nullary operator. We define the weight $w(\alpha)$ of an arbitrary word to be the minimum weight of all pure words that have the form of α:

$$w(\alpha) = w_0 \sum_{j \geq 1} n(v_j, \alpha) + \sum_{j \geq 1} w_j n(f_j, \alpha). \tag{2.3}$$

We now extend the ">" relation, which was defined only for pure words in §1, to words involving variables. Let us say that $\alpha > \beta$ if and only if one of the following three conditions is satisfied:

1) $w(\alpha) > w(\beta)$ and $n(v_i, \alpha) \geq n(v_i, \beta)$ for all $i \geq 1$;

2) $w(\alpha) = w(\beta)$ and $n(v_i, \alpha) = n(v_i, \beta)$ for all $i \geq 1$, where we have $\alpha = f_j \alpha_1 \ldots \alpha_{d_j}$ and $\beta = f_k \beta_1 \ldots \beta_{d_k}$ and either
   2a) $j > k$; or
   2b) $j = k$, $\alpha_1 = \beta_1$, …, $\alpha_{t-1} = \beta_{t-1}$, and $\alpha_t > \beta_t$ for some t with $1 \leq t \leq d_j$.
3) $w(\alpha) = w(\beta)$ and $n(v_i, \alpha) = n(v_i, \beta)$ for all $i \geq 1$ and $\alpha \neq \beta = v_k$.

Case (3) can occur only if α has the form $f_N \ldots f_N v_k$, with at least one occurrence of f_N.

It is not difficult to design a relatively simple algorithm that determines, given words α and β, whether $\alpha < \beta$, or $\alpha = \beta$, or $\alpha > \beta$, or $\alpha \# \beta$, where the latter condition means that α and β are unrelated. When α and β are pure words, the situation $\alpha \# \beta$ is impossible; but when variables are involved, we can have unrelated words such as

$$f_5 v_1 v_1 v_1 v_1 v_2 \# f_3 v_2 v_2 v_1, \tag{2.4}$$

$$f_2 v_1 v_2 \# f_2 v_2 v_1, \tag{2.5}$$

where f_2, f_3, and f_5 are operators of degrees 2, 3, and 5, respectively.

The principal motivation for the given definition of $\alpha > \beta$ is the following fact:

Theorem 2. *If $\alpha > \beta$ then $S(\theta_1, \theta_2, \ldots, \theta_n; \alpha) > S(\theta_1, \theta_2, \ldots, \theta_n; \beta)$, for all words θ_1, …, θ_n.*

Proof. Let $\alpha' = S(\theta_1, \theta_2, \ldots, \theta_n; \alpha)$ and $\beta' = S(\theta_1, \theta_2, \ldots, \theta_n; \beta)$. If condition (1) holds for α and β, then it must hold also for α' and β'. For in the first place, every word has weight $\geq w_0$, so

$$\begin{aligned} w(\alpha') &= w(\alpha) + \sum_{j \geq 1} n(v_j, \alpha)(w(\theta_j) - w_0) \\ &> w(\beta) + \sum_{j \geq 1} n(v_j, \beta)(w(\theta_j) - w_0) = w(\beta'). \end{aligned}$$

Secondly, $n(v_i, \alpha') = \sum_{j \geq 1} n(v_j, \alpha) n(v_i, \theta_j) \geq \sum_{j \geq 1} n(v_j, \beta) n(v_i, \theta_j) = n(v_i, \beta')$.

If condition (2) holds for α and β, then similarly we find $w(\alpha') = w(\beta')$ and $n(v_i, \alpha') = n(v_i, \beta')$ for all i, and $\alpha' = f_j \alpha'_1 \ldots \alpha'_{d_j}$, $\beta' = f_k \beta'_1 \ldots \beta'_{d_k}$ where $\alpha'_r = S(\theta_1, \ldots, \theta_n; \alpha_r)$ and $\beta'_r = S(\theta_1, \ldots, \theta_n; \beta_r)$ for all r. Hence we have either $j > k$, or an inductive argument based on the length of α will complete the proof.

Finally if condition (3) holds for α and β, we have $\alpha' = f_N^t \theta_k$ and $\beta' = \theta_k$ for some $t \geq 1$. Then $\alpha' > \beta'$ because α' begins with more occurrences of f_N than β' does. □

Corollary. *There is no infinite sequence of words such that* $\alpha_1 > \alpha_2 > \alpha_3 > \cdots$.

Proof. If there were such a sequence, we could substitute a nullary operator f for each variable v_j. But this would give an infinite descending sequence of pure words, by Theorem 2, contradicting Theorem 1. □

Theorem 2 and its corollary are of great importance for the method that will be explained in detail below. Indeed, the fact that $\alpha \# \beta$ can occur for certain words α and β is a serious restriction on the present applicability of the method. The authors believe that further theory can be developed to ameliorate this restriction, but such research will have to be left for later investigations.

It may seem curious that $f_5v_1v_1v_1v_1v_2 \# f_3v_2v_2v_1$; surely the word $f_5v_1v_1v_1v_1v_2$ appears to be much "bigger" than $f_3v_2v_2v_1$. But if we substitute a short formula for v_1 and a long formula for v_2, we will find that $f_3v_2v_2v_1$ is actually longer than $f_5v_1v_1v_1v_1v_2$.

Theorem 2 is not quite the best possible result of its kind: There are words α and β for which we have $\alpha \# \beta$ yet $S(\theta_1, \theta_2, \ldots, \theta_n; \alpha) > S(\theta_1, \theta_2, \ldots, \theta_n; \beta)$ for all pure words $\theta_1, \ldots, \theta_n$. For example, consider

$$f_3v_1 \# f_2f_1 \tag{2.6}$$

where f_3 and f_2 are unary operators and f_1 is a nullary operator, all of weight 1. If we substitute for v_1 a pure word θ of weight 1, we have $f_3\theta > f_2f_1$ by case (2a); but if we substitute for v_1 any word θ of weight greater than one, we get $f_3\theta > f_2f_1$ by case (1). We could therefore have made the methods of this paper slightly more powerful if we had been able to define $f_3v_1 > f_2f_1$; but an effort to extend Theorem 2 along these lines appears to lead to such a complicated definition of the relation $\alpha > \beta$ that the comparatively simple definition given here is preferable. No situation such as (2.6) has occurred so far in practice.

Let α and β be words with $v(\alpha) \le n$ and $v(\beta) \le n$. In the following discussion we will be interested in the general solution of the equation

$$S(\theta_1, \ldots, \theta_n; \alpha) = S(\theta_1, \ldots, \theta_n; \beta) \tag{2.7}$$

in words $\theta_1, \ldots, \theta_n$. Such an equation can always be treated in a reasonably simple manner:

Theorem 3. *If equation* (2.7) *has at least one solution, there is a number k with $0 \le k \le n$ and there are words $\sigma_1, \ldots, \sigma_n$, with $v(\sigma_j) \le k$ for $1 \le j \le n$ and*

$$\{v_1, v_2, \ldots, v_k\} \subseteq \{\sigma_1, \ldots, \sigma_n\}, \tag{2.8}$$

such that all solutions of (2.7) *have the form*

$$\theta_j = S(\varphi_1, \ldots, \varphi_k; \sigma_j), \quad 1 \le j \le n. \tag{2.9}$$

Moreover, there is an algorithm that either finds σ_1, ..., σ_n *or determines that* (2.7) *is unsolvable.*

(This theorem provides the general solution of (2.7). The significance of relation (2.8) is that the simple words v_1, v_2, ..., v_k are included among the σ's; thus some k of the θ's may be selected arbitrarily and the other $n-k$ θ's must have a specified relationship to these k "independent" variables. An example appears in (2.12) below. This result is equivalent to the "unification theorem" of J. A. Robinson [10].)

Proof. Theorem 3 can be proved by induction on n, and for fixed n by induction on the length of $\alpha\beta$, as follows.

Case 1, $\alpha = v_p$ and $\beta = v_q$. If $p = q$, then obviously any words σ_1, ..., σ_n will satisfy (2.7), so we may take $k = n$, $\sigma_1 = v_1$, ..., $\sigma_n = v_n$. If $p < q$, the general solution is clearly obtained by taking $k = n - 1$,

$$\sigma_1 = v_1, \ \ldots, \ \sigma_{q-1} = v_{q-1}, \ \sigma_q = v_p, \ \sigma_{q+1} = v_q, \ \ldots, \ \sigma_n = v_{n-1}.$$

Case 2, $\alpha = f_p\alpha_1 \ldots \alpha_d$ and $\beta = v_q$. If the variable v_q appears in α, the equation (2.7) has no solution, since the length of $S(\theta_1, \ldots, \theta_n; \alpha)$ is greater than the length of $\theta_q = S(\theta_1, \ldots, \theta_n; \beta)$. On the other hand if v_q does not appear in α we clearly have $k = n - 1$, $\sigma_1 = v_1$, ..., $\sigma_{q-1} = v_{q-1}$, $\sigma_q = S(v_1, \ldots, v_{q-1}, v_{q-1}, \ldots, v_{n-1}; \alpha)$, $\sigma_{q+1} = v_q$, ..., $\sigma_n = v_{n-1}$ as the general solution.

Case 3, $\alpha = v_p$ and $\beta = f_q\beta_1 \ldots \beta_d$. This is case 2 with α and β interchanged.

Case 4, $\alpha = f_p\alpha_1 \ldots \alpha_d$ and $\beta = f_q\beta_1 \ldots \beta_{d'}$. Here there is no solution of (2.7) unless $p = q$, so we may assume that $p = q$ and $d = d'$. Now (2.7) is equivalent to the system of d simultaneous equations

$$S(\theta_1, \ldots, \theta_n; \alpha_j) = S(\theta_1, \ldots, \theta_n; \beta_j) \tag{2.10}$$

for $1 \le j \le d$. If $d = 0$, the general solution is of course to take $k = n$, $\sigma_1 = v_1$, ..., $\sigma_n = v_n$. Suppose we have obtained the general solution of the system (2.10) for $1 \le j \le r$, where $0 \le r < d$; we will show how to extend this to a solution of (2.10) for $1 \le j \le r + 1$: If (2.10), for $1 \le j \le r$, has no solution, then (2.10) certainly has no solution for $1 \le j \le r + 1$. Otherwise let the general solution to (2.10), for

$1 \le j \le r$, be given by k, σ_1, ..., σ_n. Now the general solution to (2.10) for $1 \le j \le r+1$ is obtained by setting $\theta_j = S(\varphi_1, \ldots, \varphi_k; \sigma_j)$ for $1 \le j \le n$ and $S(\theta_1, \ldots, \theta_n; \alpha_{r+1}) = S(\theta_1, \ldots, \theta_n; \beta_{r+1})$. By (2.2) this requires solving

$$S(\varphi_1, \ldots, \varphi_k; S(\sigma_1, \ldots, \sigma_n; \alpha_{r+1})) = S(\varphi_1, \ldots, \varphi_k; S(\sigma_1, \ldots, \sigma_n; \beta_{r+1})). \tag{2.11}$$

The general solution of this equation can be obtained by induction, if $k < n$. And induction also applies if $k = n$, since $\{\sigma_1, \ldots, \sigma_n\} = \{v_1, \ldots, v_n\}$ in that case and $S(\sigma_1, \ldots, \sigma_n; \alpha_{r+1}) S(\sigma_1, \ldots, \sigma_n; \beta_{r+1})$ is shorter than $\alpha\beta$. If (2.11) has the general solution k', σ'_1, ..., σ'_k, then (2.10) for $1 \le j \le r+1$ has the general solution k', $S(\sigma'_1, \ldots, \sigma'_k; \sigma_1)$, ..., $S(\sigma'_1, \ldots, \sigma'_k; \sigma_n)$. The latter strings include $\{v_1, \ldots, v_{k'}\}$, since $\{\sigma'_1, \ldots, \sigma'_k\} \supseteq \{v_1, \ldots, v_{k'}\}$ and $\{\sigma_1, \ldots, \sigma_n\} \supseteq \{v_1, \ldots, v_k\}$. This inductive process ultimately allows us to solve (2.10) for $1 \le j \le d$, as required.

We have now proved that a general solution (2.8), (2.9) to the equation (2.7) can always be obtained, and it is evident that our proof is equivalent to a recursive algorithm for obtaining the solution. □

As an example of the process used in the proof of Theorem 3, let $n = 7$, $d_1 = 1$, $d_2 = 2$, and

$$\begin{aligned} \alpha &= f_2 f_1 f_2 f_1 v_4 f_2 v_3 f_1 f_2 v_2 v_2 f_2 v_1 f_2 v_3 f_1 v_1, \\ \beta &= f_2 f_1 f_2 v_5 f_2 v_5 v_6 f_2 v_7 f_2 f_1 v_6 f_1 f_2 v_5 v_6. \end{aligned} \tag{2.12}$$

We wish to determine what formulas can be obtained by a common substitution in α and β, which is essentially saying that we want to solve the equation $\alpha = \beta$ for v_1, ..., v_7. This problem reduces, first, to solving the simultaneous equations

$$f_1 f_2 f_1 v_4 f_2 v_3 f_1 f_2 v_2 v_2 = f_1 f_2 v_5 f_2 v_5 v_6, \tag{2.13}$$

$$f_2 v_1 f_2 v_3 f_1 v_1 = f_2 v_7 f_2 f_1 v_6 f_1 f_2 v_5 v_6. \tag{2.14}$$

To solve (2.13), we first remove the common f_1 at the left, then solve the system $f_1 v_4 = v_5$, $f_2 v_3 f_1 f_2 v_2 v_2 = f_2 v_5 v_6$, etc., and we ultimately obtain the conditions

$$v_3 = v_5 = f_1 v_4, \quad v_6 = f_1 f_2 v_2 v_2. \tag{2.15}$$

Substituting these into (2.14) gives the equation

$$f_2 v_1 f_2 f_1 v_4 f_1 v_1 = f_2 v_7 f_2 f_1 f_1 f_2 v_2 v_2 f_1 f_2 f_1 v_4 f_1 f_2 v_2 v_2;$$

and to make a long story short this equation in the variables v_1, v_2, v_4, v_7 ultimately implies that

$$v_4 = f_1 f_2 v_2 v_2, \quad v_1 = v_7 = f_2 f_1 f_1 f_2 v_2 v_2 f_1 f_2 v_2 v_2.$$

Finally, in connection with (2.15), we have found that every word obtainable by a common substitution of words into α and β is obtained by substituting some word for v_2 in the long word

$$\begin{aligned}&f_2 f_1 f_2 f_1 f_1 f_2 v_2 v_2 f_2 f_1 f_1 f_2 v_2 v_2 f_1 f_2 v_2 v_2\\&\quad f_2 f_2 f_1 f_1 f_2 v_2 v_2 f_1 f_2 v_2 v_2 f_2 f_1 f_1 f_2 v_2 v_2 f_1 f_2 f_1 f_1 f_2 v_2 v_2 f_1 f_2 v_2 v_2.\end{aligned}$$

Stating this in the more formal language of Theorem 3 and its proof, the general solution to (2.7) and (2.12) is given by

$$\begin{aligned}k = 1, \quad \sigma_1 = \sigma_7 = f_2 f_1 f_1 f_2 v_1 v_1 f_1 f_2 v_1 v_1, \quad \sigma_2 = v_1,\\ \sigma_3 = \sigma_5 = f_1 f_1 f_2 v_1 v_1, \quad \sigma_4 = \sigma_6 = f_1 f_2 v_1 v_1.\end{aligned}$$

3. The Word Problem

Given a set $R = \{(\lambda_1, \varrho_1), \ldots, (\lambda_m, \varrho_m)\}$ of pairs of words, called "relations," we can define a corresponding equivalence relation (in fact, a congruence relation) between words in a natural manner, by regarding the relations as "axioms,"

$$\lambda_k \equiv \varrho_k \pmod{R}, \quad 1 \le k \le m, \tag{3.1}$$

where the variables range over the set of all words. This relation "$\equiv$" is to be extended to the smallest congruence relation containing (3.1).

For our purposes it is most convenient to define the congruence relations in the following more precise manner: Let β be a subword of α, so that α has the form $\varphi\beta\psi$ for some strings φ, ψ. Assume that there is a relation (λ, ϱ) in R such that β has the form of λ; in other words, $\beta = S(\theta_1, \ldots, \theta_n; \lambda)$ for some $\theta_1, \ldots, \theta_n$ where $n \ge v(\lambda)$ and $n \ge v(\varrho)$. Let $\beta' = S(\theta_1, \ldots, \theta_n; \varrho)$, so that β and β' are obtained from λ and ϱ by means of the same substitutions. Let $\alpha' = \varphi\beta'\psi$ be the word α with its component β replaced by β'. Then we say that α *reduces* to α' with respect to R, and we write

$$\alpha \to \alpha' \pmod{R}. \tag{3.2}$$

Finally, we say that

$$\alpha \equiv \beta \pmod{R} \tag{3.3}$$

if there is a sequence of words $\alpha_0, \alpha_1, \ldots, \alpha_n$ for some $n \geq 0$ such that $\alpha = \alpha_0$, $\alpha_n = \beta$, and for $0 \leq j < n$ we have either $\alpha_j \to \alpha_{j+1} \pmod{R}$ or $\alpha_{j+1} \to \alpha_j \pmod{R}$. (Note: When the set R is understood from the context, the "(mod R)" may be omitted from notations (3.2) and (3.3).)

The *word problem* is the problem of deciding whether or not $\alpha \equiv \beta \pmod{R}$, given two words α and β and a set of relations R. Although the word problem is known to be quite difficult (indeed, unsolvable) in general, we present here a method for solving certain word problems that are general enough to be of wide interest.

The principal restriction is that we require all of the relations to be comparable in the sense of §2: We require that

$$\lambda > \varrho \tag{3.4}$$

for each relation in R. In such a case we say that R is a set of *reductions*. It follows from Theorem 2 that

$$\alpha \to \alpha' \pmod{R} \quad \text{implies} \quad \alpha > \alpha'. \tag{3.5}$$

4. The Completeness Theorem

Let R be a set of reductions. We say that a word α is *irreducible* with respect to R if there is no α' such that $\alpha \to \alpha' \pmod{R}$.

It is not difficult to design an algorithm that determines whether or not a given word is irreducible with respect to R. If $R = \{(\lambda_1, \varrho_1), \ldots, (\lambda_m, \varrho_m)\}$, we simply must verify that no subword of α has the form of λ_1, or λ_2, ..., or λ_m.

If α is reducible with respect to R, the algorithm just outlined can be extended so that it finds some α' for which $\alpha \to \alpha'$. Now the same procedure can be applied to α', and if α' is reducible we can find a further word α'', and so on. We have $\alpha \to \alpha' \to \alpha'' \to \cdots$; so by (3.5) and the corollary to Theorem 2, this process eventually terminates.

Thus, *there is an algorithm that, given any word α and any set of reductions R, finds an irreducible word α_0 such that $\alpha \equiv \alpha_0$ with respect to R.*

It would be very pleasant if we could also show that each word is equivalent to *at most* one irreducible word; for then the algorithm above solves the word problem! Take any two words α and β, and use the given algorithm to find irreducible α_0 and β_0. If $\alpha \equiv \beta$ then $\alpha_0 \equiv \beta_0$; so by hypothesis α_0 must be equal to β_0. If $\alpha \not\equiv \beta$, then $\alpha_0 \not\equiv \beta_0$; so α_0 must be unequal to β_0. In effect, α_0 and β_0 are canonical representatives of the equivalence classes.

This pleasant state of affairs is of course not true for every set of reductions R. But we will see that it is true for surprisingly many sets, and therefore it is an important property worthy of a special name. Let us say that R is a *complete* set of reductions if no two distinct irreducible words are equivalent, with respect to R. We will show in the next section that there is an algorithm to determine whether or not a given set of reductions is complete.

First we need to characterize the completeness condition in a more useful way.

Let $\to^*$ denote the reflexive transitive completion of the relation $\to$, so that $\alpha \to^* \beta$ holds if and only if there are words $\alpha_0, \alpha_1, \ldots, \alpha_n$ for some $n \geq 0$ such that $\alpha = \alpha_0$, $\alpha_j \to \alpha_{j+1}$ for $0 \leq j < n$, and $\alpha_n = \beta$.

Theorem 4. *A set of reductions R is complete if and only if the following "lattice condition" is satisfied:*

$$\text{If } \alpha \to \alpha' \text{ and } \alpha \to \alpha'' \text{ there exists a word } \gamma \text{ such that } \alpha' \to^* \gamma \text{ and } \alpha'' \to^* \gamma. \tag{4.1}$$

Proof. If $\alpha \to \alpha'$ and $\alpha \to \alpha''$, we can find irreducible words α_0' and α_0'' such that $\alpha' \to^* \alpha_0'$ and $\alpha'' \to^* \alpha_0''$. Since $\alpha_0' \equiv \alpha_0''$, we may take $\gamma = \alpha_0' = \alpha_0''$ if R is complete.

Conversely let us assume that the lattice condition holds; we will prove that R is complete. First, we show that *if $\alpha \to^* \alpha_0$ and $\alpha \to^* \alpha_0'$ where α_0 and α_0' are irreducible, we must have $\alpha_0 = \alpha_0'$.* For if not, the set of all α that violate this property has no infinite decreasing sequence so there must be a "smallest" α (with respect to the $>$ relation) such that $\alpha \to^* \alpha_0$ and $\alpha \to^* \alpha_0' \not\equiv \alpha_0$, where both α_0 and α_0' are irreducible. Clearly α is not itself irreducible, since otherwise $\alpha_0 = \alpha = \alpha_0'$. So we must have $\alpha \not\equiv \alpha_0$, $\alpha \not\equiv \alpha_0'$, and there must be elements α_1, α_1' such that $\alpha \to \alpha_1 \to^* \alpha_0$, $\alpha \to \alpha_1' \to^* \alpha_0'$. By the lattice condition there is a word γ such that $\alpha_1 \to^* \gamma$ and $\alpha_1' \to^* \gamma$. Furthermore there is an irreducible word γ_0 such that $\gamma \to^* \gamma_0$. Now by (3.5), $\alpha > \alpha_1$; so (by the way we chose α) we must have $\alpha_0 = \gamma_0$. Similarly the relation $\alpha > \alpha_1'$ implies that $\alpha_0' = \gamma_0$. This contradicts the assumption that $\alpha_0 \neq \alpha_0'$.

Now to show that R is complete, we will prove the following fact: *If* $\alpha \equiv \beta$, $\alpha \to^* \alpha_0$, *and* $\beta \to^* \beta_0$, *where* α_0 *and* β_0 *are irreducible, then* $\alpha_0 = \beta_0$. Let the derivation of the relation $\alpha \equiv \beta$ be $\alpha = \sigma_0 \leftrightarrow \sigma_1 \leftrightarrow \cdots \leftrightarrow \sigma_n = \beta$, where "$\leftrightarrow$" denotes either "$\to$" or "$\leftarrow$". If $n = 0$, we have $\alpha = \beta$, hence $\alpha_0 = \beta_0$ by the proof in the preceding paragraph. If $n = 1$, we have either $\alpha \to \beta$ or $\beta \to \alpha$, and again the result holds by the preceding paragraph. Finally if $n > 1$, let $\sigma_1 \to^* \sigma_1'$, where σ_1' is irreducible. By induction on n, we have $\sigma_1' = \beta_0$ and also $\sigma_1' = \alpha_0$. Therefore R and the proof are both complete. □

5. The Superposition Process

Our immediate goal, in view of Theorem 4, is to design an algorithm that is capable of testing whether or not the "lattice condition" is satisfied for all words.

In terms of the definitions already given, the hypothesis in (4.1) that $\alpha \to \alpha'$ and $\alpha \to \alpha''$ has the following detailed meaning: There are subwords β_1 and β_2 of α, such that α has the form

$$\alpha = \varphi_1\beta_1\psi_1 = \varphi_2\beta_2\psi_2. \tag{5.1}$$

There also are relations (λ_1, ϱ_1), (λ_2, ϱ_2) in R, and words θ_1, ..., θ_m, σ_1, ..., σ_n such that

$$\beta_1 = S(\theta_1, \ldots, \theta_m; \lambda_1), \quad \beta_2 = S(\sigma_1, \ldots, \sigma_n; \lambda_2) \tag{5.2}$$

and

$$\alpha' = \varphi_1 S(\theta_1, \ldots, \theta_m; \varrho_1)\psi_1, \quad \alpha'' = \varphi_2 S(\sigma_1, \ldots, \sigma_n; \varrho_2)\psi_2. \tag{5.3}$$

The lattice condition will hold if we can find a word γ such that $\alpha' \to^* \gamma$ and $\alpha'' \to^* \gamma$.

Several possibilities arise, depending on the relative positions of β_1 and β_2 in (5.1). If β_1 and β_2 are disjoint (have no common symbols), then assuming φ_1 is shorter than φ_2 we have $\varphi_2 = \varphi_1\beta_1\varphi_3$ for some φ_3, and the lattice condition is trivially satisfied with

$$\gamma = \varphi_1 S(\theta_1, \ldots, \theta_m; \varrho_1)\varphi_3 S(\sigma_1, \ldots, \sigma_n; \varrho_2)\psi_2.$$

If β_1 and β_2 are not disjoint, then one must be a subword of the other, and by symmetry we may assume that β_1 is a subword of β_2. In fact we may even assume that $\alpha = \beta_2$; for the lattice condition must hold in this special case, and it will hold for $\alpha = \varphi_2\beta_2\psi_2$ if it holds for $\alpha = \beta_2$. In view of (5.2), two cases can arise:

Case 1, β_1 is a subword of one of the occurrences of σ_j, for some j. In this case, notice that there are $n(v_j, \lambda_2)$ occurrences of σ_j in α, and α' has been obtained from α by replacing one of these occurrences of σ_j by the word σ'_j, where $\sigma_j \to \sigma'_j$. If we now replace σ_j by σ'_j in each of its remaining $n(v_j, \lambda_2) - 1$ occurrences in α, we obtain the word

$$\alpha_1 = S(\sigma_1, \ldots, \sigma_{j-1}, \sigma'_j, \sigma_{j+1}, \ldots, \sigma_n; \lambda_2);$$

and it is clear that $\alpha' \to^* \alpha_1$. Therefore the lattice condition is satisfied in this case if we take

$$\gamma = S(\sigma_1, \ldots, \sigma_{j-1}, \sigma'_j, \sigma_{j+1}, \ldots, \sigma_n; \varrho_2).$$

Case 2. The only remaining possibility is that

$$\beta_1 = S(\sigma_1, \ldots, \sigma_n; \mu) \tag{5.4}$$

where μ is a nontrivial subword of λ_2. (See the definition of "nontrivial subword" in §1.) The observations above show that the lattice condition holds in all other cases, regardless of the set of reductions R; so an algorithm that tests R for completeness need only consider this case. It therefore behooves us to make a thorough investigation of this remaining possibility.

For convenience, let us write simply λ instead of λ_2. Since μ is a subword of λ we must have $\lambda = \varphi\mu\psi$, for some strings φ and ψ, and it follows from the assumptions above that

$$\varphi_1 = S(\sigma_1, \ldots, \sigma_n; \varphi), \quad \psi_1 = S(\sigma_1, \ldots, \sigma_n; \psi). \tag{5.5}$$

Theorem 5. *Let μ be a subword of the word λ, where $\lambda = \varphi\mu\psi$, and let $C(\lambda_1, \mu, \lambda)$ be the set of all words α that can be written in the form*

$$\alpha = \varphi_1 S(\theta_1, \ldots, \theta_m; \lambda_1)\psi_1 = S(\sigma_1, \ldots, \sigma_n; \lambda) \tag{5.6}$$

for words $\theta_1, \ldots, \theta_m, \sigma_1, \ldots, \sigma_n$, where φ_1 and ψ_1 are defined by (5.5). Then either $C(\lambda_1, \mu, \lambda)$ is the empty set, or there is a word $\sigma(\lambda_1, \mu, \lambda)$, the "superposition of λ_1 on μ in λ," such that $C(\lambda_1, \mu, \lambda)$ is the set of all words that have the form of $\sigma(\lambda_1, \mu, \lambda)$; that is,

$$C(\lambda_1, \mu, \lambda) = \{S(\varphi_1, \ldots, \varphi_k; \sigma(\lambda_1, \mu, \lambda)) \mid \varphi_1, \ldots, \varphi_k \text{ are words}\}. \tag{5.7}$$

Furthermore there is an algorithm that either discovers such a word $\sigma(\lambda_1, \mu, \lambda)$ or determines that $C(\lambda_1, \mu, \lambda)$ is empty.

Proof. Let $\lambda' = S(v_{n+1}, \ldots, v_{n+m}; \lambda_1)$ be the word obtained by changing all the variables v_j in λ_1 to v_{n+j}; then λ' and λ have distinct variables. Let $\sigma_{n+1} = \theta_1$, …, $\sigma_{n+m} = \theta_m$, and let $r = m + n$. Then the words σ_1, …, σ_r are solutions to the equation

$$S(\sigma_1, \ldots, \sigma_r; \lambda) = S(\sigma_1, \ldots, \sigma_r; \varphi)S(\sigma_1, \ldots, \sigma_r; \lambda')S(\sigma_1, \ldots, \sigma_r; \psi).$$

By Theorem 3, we can determine whether or not this equation has solutions; and when solutions exist, we can find a general solution k, σ'_1, …, σ'_r. Theorem 5 follows if we now define $\sigma(\lambda_1, \mu, \lambda) = S(\sigma'_1, \ldots, \sigma'_r; \lambda)$. □

Corollary. *Let R be a set of reductions; and let A be any algorithm that, given a word α, finds an irreducible word α_0 such that $\alpha \to^* \alpha_0$ with respect to R. Then R is complete if and only if the following condition holds for all pairs of reductions (λ_1, ϱ_1), (λ_2, ϱ_2) in R and all nontrivial subwords μ of λ_2 such that the superposition $\sigma(\lambda_1, \mu, \lambda_2)$ exists: Let*

$$\sigma = \sigma(\lambda_1, \mu, \lambda_2) = \varphi_1 S(\theta_1, \ldots, \theta_m; \lambda_1)\psi_1 = S(\sigma_1, \ldots, \sigma_n; \lambda_2), \quad (5.8)$$

where φ_1 and ψ_1 are defined by (5.5). Let

$$\sigma' = \varphi_1 S(\theta_1, \ldots, \theta_m; \varrho_1)\psi_1, \qquad \sigma'' = S(\sigma_1, \ldots, \sigma_n; \varrho_2), \quad (5.9)$$

and use algorithm A to find irreducible words σ'_0 and σ''_0 such that $\sigma' \to^* \sigma'_0$ and $\sigma'' \to^* \sigma''_0$. Then σ'_0 must be identically equal to σ''_0.

Proof. Since $\sigma \to \sigma'$ and $\sigma \to \sigma''$, the condition that $\sigma'_0 = \sigma''_0$ is certainly necessary if R is complete. Conversely we must show that R is complete under the stated conditions.

The condition of Theorem 4 will be satisfied for all words α unless we can find reductions $(\lambda_1, \varrho_1), (\lambda_2, \varrho_2)$ and a nontrivial subword μ of λ_2 such that, in the notation of Theorem 4, we have

$$\alpha = S(\varphi_1, \ldots, \varphi_k; \sigma),\ \alpha' = S(\varphi_1, \ldots, \varphi_k; \sigma'),\ \alpha'' = S(\varphi_1, \ldots, \varphi_k; \sigma'')$$

for some words φ_1, …, φ_k. (This must happen, because the discussion earlier in this section proves that α can be assumed to be a member of $C(\lambda_1, \mu, \lambda)$ if the condition of Theorem 4 is violated, and because Theorem 5 states that α has this form.) But now we may take $\gamma = S(\varphi_1, \ldots, \varphi_k; \sigma'_0) = S(\varphi_1, \ldots, \varphi_k; \sigma''_0)$, and the condition of Theorem 4 is satisfied. □

This corollary amounts to an algorithm for testing the completeness of any set of reductions. A computer implementation is facilitated by observing that the words $\sigma_1, \ldots, \sigma_n, \theta_1, \ldots, \theta_m$ of (5.9) are precisely the words $\sigma'_1, \ldots, \sigma'_r$ obtained during the construction of $\sigma(\lambda_1, \mu, \lambda)$ in the proof of Theorem 5.

As an example of this corollary, let us consider the case when R contains the single reduction

$$(\lambda, \varrho) = (f_2 f_2 v_1 v_2 v_3, f_2 v_1 f_2 v_2 v_3).$$

Here f_2 is a binary operator, and the relation $\lambda \to \varrho$ is the well-known associative law, $((v_1 \cdot v_2) \cdot v_3) \to (v_1 \cdot (v_2 \cdot v_3))$, if we write $(v_1 \cdot v_2)$ for $f_2 v_1 v_2$. (Notice that $\lambda > \varrho$, by the ordering defined in §2.)

Since $f_2 f_2 v_1 v_2 v_3$ has two nontrivial subwords, the corollary in this case requires us to test $\sigma(\lambda, \lambda, \lambda)$ and $\sigma(\lambda, f_2 v_1 v_2, \lambda)$. In the former case we obviously have a very uninteresting situation where $\sigma' = \sigma''$, so the condition is clearly fulfilled. In the latter case, we may take

$$\begin{aligned} \sigma = \sigma(\lambda, f_2 v_1 v_2, \lambda) &= f_2 f_2 f_2 v_1 v_2 v_3 v_4, \\ \sigma' = f_2 f_2 v_1 f_2 v_2 v_3 v_4, &\quad \sigma'' = f_2 f_2 v_1 v_2 f_2 v_3 v_4. \end{aligned}$$

Both of the latter reduce to $f_2 v_1 f_2 v_2 f_2 v_3 v_4$, so the associative law by itself is a "complete" reduction.

The argument just given amounts to the traditional theorem (found in the early pages of most algebra textbooks) that, as a consequence of the associative law, any two ways of parenthesizing a formula are equal when the variables appear in the same order from left to right.

In general, the testing procedure in the corollary may be simplified by omitting the case when $\lambda_1 = \lambda_2 = \mu$, since $\sigma' = \sigma''$. Furthermore we may omit the case when μ is simply a nullary operator f_q, since in that case we must have $\lambda_1 = f_q$, and both σ' and σ'' reduce to the common word γ obtained by replacing all occurrences of f_q in ϱ_2 by ϱ_1. (The argument is essentially the same as the argument of "Case 1" at the beginning of this section.)

6. Extension to a Complete Set

When a set of reductions is incomplete, we may be able to add further reductions to obtain a complete set. In this section we will show how the procedure of the corollary to Theorem 5 can be extended so that a complete set may be obtained in many cases.

First we note that if R is a set of reductions and if $R_1 = R \cup \{(\lambda, \varrho)\}$ where $\lambda \equiv \varrho \pmod{R}$, then R_1 and R generate the same equivalence relation:

$$\alpha \equiv \beta \pmod{R} \qquad \text{if and only if} \qquad \alpha \equiv \beta \pmod{R_1}. \tag{6.1}$$

For if $\alpha \equiv \beta \pmod{R}$ we certainly have $\alpha \equiv \beta \pmod{R_1}$. Conversely, if $\theta \to \varphi \pmod{R_1}$ using the relation (λ, ϱ), it follows from $\lambda \equiv \varrho \pmod{R}$ that $\theta \equiv \varphi \pmod{R}$; and this fact suffices to prove (6.1) since all applications of the extra reduction (λ, ϱ) can be replaced by sequences of reductions using R alone.

Now if $R_1 = R \cup \{(\lambda, \varrho)\}$ and $R_2 = R \cup \{(\lambda', \varrho')\}$, where

$$\lambda \equiv \varrho \pmod{R_2} \quad \text{and} \quad \lambda' \equiv \varrho' \pmod{R_1}, \tag{6.2}$$

we can prove that R_1 and R_2 are equivalent sets of reductions, in the sense that

$$\alpha \equiv \beta \pmod{R_1} \quad \text{if and only if} \quad \alpha \equiv \beta \pmod{R_2}. \tag{6.3}$$

For (6.1) tells us that both of these relations are equivalent to the condition $\alpha \equiv \beta \pmod{R_1 \cup R_2}$.

Because of (6.3), we may assume that, for each reduction (λ, ϱ) in R, both λ and ϱ are irreducible with respect to the other reductions of R.

The following procedure may now be used, in an attempt to complete a given set R of reductions:

Apply the tests of the corollary to Theorem 5, for all λ_1, λ_2, and μ. If in every case $\sigma_0' = \sigma_0''$, R is complete and the procedure terminates. If some choice of $\lambda_1, \lambda_2, \mu$ leads to $\sigma_0' \neq \sigma_0''$ then we have either $\sigma_0' > \sigma_0''$ or $\sigma_0'' > \sigma_0'$ or $\sigma_0' \# \sigma_0''$. In the latter case, the process terminates unsuccessfully, having derived an equivalence $\sigma_0' \equiv \sigma_0'' \pmod{R}$ for which no reduction (as defined in this paper) can be used. In the former cases, we add a new reduction (σ_0', σ_0'') or (σ_0'', σ_0'), respectively, to R, and begin the procedure again.

Whenever a new reduction (λ', ϱ') is added to R, the entire new set R is checked to make sure that it contains only irreducible words. This means, for each reduction (λ, ϱ) in R, we find irreducible λ_0 and ϱ_0 such that $\lambda \to^* \lambda_0$ and $\varrho \to^* \varrho_0$ with respect to $R \setminus \{(\lambda, \varrho)\}$. Here it is possible that $\lambda_0 = \varrho_0$, in which case by (6.1) we may remove (λ, ϱ) from R. Otherwise we might have $\lambda_0 > \varrho_0$ or $\varrho_0 > \lambda_0$; then (λ, ϱ) may be replaced by (λ_0, ϱ_0) or (ϱ_0, λ_0), respectively, by (6.3). We might also find that $\lambda_0 \# \varrho_0$, in which case the process terminates unsuccessfully as above.

Several examples of experiments with this procedure appear in the remainder of this paper. It was found to be most useful to test short reductions first (that is, to consider first those λ_1 and λ_2 that have small weight or short length). Shorter words are more likely to lead to interesting consequences, by which relations that involve longer words can be reduced or even eliminated.

In practice, when equivalent words α and β are found such that $\alpha \# \beta$, it is often possible to continue the process by introducing a new operator into the system, as shown in the examples of the next section.

7. Computational Experiments

In this section we will make free use of more familiar "infix" notations, such as $\alpha \cdot \beta$, in place of the prefix notation $f_j\alpha\beta$ that was more convenient for a formal development of the theory. Furthermore the word "axiom" will often be used instead of "reduction," and the letters a, b, c, d will be used in place of the variables v_1, v_2, v_3, v_4.

The computational procedure explained in §6 was programmed in FORTRAN IV for an IBM 7094 computer, making use of standard techniques of tree structure manipulation. The running times quoted below could be improved somewhat, perhaps by an order of magnitude, (a) by recoding the most extensively used subroutines in assembly language; (b) by keeping more detailed records of which pairs (λ_1, λ_2) have already been tested against each other; and (c) by keeping more detailed records of those pairs (α, λ) of words for which we have already verified that α does not have the form of λ. Such improvements have not been made at the time of writing, because of the experimental nature of the algorithm.

Example 1. Group theory I. The first example on which this method was tried was the traditional definition of an abstract group. Here we have three operators: a binary operator $f_2 = \cdot$, of weight zero; a unary operator $f_3 = {}^-$, of weight zero; and a nullary operator $f_1 = e$, of weight one. These operators are supposed to obey three given axioms:

1. $e \cdot a \to a$ ("There exists a left identity, e.")
2. $a^- \cdot a \to e$ ("Each element has a left inverse with respect to e.")
3. $(a \cdot b) \cdot c \to a \cdot (b \cdot c)$ ("Multiplication is associative.")

The procedure was first carried out by hand, to see if it would succeed in deriving identities such as $a \cdot e = a$, $a^{--} = a$, etc., without making use of any more ingenuity than can normally be expected of a computer's brain. The success of this hand-computation experiment provided the initial incentive to create the computer program, so that experiments on other axiom systems could be performed.

When the computer program was finally completed, the machine treated the three axioms above as follows: First axioms 1 and 2 were found to be complete, by themselves. But when $\lambda_1 = a^- \cdot a$ of axiom 2 was superposed on the subword $\mu = a \cdot b$ of $\lambda_2 = (a \cdot b) \cdot c$ in axiom 3, the resulting formula $(a^- \cdot a) \cdot b$ could be reduced in two ways as

$$(a^- \cdot a) \cdot b \;\to\; a^- \cdot (a \cdot b)$$

and

$$(a^- \cdot a) \cdot b \;\to\; e \cdot b \;\to\; b.$$

Therefore a new axiom was added:

4. $a^- \cdot (a \cdot b) \to b$

Axiom 1 was superposed on the subword $a \cdot b$ of this new axiom, and another new axiom resulted:

5. $e^- \cdot a \to a$

The computation continued as follows:

6. $a^{--} \cdot e \to a$ — from 2 and 4.
7. $a^{--} \cdot b \to a \cdot b$ — from 6 and 3.

Now axiom 6 was no longer irreducible, and it could be replaced by a better one:

8. $a \cdot e \to a$

Thus, the computer had found a proof that e is a *right* identity. Its proof was essentially the following, if reduced to applications of axioms 1, 2, and 3:

$$\begin{aligned} a \cdot e &\equiv (e \cdot a) \cdot e \equiv ((a^{--} \cdot a^-) \cdot a) \cdot e \equiv (a^{--} \cdot (a^- \cdot a)) \cdot e \\ &\equiv (a^{--} \cdot e) \cdot e \equiv a^{--} \cdot (e \cdot e) \equiv a^{--} \cdot e \equiv a^{--} \cdot (a^- \cdot a) \\ &\equiv (a^{--} \cdot a^-) \cdot a \equiv e \cdot a \equiv a. \end{aligned}$$

This ten-step proof may well be the shortest possible.

The computation continued further:

9. $e^- \to e$ — from 2 and 8.

(At this point axiom 5 disappeared.)

10. $a^{--} \to a$ — from 7 and 8.

(Hence axiom 7 also disappeared.)

11. $a \cdot a^- \to e$ from 10 and 2.
12. $a \cdot (b \cdot (a \cdot b)^-) \to e$ from 3 and 11.
13. $a \cdot (a^- \cdot b) \to b$ from 11 and 3.

So far, the computation was done almost as a professional mathematician would have performed things. The axioms present at this point were $\{1, 2, 3, 4, 8, 9, 10, 11, 12, 13\}$; they do not form a complete set. The ensuing computation reflected the computer's groping for a suitable way to complete it:

14. $(a \cdot b)^- \cdot (a \cdot (b \cdot c)) \to c$ from 3 and 4.
15. $b \cdot (c \cdot ((b \cdot c)^- \cdot a)) \to a$ from 13 and 3.
16. $b \cdot (c \cdot (a \cdot (b \cdot (c \cdot a))^-)) \to e$ from 12 and 3.
17. $a \cdot (b \cdot a)^- \to b^-$ from 12 and 4, using 8.

(Now axiom 12 disappeared.)

18. $b \cdot ((a \cdot b)^- \cdot c) \to a^- \cdot c$ from 17 and 3.

(Now axiom 15 disappeared.)

19. $b \cdot (c \cdot (a \cdot (b \cdot c))^-) \to a^-$ from 17 and 3.

(Now axiom 16 disappeared.)

20. $(a \cdot b)^- \to b^- \cdot a^-$ from 17 and 4.

Success! Axioms 17, 14, 18, and 19 now disappeared, and the resulting set of axioms was complete:

1. $e \cdot a \to a$	9. $e^- \to e$
2. $a^- \cdot a \to e$	10. $a^{--} \to a$
3. $(a \cdot b) \cdot c \to a \cdot (b \cdot c)$	11. $a \cdot a^- \to e$
4. $a^- \cdot (a \cdot b) \to b$	13. $a \cdot (a^- \cdot b) \to b$
8. $a \cdot e \to a$	20. $(a \cdot b)^- \to b^- \cdot a^-$

A study of these ten reductions confirms that they suffice to solve the word problem for free groups with no relations. Two words formed with the operators $\cdot$, $^-$, and e can be proved equivalent as a consequence of axioms $\{1, 2, 3\}$ if and only if they reduce to the same irreducible word, when the ten reductions above are applied in any order.

The computer took 30 seconds for this calculation. Notice that, of the 17 axioms derived during the process, axioms 5, 14, 15, 16, 18, 19 never took part in the derivations of the final complete set; so we can give the machine an "efficiency rating" of $11/17 \approx 65\%$, if we consider how many of its attempts were along fruitful lines. This would seem to compare favorably with the behavior of most novice students of algebra, who do not have the benefit of the corollary to Theorem 5 to show them which combinations of axioms can possibly lead to new results.

Example 2. Group theory II. In the previous example, the unary operator $^-$ was assigned weight zero. In §1 we observed that a unary operator may be assigned weight zero only in exceptional circumstances (at least under the well-ordering we are considering), so it may be interesting to consider what would have happened if we had attempted to complete the group theory axioms of Example 1 with a "slight" change, giving the operator $^-$ a positive weight.

From the description of Example 1, it is clear that the computation would proceed in exactly the same manner, regardless of the weight of $^-$, until reaching step 20. But then the axiom would be reversed:

20. $b^- \cdot a^- \to (a \cdot b)^-$

Thus, $(a \cdot b)^- = f_3 f_2 ab$ would be considered as a "reduction" of the word $b^- \cdot a^- = f_2 f_3 b f_3 a$; and this is apparently quite a reasonable idea, because $(a \cdot b)^-$ is in fact a shorter formula.

But if axiom 20 is written in this way, the computation will never terminate, and no complete set of axioms will ever be produced!

Theorem 6. *If the operator* $^-$ *is assigned a positive weight, no finite complete set of reductions is equivalent to the group theory axioms*

$$e \cdot a \to a, \quad a^- \cdot a \to e, \quad (a \cdot b) \cdot c \to a \cdot (b \cdot c).$$

Proof. Consider the two words

$$\alpha_n = v_{n+1} \cdot (v_1 \cdot (v_2 \cdot \ldots \cdot (v_n \cdot v_{n+1}) \ldots))^-,$$
$$\beta_n = (v_1 \cdot (v_2 \cdot \ldots \cdot (v_{n-1} \cdot v_n) \ldots))^-,$$

for $n \geq 2$. It is obvious that β_n is not equivalent to any lesser word in the well-ordering, since all words equivalent to β_n have at least one occurrence of each variable $v_1, \ldots, v_n$, plus at least $n-1$ multiplication operators, plus at least one $^-$ operator. Since α_n is equivalent to β_n, any complete set R of reductions must include some (λ, ϱ) that reduces α_n. But no subword of α_n, except α_n itself, can be reduced, since each of its smaller subwords is the least in its equivalence class. Therefore α_n itself must have the form of λ; we must have $\alpha_n = S(\theta_1, \ldots, \theta_m; \lambda)$ for some words $\theta_1, \ldots, \theta_m$. It is easy to see that this condition leaves only a few possibilities for the word λ. Now the word

$$\alpha_n' = v_{n+2} \cdot (v_1 \cdot (v_2 \cdot \ldots \cdot (v_n \cdot v_{n+1}) \ldots))^-$$

is *not* equivalent to any lesser word in the well-ordering, so α'_n cannot have the form of λ. We conclude that $\lambda = \alpha_n$, except perhaps for permutation of variables; so R must contain infinitely many reductions. □

Example 3. Group theory III. Suppose we start as in Example 1 but with left identity and left inverse replaced by right identity and right inverse:

1. $a \cdot e \to a$
2. $a \cdot a^- \to e$
3. $(a \cdot b) \cdot c \to a \cdot (b \cdot c)$

It should be emphasized that the computational procedure of §6 is *not* symmetrical between right and left, due to the nature of the well-ordering; therefore this is quite a different problem from Example 1. For example, the new axiom 1 combined with axiom 3 generates the reduction "$a \cdot (e \cdot b) \to a \cdot b$," which has no analog in the system of Example 1.

The computer found this system slightly more difficult than the system of Example 1; 24 axioms were generated during the computation, of which 8 did not participate in the derivation of the final set. This gave an "efficiency rating" of 67%, roughly the same as before. The computation required 40 seconds, compared with 30 seconds in the former case. The same set of ten reductions was obtained as the final answer.

Example 4. An inverse property. Suppose we have only two operators $\cdot$ and $^-$ as in the previous examples, and suppose that only the single axiom

1. $a^- \cdot (a \cdot b) \to b$

is given. No associative law, etc., is assumed. There is no nullary operator; but nullary operators were needed only to establish the theory of §§1–5, they need not be present when we study general word problems.

This example can be worked by hand: First we superpose $a^- \cdot (a \cdot b)$ onto its component $(a \cdot b)$, obtaining the word $a^{--} \cdot (a^- \cdot (a \cdot b))$, which can be reduced both to $a \cdot b$ and to $a^{--} \cdot b$. This gives us a second axiom

2. $a^{--} \cdot b \to a \cdot b$

as a consequence of axiom 1.

Now $a^- \cdot (a \cdot b)$ can be superposed onto $a^{--} \cdot b$; we obtain the word $a^{--} \cdot (a^- \cdot b)$, which reduces to b by axiom 1, and to $a \cdot (a^- \cdot b)$ by axiom 2. Thus, a third axiom

3. $a \cdot (a^- \cdot b) \to b$

is generated. It is interesting (and not well known) that axiom 3 follows from axiom 1 and no other hypotheses; this fact can be used to simplify several proofs that appear in the literature, for example in the algebraic structures associated with projective geometry.

A rather tedious further consideration of about ten more cases shows that axioms $\{1, 2, 3\}$ form a complete set. Thus, we can show that the identity $a^{--} \cdot b \equiv a \cdot b$ is a consequence of axiom 1, but we cannot prove that $a^{--} \equiv a$ without further assumptions.

A similar process shows that axioms 1 and 2 follow from axiom 3.

Example 5. Group theory IV. The axioms in Example 1 are slightly stronger than the "classical" definition (see, for example, Dickson [3]), which states that multiplication is associative, there is at least one left identity, and that *for each left identity* there exists a left inverse of each element. Our axioms of Example 1 just state that there is a left inverse for the left identity e.

Consider the five axioms

1. $(a \cdot b) \cdot c \to a \cdot (b \cdot c)$
2. $e \cdot a \to a$
3. $f \cdot a \to a$
4. $a^{-} \cdot a \to e$
5. $a^{\sim} \cdot a \to f$

where e and f are nullary operators, $^{-}$ and $^{\sim}$ are unary operators, and $\cdot$ is a binary operator. Here we are postulating two left identities, and a left inverse for each one. The computer, when presented with these axioms, found a complete set of reductions in 50 seconds, namely the two reductions

$$f \to e$$
$$a^{\sim} \to a^{-}$$

together with the ten reductions in Example 1. As a consequence, it is clear that the identity and inverse functions are unique.

The derivation of "$f \to e$" was achieved quickly in a rather simple way, by first deriving "$a^{-} \cdot (a \cdot b) \to b$" as in Example 1, then deriving "$f^{-} \cdot b \to b$" by setting $a = f$, and finally deriving "$f \to e$" by setting $b = f$.

Example 6. Central groupoids I. An interesting algebraic system has recently been described by Trevor Evans [5]. There is one binary operator $\cdot$ and one axiom:

1. $(a \cdot b) \cdot (b \cdot c) \to b$

Let us call this a "central groupoid," since the product $(a \cdot b) \cdot (b \cdot c)$ reduces to its central element b. As in Example 4, the computational procedure of §6 can be carried out easily by hand, and we obtain two further axioms

2. $a \cdot ((a \cdot b) \cdot c) \to a \cdot b$
3. $(a \cdot (b \cdot c)) \cdot c \to b \cdot c$

which complete the set.

Evans [5] has shown that every finite central groupoid has n^2 elements for some nonnegative integer n. It is also possible to show [7] that every finite central groupoid with n^2 elements has exactly n idempotent elements, i.e., elements with $a \cdot a = a$. On the other hand one can show (by virtue of the fact that the three axioms above form a complete set) that the *free* central groupoid on any number of generators has no idempotents at all. For if there is an idempotent, consider the least word α in the well-ordering such that $\alpha \equiv \alpha \cdot \alpha$. Clearly α is not a generator, and so α must have the form $\alpha = \beta \cdot \gamma$ where α, β, and γ are irreducible. Thus $(\beta \cdot \gamma) \cdot (\beta \cdot \gamma)$ must be reducible; but this is possible only if $\gamma = \beta$, and then $\beta \cdot \beta = \alpha = \alpha \cdot \alpha = \beta$ is not irreducible after all. (This proof was communicated to the authors by Professor Evans in 1966.)

Example 7. A "random" axiom. Experiments on several axioms that were more or less selected at random show that the resulting systems often degenerate. For example, suppose we have a ternary operator denoted by (x, y, z), which satisfies the following axiom:

1. $(a, (b, c, a), d) \to c$

Superposing the left-hand side onto (b, c, a) gives the word

$$(b, (a, (e, c, a), b), d),$$

and this reduces both to (e, c, a) and to (b, c, d). Hence we find

$$(e, c, a) \equiv (b, c, d).$$

Now the computational method described in §6 will stop, since

$$(e, c, a) \# (b, c, d).$$

But there is an obvious way to proceed: Since $(e, c, a) \equiv (b, c, d)$, clearly (e, c, a) is a function of c only. So we may introduce a new unary operator $'$ and a new axiom:

2. $(a, b, c) \to b'$

Now axiom 1 may be replaced by

3. $a'' \to a$

and we're done(!), because axioms 2 and 3 form a complete set.

We've proved that two words involving the ternary operator are equivalent as a consequence of axiom 1 if and only if they reduce to the same word by applying reductions 2 and 3. The free system on n generators has $2n$ elements.

Example 8. Another "random" axiom. If we start with

1. $(a \cdot b) \cdot (c \cdot (b \cdot a)) \to b$

where $\cdot$ is a binary operator, the computer deduces the equivalences

$$c \equiv ((b \cdot a) \cdot c) \cdot ((a \cdot b) \cdot (c \cdot (b \cdot a))) \equiv ((b \cdot a) \cdot c) \cdot b;$$

so $((b \cdot a) \cdot c) \cdot b \to c$. This simpler axiom implies that

$$b \equiv (((b \cdot a) \cdot c) \cdot b) \cdot (b \cdot a) \equiv c \cdot (b \cdot a),$$

and the original axiom now becomes

$$c \equiv b.$$

Clearly this is a totally degenerate system. Following the general procedure outlined above, we introduce a new nullary operator e, and we are left with the axiom

$$a \to e.$$

The free system on n generators has one element.

Example 9. The cancellation law. The previous two examples illustrate how to introduce new operators in order to apply our reduction method to axioms that the method cannot handle directly. A similar technique can be used to take the place of axioms that cannot be expressed directly in terms of "identities."

Our axioms up to now have always been formulas that are identically equal under all substitutions. For example, the reduction $(a \cdot b) \cdot c \to a \cdot (b \cdot c)$ is essentially the statement that

$$\text{for all words } a, b, c, \quad (a \cdot b) \cdot c \equiv a \cdot (b \cdot c).$$

A general reduction $\alpha \to \beta$ means that $\alpha \equiv \beta$ for all values of the variables appearing in α and β. But many mathematical axioms are not simply identities; one common example is the left cancellation law, which states that

$$\text{for all words } a, b, c, \quad \text{if} \quad a \cdot b \equiv a \cdot c \quad \text{then} \quad b \equiv c. \tag{7.1}$$

The left cancellation law can be represented as an identity in the following way. Consider a function $f(x, y)$ that satisfies the identity

$$f(a, a \cdot b) \to b. \tag{7.2}$$

If $\mathcal{S}$ represents any set of axioms, let $\mathcal{S}'$ be the set of axioms obtained by adding the left cancellation law (7.1) to $\mathcal{S}$, and let $\mathcal{S}''$ be the set of axioms obtained by adding the reduction (7.2) to $\mathcal{S}$, where f is a binary operator that does not appear in $\mathcal{S}$. Then any two words *not* involving f that can be proved equivalent in $\mathcal{S}'$ can be proved equivalent in $\mathcal{S}''$. For whenever (7.1) is used, we must have already proved that $a \cdot b \equiv a \cdot c$; hence $f(a, a \cdot b) \equiv f(a, a \cdot c)$; hence $b \equiv c$, by (7.2). Conversely, any two words α and β not involving f that can be proved equivalent in $\mathcal{S}''$ can be proved equivalent in $\mathcal{S}'$: For if (7.1) holds, there exists a binary operator f satisfying (7.2); one such binary operator can, for example, be defined by letting $f(x, y)$ equal z if y can be written in the form $x \cdot z$ (here z is unique by (7.1)), and by letting $f(x, y)$ equal x otherwise. This function f has properties that are in fact somewhat stronger than (7.2) asserts; if we can prove $\alpha \equiv \beta$ under the weaker hypotheses $\mathcal{S}''$, we can surely prove $\alpha \equiv \beta$ with $\mathcal{S}'$.

(The argument just given seems to rely on rules of inference that aren't admissible in some logical systems. Another argument that systematically removes all appearances of f from a proof of $\alpha \to \beta$ in the system $\mathcal{S}''' \equiv \mathcal{S} \cup \{(7.1), (7.2)\}$ can be given, but it will be omitted here; we will content ourselves with the validity of the more intuitive but less intuitionistic argument above.)

A system that has a binary operation $\cdot$ and both left and right cancellation laws, but no further axioms, can be defined by two axioms:

1. $f(a, a \cdot b) \to b$
2. $g(a \cdot b, b) \to a$

Here f and g are two new binary operators. Axioms 1 and 2 are complete by themselves, so they suffice to solve the word problem for any words involving f, $\cdot$, and g. In particular, two words involving only $\cdot$ are equivalent if and only if they are equal.

Suppose we add a unit element, namely a nullary operator e that satisfies two special properties:

3. $e \cdot a \to a$
4. $a \cdot e \to a$

Then the computer will complete the set by adding four more reductions:

5. $f(a, a) \to e$
6. $f(e, a) \to a$
7. $g(a, a) \to e$
8. $g(a, e) \to a$

Example 10. Loops. Consider the axiom "for all a and b there exists c such that $a \cdot c \equiv b$." This amounts to saying that there is a binary operation $\backslash$ such that $c = a \backslash b$; in other words, we have $a \cdot (a \backslash b) \equiv b$. This law is a companion to the cancellation law (7.1) of Example 9, which asserts that at *most* one such c exists.

In the mathematical system known as an abstract loop, we have both the law above and its left-right dual; so there are three binary operators $\cdot$, $\backslash$, and $/$, together with two axioms:

1. $a \cdot (a \backslash b) \to b$
2. $(a \,/\, b) \cdot b \to a$

There also is a unit element:

3. $e \cdot a \to a$
4. $a \cdot e \to a$

When presented with these axioms, a computer will generate two more:

5. $e \backslash a \to a$
6. $a \,/\, e \to a$

Axioms 1 through 6 form a complete set, but they do not define a loop; two important axioms have been left out of the discussion above, namely the left and right cancellation laws. After we postulate two further binary operators f and g as in Example 9, together with two further axioms

7. $f(a, a \cdot b) \to b$
8. $g(a \cdot b, b) \to a$

the computer will generate

9. $f(a, b) \to a \backslash b$
10. $g(a, b) \to a \,/\, b$
11. $a \backslash (a \cdot b) \to b$
12. $(a \cdot b) \,/\, b \to a$
13. $a \,/\, a \to e$
14. $a \backslash a \to e$
15. $a \,/\, (b \backslash a) \to b$
16. $(a \,/\, b) \backslash a \to b$

Axioms $\{1, 2, \ldots, 6, 9, 10, \ldots, 16\}$ form a complete set of reductions; and if we remove axioms 9 and 10 (which merely serve to eliminate the auxiliary functions f and g) we obtain reductions for a free loop. This is a special case of the complete set given by Evans [4], who also added relations between generators (i.e., between additional nullary operators).

Notice that in Example 9 the cancellation laws had no effect on the word problem, while in this case the rules 11 through 16 could not be obtained from 1 through 4 without postulating the cancellation laws. On the other hand, when the mathematical system is known to be finite, the existence of a solution c to the equation $a \cdot c \equiv b$, for all a and b, implies the uniqueness of that solution. Thus laws 11 through 16 can be deduced from 1 through 4 in a finite system, but not in a free system on a finite number of generators.

The process of generating the complete set above, starting from $\{1, 2, 3, 4, 7, 8\}$, took 20 seconds. Axiom 9 was found quickly since

$$b \setminus a \equiv f(b, b \cdot (b \setminus a)) \equiv f(b, a).$$

Example 11. Group theory V. An interesting way to define a group with axioms even weaker than the classical axioms in Example 5 has been pointed out by O. Taussky [11]. Besides the associative law

1. $(a \cdot b) \cdot c \to a \cdot (b \cdot c)$

we postulate the existence of an idempotent element e:

2. $e \cdot e \to e$

Furthermore, each element has *at least one right inverse* with respect to e, that is, there is a unary operator $^-$ such that

3. $a \cdot a^- \to e$

holds. Finally, we postulate that each element has *at most one left inverse* with respect to e. This last assertion is equivalent to a very special type of cancellation law, which is more difficult to handle than (7.1):

$$\text{for all } a, b, c, \quad \text{if} \quad b \cdot a \equiv c \cdot a \equiv e \quad \text{then} \quad b \equiv c. \tag{7.3}$$

Axiom (7.3) can be replaced, as in Example 9, by identities involving new operators. Let f be a ternary operator and g a binary operator, and postulate the following:

4. $f(e, a, b) \to a$
5. $f(a \cdot b, a, b) \to g(a \cdot b, b)$

It is easy to see that these axioms imply (7.3). Conversely, (7.3) implies the existence of such functions f and g, since we may define for example

$$f(x, y, z) = \begin{cases} y, & \text{if } x \equiv e; \\ x, & \text{if } x \not\equiv e; \end{cases}$$

$$g(x, y) = \begin{cases} z, & \text{if } x \equiv e \text{ and } z \cdot y \equiv e; \\ x, & \text{if } x \not\equiv e \text{ or if there is no } z \text{ such that } z \cdot y \equiv e. \end{cases}$$

The latter function g is well defined when (7.3) holds.

Thus, axioms 4 and 5 may be regarded as equivalent to (7.3), just as in Examples 9 and 10, if we consider the word problem for words that do not involve f and g. (A binary operation $f(x, y)$ could actually have been used; but we introduced a ternary operation so that axiom 5 could be considered as a reduction, because $f(a \cdot b, a) \# g(a \cdot b, b)$.)

The computer was presented with axioms 1 through 5, and an interesting sequence of computations began. One of the consequences of axioms 1 and 3 alone is that

$$e \cdot a^{--} \equiv (a \cdot a^{-}) \cdot a^{--} \equiv a \cdot (a^{-} \cdot a^{--}) \equiv a \cdot e. \tag{7.4}$$

After 2 minutes and 15 seconds, the procedure derived its 29th consequence of axioms 1 through 5, namely that $a^{--} \to a$. This meant that (7.4) became

$$e \cdot a \equiv a \cdot e;$$

and the computer stopped, since the definitions of §2 imply that $e \cdot a \# a \cdot e$. It was sensible to balk here; for if we were to say $e \cdot a \to a \cdot e$, the machine would loop indefinitely trying to reduce the word $e \cdot e$.

Now we restarted the process as in Examples 7 and 8, by introducing a new unary operator $'$ with $e \cdot a \to a'$. The axioms currently in the system at that time were thereby transformed to include the following, among others:

$$\begin{aligned}
&e' \to e\\
&a'' \to a'\\
&a \cdot e \to a'\\
&e \cdot a \to a'\\
&(a \cdot b)' \to a \cdot b'\\
&g(e, a') \to a^{-}
\end{aligned}$$

In order to make the well-ordering come out correctly for these reductions, the authors changed the weight of $\cdot$ from zero to one, changed the weight of f from one to two, and made $'$ a unary operator of weight one that was higher than $\cdot$ in the ordering of operators.

Another axiom in the system at this time, which had been derived quite early by superposing 3 onto 5 and applying 4, was

$$g(e, a^{-}) \to a.$$

This now was combined with the rule $a^{--} \to a$ to derive

$$g(e, a) \to a^{-}.$$

The reduction $g(e, a') \to a^-$ was consequently transformed to

$$a'^- \to a^-$$

and, with the law $a^{--} \to a$, this became

$$a' \to a.$$

Thus, the $'$ operator disappeared, and the traditional group axioms were obtained immediately. After approximately 3 minutes of computer time from the beginning of the computations, all ten reductions of Example 1 had been derived.

Actually it is not hard to see that, as in the discussion of Example 2, axioms 1 through 5 cannot be completed to a finite set. After $4\frac{1}{2}$ minutes of execution time, the computer was deriving esoteric reductions such as

$$f(c, c \cdot (a^- \cdot b^-), b \cdot a) \to g(c, b \cdot a).$$

Since the process would never terminate, there was perhaps a logical question remaining whether any new reductions would be derived (besides the 10 in the final set of Example 1) that would give us *more* than a group. Group theory tells us that this could not happen, but the authors wanted a direct way to reach this conclusion as a consequence of axioms 1 through 5. This can be done in fact, by adding new axioms

$$g(a, b) \to a \cdot b^-$$
$$f(a, b, c) \to b$$

to the long list of axioms derived by the machine after 3 minutes. The new axioms are even stronger than 4 and 5, and together with the ten final axioms of Example 1 they form a complete set of twelve reductions. Thus we can be sure that axioms 1 through 5 do not prove any more about words in $\cdot$, $^-$, and e that could not be proved in groups.

The computer's derivation of the laws of group theory from axioms 1, 2, 3, and (7.3) may be reformulated as follows, if we examine the computations and remove references to f and g:

"We have $e \cdot a^{--} \equiv a \cdot e$, as in (7.4), hence

$$a \cdot e \equiv e \cdot a^{--} \equiv (e \cdot e) \cdot a^{--} \equiv e \cdot (e \cdot a^{--}) \equiv e \cdot (a \cdot e).$$
$$\therefore\ a^{--} \cdot e \equiv e \cdot (a^{--} \cdot e) \equiv (e \cdot a^{--}) \cdot e \equiv (a \cdot e) \cdot e \equiv a \cdot (e \cdot e) \equiv a \cdot e.$$
$$\therefore\ a^- \cdot (a \cdot e) \equiv a^- \cdot (a^{--} \cdot e) \equiv (a^- \cdot a^{--}) \cdot e \equiv e \cdot e \equiv e.$$

So, by (7.3), a^- is the left inverse of $a \cdot e$, and similarly a^{---} is the left inverse of $a^{--} \cdot e \equiv a \cdot e$. Hence

$$a^{---} \equiv a^-.$$

But now a is the left inverse of a^- by (7.3) and axiom 3, and so a^{--} is the left inverse of $a^{---} \equiv a^-$, so

$$a^{--} \equiv a.$$

This implies that a^- is the left inverse of $a \equiv a^{--}$, so each element has a unique left inverse. The left inverse of $a \cdot e$ is $(a \cdot e)^-$, and we have seen that the left inverse of $a \cdot e$ is a^-; hence $(a \cdot e)^- = a^-$. Now, taking inverses of both sides, we see that $a \cdot e = a$, and the rest of the properties of group theory follow as usual."

A simpler proof results if we start by observing that $(e \cdot a) \cdot a^- \equiv e \cdot (a \cdot a^-) \equiv e \cdot e \equiv e \equiv a \cdot a^-$; hence, by (7.3), $e \cdot a \equiv a$. Now $(a \cdot e) \cdot a^- \equiv a \cdot (e \cdot a^-) \equiv a \cdot a^- \equiv e$; hence, by (7.3), $a \cdot e \equiv a$.

The computer's proof is longer, but interesting in that it does not require application of (7.3) until after several consequences of axioms $\{1, 2, 3\}$ alone are derived.

Example 12. (l, r) systems I. It is interesting to ask what happens if we modify the axioms of group theory slightly, postulating a *left* identity element and a *right* inverse. (Compare with Examples 1 and 3.) This modification leads to an algebraic system that apparently was first discussed by A. H. Clifford [1]. H. B. Mann [8] considered this question independently, and called the systems "(l, r) systems." They are also called "left groups" [2].

Starting with the axioms

1. $e \cdot a \to a$
2. $a \cdot a^- \to e$
3. $(a \cdot b) \cdot c \to a \cdot (b \cdot c)$

the computer extended them to the following complete set of reductions:

4. $e^- \to e$
6. $a \cdot (a^- \cdot b) \to b$
8. $a \cdot e \to a^{--}$
10. $a^{--} \cdot b \to a \cdot b$
16. $a^- \cdot (a \cdot b) \to b$
18. $a^{---} \to a^-$
29. $(a \cdot b)^- \to b^- \cdot a^-$

(The numbers 4, 6, 8, ... that appear here reflect the order of "discovery" of these reductions. The computation took 110 seconds. Of 26 axioms generated, 14 were never used to derive members of the final set, so the "efficiency ratio" in this case was 46%.) These ten reductions solve the word problem for free (l, r) systems defined by axioms 1, 2, and 3.

Example 13. (r,l) systems. Similarly, we can postulate a right identity and a left inverse. This leads to an algebraic system dual to the system of Example 12; so it is not essentially different, from a theoretical standpoint. But the method of §6 is not symmetrical between left and right, so a test of these axioms was worthwhile as a further test of the algorithm's usefulness.

This set of axioms was substantially more difficult for the computer to resolve, apparently because the derivation of the law $(a\cdot b)^- \equiv b^-\cdot a^-$ in this case requires the use of a fairly complex intermediate reduction, $(a\cdot b)^-\cdot(a\cdot(b\cdot c))\to c^{--}$, which will not be examined until all simpler possibilities have been explored. When the roles of left and right are interchanged as in Example 12, the steps leading to $(a\cdot b)^- \equiv b^-\cdot a^-$ are much less complicated.

After $2\frac{1}{2}$ minutes of computation, the identity

$$b^{--}\cdot(a\cdot b)^- \equiv (c\cdot a)^-\cdot c$$

was derived, and computation ceased because $b^{--}\cdot(a\cdot b)^- \,\#\, (c\cdot a)^-\cdot c$. However, it is plain that this quantity is a function of a alone; so the authors introduced a new unary operator $'$ and the rule $(c\cdot a)^-\cdot c\to a'$. After another $2\frac{1}{2}$ minutes of computation the following complete set of 12 reductions for (r,l) systems was obtained:

$a\cdot e \to a$	$b\cdot(a\cdot a^-) \to b$
$a^-\cdot a \to e$	$b\cdot(a\cdot(a^-\cdot c)) \to b\cdot c$
$(a\cdot b)\cdot c \to a\cdot(b\cdot c)$	$a^-\cdot(a\cdot b) \to b^{--}$
$e^- \to e$	$b\cdot(a^{--}\cdot c) \to b\cdot(a\cdot c)$
$e\cdot a \to a^{--}$	$a^{---} \to a^-$
$a\cdot b^{--} \to a\cdot b$	$(a\cdot b)^- \to b^-\cdot a^-$

plus the further reduction $a'\to a^-$ which was, of course, discarded.

Example 14. (l,r) systems II. If we introduce two left identity elements and two corresponding right inverse operators, we have five axioms instead of three:

1. $(a\cdot b)\cdot c\to a\cdot(b\cdot c)$
2. $e\cdot a\to a$
3. $f\cdot a\to a$
4. $a\cdot a^-\to e$
5. $a\cdot a^\sim\to f$

(Compare with Examples 5 and 12.) After 2 minutes of computation, the computer was only slowly approaching a solution to the complete set; at that point 35 different axioms were in the system, including things such as $a^{----} \to a^{--}$, $a^{---\sim} \to a^{-\sim}$, $a \cdot a^{\sim--} \to e$, etc.; the reduction $a^{-\sim-} \cdot b \to a^{-} \cdot b$ was generated just before we terminated the computation manually.

It was apparent that more efficient use could be made of the computer time if the machine was also presented with the information it had already derived in Example 12. Axioms 1, 2, and 4 by themselves generate a complete set of ten axioms as listed in Example 12, and axioms 1, 3, 5 generate an analogous set of ten with e and $^-$ replaced by f and $^\sim$. Therefore the calculation was restarted with 19 initial axioms in place of the 5 above.

(In general, it seems worthwhile to apply the computational method to subsets of a given set of axioms first, and later to add the consequences of those subsets to the original set, since the computation time depends critically on the number of axioms currently being considered.)

With those 19 well-chosen axioms as a starting point, a complete set of consequences of axioms 1 through 5 was obtained after $2\frac{1}{2}$ minutes of calculation. This complete set consisted of the following 21 reductions:

$$\begin{array}{rl}
e^{-} \to e & f^{\sim} \to f \\
e^{\sim} \to f & f^{-} \to e \\
e \cdot a \to a & f \cdot a \to a \\
a \cdot a^{-} \to e & a \cdot a^{\sim} \to f \\
a^{\sim\sim} \to a^{-\sim} & a^{\sim-} \to a^{--} \\
a^{---} \to a^{-} & a^{--\sim} \to a^{\sim} \\
a \cdot e \to a^{--} & a \cdot f \to a^{-\sim}
\end{array}$$

$$(a \cdot b) \cdot c \to a \cdot (b \cdot c)$$

$$a^{\sim} \cdot b \to a^{-} \cdot b$$

$$\begin{array}{rl}
(a \cdot b)^{-} \to b^{-} \cdot a^{-} & (a \cdot b)^{\sim} \to b^{\sim} \cdot a^{\sim} \\
a^{-} \cdot (a \cdot b) \to b & a \cdot (a^{-} \cdot b) \to b
\end{array}$$

$$a^{--} \cdot b \to a \cdot b$$

It is clear from this set what would be obtained if additional left inverse and right identity functions were supplied. Furthermore if we were to postulate that $a^{-} \equiv a^{\sim}$, then $e \equiv f$. If we postulate that $e \equiv f$, then it follows quickly that $a^{--} \equiv a^{-\sim}$, hence $a^{-} \equiv a^{---} = a^{--\sim} \equiv a^{\sim}$.

Example 15. (l, r) systems III. Clifford's paper [1] introduced still another weakening of the group theory axioms: Besides the associative law

1. $(a \cdot b) \cdot c \to a \cdot (b \cdot c)$

and the existence of a left identity

2. $e \cdot a \to a$

he added the axiom, "For every element a there exists a left identity e and an element b such that $b \cdot a = e$." This setup was suggested by an ambiguous statement of the group theory axioms in the first edition of B. L. van der Waerden's *Moderne Algebra* [Berlin: Springer, 1930, page 15]. Following the conventions of the present paper, his third axiom is equivalent to asserting the existence of *two* unary operators, $'$ and *, with the following two axioms:

3. $a' \cdot a \to a^*$
4. $a^* \cdot b \to b$

Clifford proved the rather surprising result that this set of axioms defines an (l, r) system; and that, conversely, every (l, r) system satisfies this set. Therefore the computer was asked to work on axioms $\{1, 2, 3, 4\}$, to see what the result would be.

After 2 minutes of computation, it was apparent that the system was diverging; 32 axioms were present, including

$$e^{'''''*} \to e^{'''''}, \quad a^{*''''*} \to a^{*''''}, \quad a \cdot a^{'''''} \to a^{'''''*}$$

and others of the same nature. It was not hard to show that, as in Example 2, no finite set of reductions is complete.

But there is a "trick" that *can* be used to solve the word problem for all words composed of the operators in axioms 1–4: Introduce two further unary operators $^\sharp$ and $^\flat$, such that $a' \cdot e \equiv a^\sharp$ and $a \cdot a' \equiv a^\flat$. One of the consequences that the machine had derived very quickly from axioms $\{1, 2, 3, 4\}$ was that $a \cdot (a' \cdot b) \to b$; so, putting $b \equiv e$, we have $a \cdot a^\sharp \equiv e$. Similarly the law $a' \cdot (a \cdot b) \to b$ had been derived, and it follows that $a' \equiv a' \cdot (a \cdot a') \equiv a' \cdot a^\flat \equiv a' \cdot (e \cdot a^\flat) \equiv (a' \cdot e) \cdot a^\flat \equiv a^\sharp \cdot a^\flat$.

Therefore if we take any word involving e, $\cdot$, $'$, and *, we can replace each component of the form α' by $\alpha^\sharp \cdot \alpha^\flat$. Then we have a word in the operators e, $\cdot$, *, $^\sharp$, and $^\flat$. For this new system, the rule

3′. $a^\sharp \cdot (a^\flat \cdot a) \to a^*$

replaces axiom 3. We also know from the discussion above that the rule

5. $a \cdot a^\sharp \to e$

is a legitimate consequence of axioms 1, 2, 3, and 4. Since axioms 1, 2, and 5 define an (l, r) system, we can add their consequences

6. $a \cdot e \to a^{\sharp\sharp}$
7. $a^{\sharp\sharp\sharp} \to a^{\sharp}$

etc., as determined in Example 12. Thus the computer could obtain the following complete set of 21 reductions for words in e, $\cdot$, *, ${}^\sharp$, and ${}^\flat$:

$$(a \cdot b) \cdot c \to a \cdot (b \cdot c)$$
$$e \cdot a \to a \qquad a \cdot a^{\sharp} \to e \qquad a \cdot e \to a^{\sharp\sharp}$$
$$a^{\sharp\sharp\sharp} \to a^{\sharp} \qquad a^{\sharp\sharp} \cdot b \to a \cdot b$$
$$a \cdot (a^{\sharp} \cdot b) \to b \qquad a^{\sharp} \cdot (a \cdot b) \to b$$
$$(a \cdot b)^{\sharp} \to b^{\sharp} \cdot a^{\sharp}$$
$$e^{\sharp} \to e \qquad e^* \to e$$
$$a^* \cdot b \to b \qquad a^{\flat} \cdot b \to b$$
$$a^{\sharp} \cdot a \to a^* \qquad a \cdot a^* \to a$$
$$a^{**} \to a^* \qquad (a \cdot b)^* \to b^*$$
$$a^{\flat\sharp} \to e \qquad a^{*\sharp} \to e \qquad a^{\sharp *} \to e \qquad a^{\flat *} \to a^{\flat}$$

This complete set can be used to solve the original word problem presented by axioms 1, 2, 3, and 4.

Note that although, as Clifford showed, systems satisfying axioms $\{1, 2, 3, 4\}$ are equivalent to (l, r) systems, the *free* systems are quite different. The free system on n generators $\{g_1, \ldots, g_n\}$ defined by the three axioms of Example 12 has exactly $n + 1$ idempotent elements, namely e, $g_1^- \cdot g_1$, $\ldots$, $g_n^- \cdot g_n$; the free system on one generator defined by axioms $\{1, 2, 3, 4\}$ of the present example has infinitely many idempotent elements, including $\alpha^{\flat}$ for each irreducible word α.

Example 16. Central groupoids II. (Compare with Example 6.) A natural model of a central groupoid with n^2 elements is obtained by considering the set S of ordered pairs $\{(x_1, x_2) \mid x_1, x_2 \in S_0\}$, where S_0 is a set of n elements. If we define the product $(x_1, x_2) \cdot (y_1, y_2) = (x_2, y_1)$, we find that the basic identity $(a \cdot b) \cdot (b \cdot c) = b$ is satisfied.

Any ordered pair $x = (x_1, x_2)$ is the product of two idempotent elements $(x_1, x_1) \cdot (x_2, x_2)$ according to these conventions. We have $(x_1, x_1) = (x \cdot x) \cdot x$, and $(x_2, x_2) = x \cdot (x \cdot x)$; this observation suggests that we define, in a central groupoid, two unary functions denoted by subscripts 1 and 2. Namely, we postulate

1. $(a \cdot a) \cdot a \to a_1$
2. $a \cdot (a \cdot a) \to a_2$

in addition to the basic axiom

3. $(a \cdot b) \cdot (b \cdot c) \to b$

that defines a central groupoid.

For reasons that are explained in detail in [7], it is especially interesting to add the further axiom

4. $a_2 \cdot b \to a \cdot b$

(which is valid in the "natural" central groupoids but not in all central groupoids), and to see if this rather weak axiom implies that we must have a "natural" central groupoid.

This conjecture is, in fact, true, although previous investigations by hand had been unable to derive the result. The computer started with axioms 1, 2, 3, 4, and after 9 minutes the following complete set of 13 reductions was found:

$$a_{11} \to a_1 \qquad a_{12} \to a_1 \qquad a_{21} \to a_2 \qquad a_{22} \to a_2$$
$$(a \cdot b)_1 \to a_2 \qquad (a \cdot b)_2 \to b_1$$
$$a \cdot (b \cdot c) \to a \cdot b_2 \qquad (a \cdot b) \cdot c \to b_1 \cdot c$$
$$a_2 \cdot b \to a \cdot b \qquad a \cdot b_1 \to a \cdot b$$
$$a \cdot a_2 \to a_2 \qquad a_1 \cdot a \to a_1 \qquad a_1 \cdot a_2 \to a$$

The computation process generated 54 axioms, of which 24 were used in the derivation of the final set, so the "efficiency rating" was 44%. This is the most difficult problem solved by the computer program so far.

As a consequence of the reduction rules above, the free system on n generators has $4n^2$ elements.

Example 17. Central groupoids III. If we start with only axioms 1, 2, and 3 of Example 16, the resulting complete set has 25 reductions:

$$(a \cdot a) \cdot a \to a_1 \qquad a \cdot (a \cdot a) \to a_2$$
$$a_1 \cdot a_2 \to a \qquad a_2 \cdot a_1 \to a \cdot a$$
$$a \cdot a_1 \to a \cdot a \qquad a_2 \cdot a \to a \cdot a$$
$$(a \cdot a)_1 \to a_2 \qquad (a \cdot a)_2 \to a_1$$
$$a_{11} \cdot a \to a_1 \qquad a \cdot a_{22} \to a_2$$
$$a_1 \cdot (a \cdot b) \to a \qquad (a \cdot b) \cdot b_2 \to b$$
$$(a \cdot b)_1 \cdot b \to a \cdot b \qquad a \cdot (a \cdot b)_2 \to a \cdot b$$
$$(a \cdot b_1) \cdot b \to b_1 \qquad a \cdot (a_2 \cdot b) \to a_2$$
$$(a \cdot a) \cdot a_{12} \to a_1 \qquad a_{21} \cdot (a \cdot a) \to a_2$$
$$(a \cdot a) \cdot (a_1 \cdot b) \to a_1 \qquad (a \cdot b_2) \cdot (b \cdot b) \to b_2$$
$$(a \cdot (b \cdot b)) \cdot b_1 \to b \cdot b \qquad a_2 \cdot ((a \cdot a) \cdot b) \to a \cdot a$$
$$(a \cdot b) \cdot (b \cdot c) \to b$$
$$a \cdot ((a \cdot b) \cdot c) \to a \cdot b \qquad (a \cdot (b \cdot c)) \cdot c \to b \cdot c$$

Of course these 25 reductions say no more than the three reductions of Example 6, if we replace a_1 by $(a \cdot a) \cdot a$ and a_2 by $a \cdot (a \cdot a)$ everywhere, so they have little mathematical interest. They have been included here merely as an indication of the speed of our experimental implementation. If these 25 axioms are presented to our program, it requires almost exactly 2 minutes to prove that they form a complete set.

Example 18. Some unsuccessful experiments. The major restriction of the present algorithm is that it cannot handle systems in which there is a commutative binary operator, where

$$a \circ b \equiv b \circ a.$$

We have no way of deciding in general how to construe this as a "reduction," so the method must be supplemented with additional techniques to cover this case. Presumably an approach could be worked out in which we use *two* reductions

$$\alpha \to \beta \text{ and } \beta \to \alpha$$

whenever we find that $\alpha \equiv \beta$ but $\alpha \# \beta$, and to make sure that no infinite looping occurs when reducing words to a new kind of "irreducible" form. At any rate it is clear that the methods of this paper ought to be extended to such cases, so that rings and other varieties can be studied.

We tried experimenting with Burnside groups, by adding the axiom $a \cdot (a \cdot a) \to e$ to the set of ten reductions of Example 1. The computer almost immediately derived

$$a \cdot (b' \cdot a) \equiv b \cdot (a' \cdot b)$$

in which each side is a commutative binary function of a and b. Therefore no more could be done by our present method.

Another type of axiom we do not presently know how to handle is a rule of the following kind:

$$\text{if } a \not\equiv 0 \text{ then } a \cdot a' \to e.$$

Thus, division rings would seem to be out of the scope of this present study even if we could handle the commutative law for addition.

The "Semi-Automated Mathematics" system of Guard, Oglesby, Bennett, and Settle [6] illustrates the fact that the superposition techniques used here lead to efficient procedures in the more general situation where axioms involving quantifiers and other logical connectives are allowed as well. That system generates "interesting" consequences of axioms it is given, by trial and error. Its techniques are related to but not identical to the methods described in this paper, since it uses both "expansions" and "reductions" separately, and it never terminates unless it has been asked to prove or disprove a specific result.

8. Conclusions

The long list of examples in the preceding section shows that the computational procedure of §6 can give useful results for many interesting and important algebraic systems. The methods of Evans [4] have essentially been extended so that the associative law can be treated, but not yet the commutative law. On small systems, the computations can be done by hand, and the method is a powerful tool for solving algebraic problems of the types described in Examples 4 and 6. On larger problems, a computer can be used to derive consequences of axioms that would be very difficult to do without machine aid. Although we deal only with "identities," other axioms such as cancellation laws can be treated as shown in Examples 9 and 11.

The method described here ought to be extended so that it can handle the commutative law and other systems discussed under Example 18. Another modification worth considering is to change the definition of the well-ordering so that it evaluates the weights of subwords differently depending on the operators that operate on these subwords. Thus, in Example 11 we would have liked to write

$$f(a \cdot b, a) \to g(a \cdot b, b),$$

and in Example 15 we would have liked to write

$$a' \to a^{\sharp} \cdot a^{\flat}.$$

These options were not allowed by the present definition of well-ordering, but other well-orderings exist in which such rules are reductions no matter what is substituted for a and b.

The work reported in this paper was supported in part by the U.S. Office of Naval Research.

References

[1] A. H. Clifford, "A system arising from a weakened set of group postulates," *Annals of Mathematics* (2) **34** (1933), 865–871.

[2] A. H. Clifford and G. B. Preston, *The Algebraic Theory of Semigroups*, Mathematical Surveys No. 7 (Providence, Rhode Island: American Mathematical Society, 1961).

[3] Leonard Eugene Dickson, "Definitions of a group and a field by independent postulates," *Transactions of the American Mathematical Society* **6** (1905), 198–204.

[4] Trevor Evans, "On multiplicative systems defined by generators and relations, I. Normal form theorems," *Proceedings of the Cambridge Philosophical Society* **47** (1951), 637–649.

[5] Trevor Evans, "Products of points — some simple algebras and their identities," *American Mathematical Monthly* **74** (1967), 362–372.

[6] J. R. Guard, F. C. Oglesby, J. H. Bennett, and L. G. Settle, "Semi-automated mathematics," *Journal of the Association for Computing Machinery* **16** (1969), 49–62.

[7] Donald E. Knuth, "Notes on central groupoids," *Journal of Combinatorial Theory* **8** (1970), 376–390. [Reprinted as Chapter 22 of *Selected Papers on Discrete Mathematics*, CSLI Lecture Notes 106 (Stanford, California: Center for the Study of Language and Information, 2003), 357–375.]

[8] Henry B. Mann, "On certain systems which are almost groups," *Bulletin of the American Mathematical Society* **50** (1944), 879–881.

[9] M. H. A. Newman, "On theories with a combinatorial definition of 'equivalence'," *Annals of Mathematics* (2) **43** (1942), 223–243.

[10] J. A. Robinson, "A machine-oriented logic based on the resolution principle," *Journal of the Association for Computing Machinery* **12** (1965), 23–41.

[11] O. Taussky, "Zur Axiomatik der Gruppen," *Ergebnisse eines mathematischen Kolloquiums* **4** (1933), 2–3.

Addendum

The ideas and methods of this chapter have been extended in many directions, too numerous to list here. Nachum Dershowitz has given an excellent survey of the earliest developments in his paper "Termination of rewriting," *Journal of Symbolic Computation* **3** (1987), 69–115; **4** (1987), 409–410.

Chapter 20

Efficient Representation of Perm Groups

[Dedicated to the memory of Marshall Hall. Originally published in Combinatorica **11** *(1991), 33–43.]*

This note presents an elementary version of Sims's algorithm for computing strong generators of a given perm group, together with a proof of correctness and some notes about appropriate low-level data structures. Upper and lower bounds on the running time are also obtained. (Following a suggestion of Vaughan Pratt, we adopt the convention that perm = permutation, perhaps thereby saving millions of syllables in future research.)

1. A Data Structure for Perm Groups

A "perm," for the purposes of this paper, is a one-to-one mapping of a set onto itself. If α and β are perms such that α takes $i \mapsto j$ and β takes $j \mapsto k$, the product $\alpha\beta$ takes $i \mapsto k$. We write α^- for the inverse of the perm α; hence $\alpha\beta = \gamma$ if and only if $\alpha = \gamma\beta^-$.

Let $\Pi(k)$ be the set of all perms of the positive integers that fix all points $> k$. Consider the following data structure: For $1 \le j \le k$, either $\sigma_{k,j} = \emptyset$ or $\sigma_{k,j}$ is a perm in $\Pi(k)$ that takes $k \mapsto j$. Let $\Sigma(k)$ be the set of all non-$\emptyset$ perms $\sigma_{k,j}$. We assume that $\sigma_{k,k}$ is the identity perm; hence $\Sigma(k)$ is always nonempty.

We write $\Gamma(k)$ for the set of all perms that can be written as products of the form $\sigma_1 \ldots \sigma_k$ where each σ_i is in $\Sigma(i)$. There is an easy way to test if a given perm $\pi \in \Pi(k)$ is a member of $\Gamma(k)$: Let π take $k \mapsto j$. Then if $\sigma_{k,j} = \emptyset$ we have $\pi \notin \Gamma(k)$; otherwise if $k = 1$ we have $\pi \in \Gamma(k)$; otherwise $\pi \in \Gamma(k)$ if and only if $\pi\sigma_{k,j}^- \in \Gamma(k-1)$.

The data structure also includes a set $T(k) \subseteq \Pi(k)$ with the invariant property that each element of $\Gamma(k)$ can be written as a product of elements of $T(k)$. In other words, $\Gamma(k)$ will be a subset of the group

$\langle T(k)\rangle$ generated by $T(k)$, for all k, throughout the course of the algorithm to be described. (Since all elements π of $\Pi(k)$ are finite perms, we have $\pi^- = \pi^r$ for some $r > 0$; hence closure under multiplication implies closure under inversion.)

The data structure is said to be up-to-date of order n if $\Gamma(k) = \langle T(k)\rangle$ for $1 \le k \le n$; equivalently, $\Gamma(k) \supseteq T(k)$ and $\Gamma(k)$ is closed under multiplication, for $1 \le k \le n$. In that case we say that the perms $\bigcup_{k=1}^{n} \Sigma(k)$ form a *transversal system* of $\Gamma(n)$, and that the perms $\bigcup_{k=1}^{n} T(k)$ are *strong generators* of $\Gamma(n)$. Having a transversal system makes it easy to determine what perms are generated by a given set of perms $T(n)$.

2. Maintaining the Data Structure

Let us now discuss two algorithms that can be used to transform the data structure when a new perm is introduced into $T(k)$. We will first look at the algorithms, then discuss why they are valid.

Algorithm $A_k(\pi)$. Assuming that the data structure is up-to-date of order k, and that $\pi \in \Pi(k)$ but $\pi \notin \Gamma(k)$, this procedure appends π to $T(k)$ and brings the data structure back up-to-date so that $\Gamma(k)$ will equal the new $\langle T(k)\rangle$.

Step A1. Insert π into the set $T(k)$.

Step A2. Perform Algorithm $B_k(\sigma\tau)$ for all $\sigma \in \Sigma(k)$ and $\tau \in T(k)$ such that $\sigma\tau$ is not already known to be a member of $\Gamma(k)$. (Algorithm B_k may increase the size of $\Sigma(k)$; any new perms σ that are added to $\Sigma(k)$ must also be included in this step. Implementation details are discussed in Section 3 below.) □

Algorithm $B_k(\pi)$. Assuming that the data structure is up-to-date of order $k-1$, and that $\pi \in \langle T(k)\rangle$, this procedure ensures that π is in $\Gamma(k)$ and that the data structure remains up-to-date of order $k-1$. (The value of k will always be greater than 1.)

Step B1. Let π take $k \mapsto j$.

Step B2. If $\sigma_{k,j} = \emptyset$, set $\sigma_{k,j} \leftarrow \pi$ and terminate the algorithm.

Step B3. If $\pi\sigma_{k,j}^- \in \Gamma(k-1)$, terminate the algorithm. (This test for membership in $\Gamma(k-1)$ has been described in Section 1 above.)

Step B4. Perform Algorithm $A_{k-1}(\pi\sigma_{k,j}^-)$. □

The correctness of these mutually recursive procedures follows readily from the stated invariant relations, except for one nontrivial fact: We must verify that $\Gamma(k)$ is closed under multiplication at the conclusion of Algorithm $A_k(\pi)$. This is obvious when $k = 1$, so we may assume that $k > 1$. Let α and β be elements of $\Gamma(k)$. By definition of $\Gamma(k)$ we can write $\alpha = \gamma\sigma$, where $\gamma \in \Gamma(k-1)$ and $\sigma \in \Sigma(k)$; and by the invariant relation $\Gamma(k) \subseteq \langle T(k) \rangle$ we can write $\beta = \tau_1 \ldots \tau_r$ where each $\tau_i \in T(k)$. We know that $\sigma\tau_1 \in \Gamma(k)$, by step A2; hence $\sigma\tau_1 = \gamma_1\sigma_1$ for some $\gamma_1 \in \Gamma(k-1)$ and some $\sigma_1 \in \Sigma(k)$. Similarly $\sigma_1\tau_2 = \gamma_2\sigma_2$, etc., and we finally obtain $\alpha\beta = \gamma\gamma_1 \ldots \gamma_r\sigma_r$. This proves that $\alpha\beta \in \Gamma(k)$, since $\gamma\gamma_1 \ldots \gamma_r$ is in $\Gamma(k-1)$ by induction.

3. Low-Level Implementation Hints

Let $s(k)$ be the cardinality of $\Sigma(k)$ and let $t(k)$ be the cardinality of $T(k)$. The algorithms of Section 2 can perhaps be implemented most efficiently in practice by keeping a linear list of the perms $\tau(k,1), \ldots, \tau\bigl(k,t(k)\bigr)$ of $T(k)$, for each k, together with an array of pointers to the representations of each $\sigma_{k,j}$ for $1 \le j < k$, using a null pointer to represent the relation $\sigma_{k,j} = \emptyset$. It is also convenient to have a linear list $j(k,1)$, $\ldots$, $j(k,s(k))$ of the indices of the non-$\emptyset$ perms $\sigma_{k,j}$, where $j(k,1) = k$. We will see below that the algorithm often completes its task without needing to make many of the sets $\Sigma(k)$ very large; thus most of the $\sigma_{k,j}$ are often $\emptyset$. Pointers can be used to avoid duplications between $T(k)$ and $T(k-1)$.

There are two fairly simple ways to handle the loop over σ and τ in step A2; one is recursive and the other is iterative. The recursive method replaces step A2 by the following operation: "Perform Algorithm $B_k(\sigma\pi)$ for all σ in the current set $\Sigma(k)$." Then step B2 is also changed: "If $\sigma_{k,j} = \emptyset$, set $\sigma_{k,j} \leftarrow \pi$ and perform $B_k(\pi\tau)$ for all τ in the current set $T(k)$, then terminate the algorithm."

The iterative method maintains an additional table, in order to remember which pairs (σ, τ) have already been tested in step A2. This table consists of counts $c(k,i)$ for each k and for $1 \le i \le s(k)$, such that the product $\sigma_{k,j(k,i)}\tau(k,l)$ is known to be in $\Gamma(k)$ for $1 \le l \le c(k,i)$. When step B2 increases the value of $s(k)$, the newly created count $c(k,s(k))$ is set to zero. Step A2 is a loop of the form

```
i ← 1;
while i ≤ s(k) do
   begin while c(k,i) < t(k) do
      begin l ← c(k,i) + 1;  B_k(σ_{k,j(k,i)} τ(k,l));
      c(k,i) ← l;
      end;
   i ← i + 1;
   end;
```

the invocation of B_k may increase $s(k)$, but it can change $t(k')$ and $c(k', i')$ only for values of k' that are less than k.

The iterative method carries out its tests in a different order from the recursive method, so it might yield a different transversal system.

It is convenient to represent each perm σ of $\Sigma(k)$ indirectly in an array q that gives inverse images, so that σ takes $q[i] \mapsto i$ for $1 \le i \le k$. All other perms π can be represented directly in an array p, with π taking $i \mapsto p[i]$ for $1 \le i \le k$. To compute the direct representation d of the product $\pi\sigma^-$, we can then simply set $d[i] \leftarrow q\bigl[p[i]\bigr]$ for $1 \le i \le k$. To compute the direct representation d of the product $\sigma\pi$, we set $d\bigl[q[i]\bigr] \leftarrow p[i]$ for $1 \le i \le k$. Thus, the elementary operations are fast.

4. Upper Bounds on the Running Time

The "inner loop" of the updating algorithms occurs in step B3, the membership test. Testing for membership of $\pi \in \Gamma(k)$ involves multiplication by some sequence of non-identity perms $\sigma^-_{k_1,j_1}, \ldots, \sigma^-_{k_r,j_r}$, where $k \ge k_1 > \cdots > k_r > 0$; so the running time is essentially proportional to $k + k_1 + \cdots + k_r$, which is $O(k^2)$ in the worst case.

The total number of executions of $B_k(\sigma\tau)$ is $s(k)t(k)$, and we have $s(k) \le k$. The value of $t(k)$ increases by 1 each time we perform $A_k(\pi)$; every time we do this, we increase $\Gamma(k)$ to a larger subgroup of $\Pi(k)$, hence $t(k)$ cannot exceed the length of the longest chain of subgroups of the symmetric group $\Pi(k)$. A straightforward upper bound is therefore $t(k) \le \theta(k!) = O(k \log\log k)$, where $\theta(N)$ is the number of prime divisors of N counting multiplicity. Babai [1] has shown that $\Pi(k)$ admits no subgroup chains of length exceeding $2k - 3$, for $k \ge 2$; hence we have the sharper estimate $t(k) = O(k)$.

It follows that Algorithm $B_k(\sigma\tau)$ is performed $O(k^2)$ times, and each occurrence of step B3 takes $O(k^2)$ units of time. Summing for $1 \le k \le n$ allows us to conclude that *a transversal system for a perm group generated by m perms of $\Pi(n)$ can be found in at most $O(n^5) + O(mn^2)$ steps.* (The term $O(mn^2)$ comes from m membership tests, which are carried out on each generator π before Algorithm $A_n(\pi)$ is applied.)

The storage requirement for each non-identity perm of $\Sigma(k)$ or $T(k)$ is $O(k)$; hence we need at most $O(k^2)$ memory cells for perms of $\Pi(k)$, and $O(n^3)$ memory cells in all.

5. A Sparse Example

Actual computations with these procedures rarely take as much time as our worst-case estimates predict. We can learn more about the true efficiency by studying particular cases in detail. Let us therefore consider first the case of a group generated by a single non-identity perm $\pi \in \Pi(n)$.

We begin, of course, with $\sigma_{k,j} = \emptyset$ for $1 \le j < k \le n$ and $T(k) = \emptyset$ for $1 \le k \le n$; the data structure is then up-to-date of order n, and we can perform $A_n(\pi)$. Suppose π takes $n \mapsto a_1 \mapsto \cdots \mapsto a_{r-1} \mapsto n$. Then $A_n(\pi)$ will set $T(n) \leftarrow \{\pi\}$ and $\sigma_{n,a_j} \leftarrow \pi^j$ for $1 \le j < r$, and it will invoke $A_{n-1}(\pi^r)$ (unless π^r is the identity perm, in which case the algorithm will terminate).

If, for example, we have

$$\pi = [1,2,3,4,5,6,7,14]\,[8,9,10,13]\,[11,12]$$

in cycle form, the algorithm will set $\sigma_{14,j} \leftarrow \pi^j$ for $1 \le j < 8$, and it will terminate with $T(14) = \{\pi\}$ and with all other $T(k)$ empty. But if we relabel points 12 and 14, obtaining the conjugate perm

$$\bar{\pi} = [1,2,3,4,5,6,7,12]\,[8,9,10,13]\,[11,14]\,,$$

the algorithm will act quite differently: The nontrivial perms $\sigma_{k,j}$ and sets $T(k)$ will now be

$$\begin{array}{lll} \sigma_{14,11} = \bar{\pi}\,, & \sigma_{13,9} = \bar{\pi}^2\,, & \sigma_{12,4} = \bar{\pi}^4\,; \\ T(14) = \{\bar{\pi}\}\,, & T(13) = \{\bar{\pi}^2\}\,, & T(12) = \{\bar{\pi}^4\}\,. \end{array}$$

When the algorithm terminates, it has produced a transversal system by which we can test if a given perm ρ is a power of π or $\bar{\pi}$, respectively. In the first case this membership test involves at most one multiplication, by $\sigma_{14,j}$ if ρ takes $14 \mapsto j$ where $j < 8$. In the second case the test will involve three multiplications if we have, say, $\rho = \bar{\pi}^7$.

These perms π and $\bar{\pi}$ are the special case $h = 4$ of an infinite family of perms of degree $n = 2^h - 2$, having cycles of lengths $2^{h-1}, 2^{h-2}, \ldots, 2^1$. In general π will cause $\sim n/2$ slots $\sigma_{k,j}$ to become nonempty, and it will terminate after performing $\sim n^2/2$ elementary machine steps, yielding a membership test whose worst-case running time is $\sim n$. The corresponding perm $\bar{\pi}$ will cause only $\sim \lg n$ slots $\sigma_{k,j}$ to become nonempty, and it will terminate after $\sim 2n \lg n$ steps, yielding a membership test whose worst-case running time is $\sim n \lg n$. Thus, the algorithm's performance can change dramatically when only two points of its input perm are relabeled.

6. A Dense Example

The algorithm needs to work harder when we wish to find the group generated by $\{\pi_2, \pi_3, \ldots, \pi_n\}$, where $\pi_k \in \Pi(k)$ takes $k \mapsto k-1$, and where the generators π_k are input in increasing order of k. Then it is not difficult to verify by induction that the algorithm will terminate with $T(k) = \{\pi_2, \ldots, \pi_k\}$ and with $\sigma_{k,j} \neq \emptyset$ for $1 \le j < k \le n$. Thus, the algorithm will fill all of the slots $\sigma_{k,j}$, thereby implicitly deducing that each $\Gamma(k)$ is the full symmetric group $\Pi(k)$.

Moreover, if the recursive method of Section 3 is being used to implement step A2, the algorithm will terminate with

$$\sigma_{k,j} = \pi_k \pi_{k-1} \ldots \pi_{j+1}\,, \qquad \text{for } 1 \le j < k \le n.$$

For after $\sigma_{k,j}$ has been defined, the modified step B2 will continue to test whether the perms

$$\sigma_{k,j}\pi_2\,, \quad \sigma_{k,j}\pi_3\,, \quad \ldots, \quad \sigma_{k,j}\pi_k$$

belong to the current $\Gamma(k)$. The first $j-2$ will succeed; then $B_k(\sigma_{k,j}\pi_j)$ will cause $\sigma_{k,j-1}$ to be defined. And by the time the recursive call on $B_k(\sigma_{k,j}\pi_j)$ returns control to $B_k(\sigma_{k,j})$, the values of $\sigma_{k,i}$ will be non-$\emptyset$ for all $i < k$; hence the remaining tests on $\sigma_{k,j}\pi_l$ for $l > j$ will succeed.

Let us examine the special case of this construction in which each π_k is the simple transposition $[k, k-1]$. How much time is taken by the $\Theta(n^3)$ membership tests $\sigma_{k,j}\pi_i \in \Gamma(k)$? We have

$$\sigma_{k,j} = [j,\, j+1,\, \ldots,\, k]\,,$$

and it follows that

$$\sigma_{k,j}\pi_i = \begin{cases} \sigma_{i,i-1}\,\sigma_{k,j}\,, & \text{if } 1 < i < j; \\ \sigma_{k,i}\,, & \text{if } i = j+1; \\ \sigma_{i-1,i-2}\,\sigma_{k,j}\,, & \text{if } i > j+1. \end{cases}$$

Each membership test therefore involves at most two multiplications by non-identity perms, and the total running time of the algorithm is $\Theta(n^4)$.

Another interesting special case occurs when each π_k is the cyclic perm $[k, k-1, \ldots, 1]$. Here we find that $\sigma_{k,j}$ takes

$$x \mapsto \begin{cases} k+1-x, & \text{if } 1 \le x \le k-j; \\ x-(k-j), & \text{if } k-j < x \le k. \end{cases}$$

It turns out that we have

$$\sigma_{k,j}\pi_i = \begin{cases} \sigma_{k-j,1}\,\sigma_{k-j+1,1}\,\sigma_{k-j+i,k-j+i-1}\,\sigma_{k,j}\,, & \text{if } i < j; \\ \sigma_{k-i,1}\,\sigma_{k-j,1}\,\sigma_{k-j+1,k-i+1}\,\sigma_{k,j-1}\,, & \text{if } 1 < j < i < k; \\ \sigma_{k-i,1}\,\sigma_{k-2,i-2}\,\sigma_{k-1,k-i+1}\,\sigma_{k,i}\,, & \text{if } j = 1 \text{ and } 2 < i < k; \\ \sigma_{k,2}\,, & \text{if } j = 1 \text{ and } i = 2; \\ \sigma_{k-1,1}\,, & \text{if } j = 1 \text{ and } i = k; \\ \sigma_{k-j,1}\,\sigma_{k-j+1,1}\,\sigma_{k,j-1}\,, & \text{if } 1 < j < i = k. \end{cases}$$

So the membership tests need at most 4 multiplications each, and again the total running time is $\Theta(n^4)$.

In both of these special cases, it turns out that the iterative implementation of step A2 will also define the same perms $\sigma_{k,j}$. Hence the running time will be $\Theta(n^4)$ under either of the implementations we have discussed.

It is interesting to analyze the algorithm in another special case, when there are just two generators

$$\sigma_n = [1, 2, \ldots, n] \qquad \text{and} \qquad \tau_n = [n-1, n].$$

Assume that the recursive implementation of A2 is used. First, Algorithm $A_n(\sigma_n)$ sets $T(n) = \{\sigma_n\}$ and performs $B_n(\sigma_n)$. Algorithm $B_n(\sigma_n)$ sets $\sigma_{n,1} \leftarrow \sigma_n$ and performs $B_n(\sigma_n^2)$, which sets $\sigma_{n,2} \leftarrow \sigma_n^2$ and performs $B_n(\sigma_n^3)$, etc. Thus $\sigma_{n,j}$ becomes σ_n^j for all j. Second, Algorithm $A_n(\tau_n)$ adds τ_n to $T(n)$ and performs $B_n(\tau_n)$, $B_n(\sigma_n\tau_n)$, …, $B_n(\sigma_n^{n-1}\tau_n)$. The first of these subroutines, $B_n(\tau_n)$, performs Algorithm $A_{n-1}(\tau_n\sigma_n)$, which is $A_{n-1}(\sigma_{n-1})$. The second subroutine, $B_n(\sigma_n\tau_n)$, performs $A_{n-1}(\sigma_n\tau_n\sigma_n^-)$, which is $A_{n-1}(\tau_{n-1})$. Therefore we can use induction on n to show that $\sigma_{k,j} = \sigma_k^j$ for all $j < k$. It is easy to verify that each membership test requires at most three nontrivial multiplications. Therefore the total running time in this special case comes to only $\Theta(n^3)$, although $\Gamma(n)$ is the full symmetric group $\Pi(n)$.

7. A Random Example

The conditions of the preceding construction allow $(k-1)!$ possibilities for each perm π_k. Let us consider the average total running time of the algorithm when each of the $1!\,2!\ldots(n-1)!$ choices of $\{\pi_2, \pi_3, \ldots, \pi_n\}$ is equally likely. On intuitive grounds it appears plausible that the average running time will be $\Theta(n^5)$, because most of the multiplications in a "random" situation will be by non-identity perms. This indeed turns out to be true, at least when the recursive implementation of step A2 is used; but the proof is a bit delicate.

As before, the running time is dominated by $\Theta(n^3)$ successful tests for membership of $\sigma_{k,j}\pi_i$ in $\Gamma(k)$, where $k > j \geq 1$ and $k \geq i > 1$ and $i \neq j$. We know that the total running time is $O(n^5)$, so we need only show that the average value is $\Omega(n^5)$; and for this purpose it will suffice to consider only the membership tests with $k > j > i$.

The membership test for $\sigma_{k,j}\pi_i$ performs the multiplications

$$\sigma_{k,j}\pi_i\sigma^-_{k,j_k}\sigma^-_{k-1,j_{k-1}}\ldots\sigma^-_{2,j_2}\,,$$

and the cost is l for each multiplication such that $j_l \neq l$. Since $j > i$, we always have $j_k = j$. Let us fix the values k, j, i, and l, where $k > j > i > 1$ and $k > l > i$, and try to determine an upper bound for the probability that $j_l = l$. The following analysis applies to any given (not necessarily random) sequence of perms $\pi_l, \ldots, \pi_2$, with $\pi_k, \ldots, \pi_{l+1}$ varying randomly.

Let $i - r$ be the number of points $\leq i$ that are fixed by the given perm π_i. By assumption, π_i takes $i \mapsto i - 1$, hence $r \geq 2$.

Our first goal is to determine the probability that we have

$$j_{k-1} = k - 1, \quad j_{k-2} = k - 2, \quad \ldots, \quad j_l = l.$$

This holds if and only if $\sigma_{k,j}\pi_i\sigma^-_{k,j} \in \Pi(l-1)$. Note that, in the recursive implementation of step A2, we have

$$\sigma_{k,j}\pi_i\sigma^-_{k,j} = \pi_k\pi_{k-1}\ldots\pi_{j+1}\pi_i\pi^-_{j+1}\ldots\pi^-_{k-1}\pi^-_k = \pi_k\rho\pi^-_k\,,$$

where ρ is a perm of $\Pi(k-2)$ that has the same cycle structure as π_i; hence ρ fixes exactly $k - 2 - r$ points $\leq k - 2$. Consider what happens to $\pi_k\rho\pi^-_k$ as π_k runs through its $(k-1)!$ possible values: We obtain a uniform distribution over all perms of $\Pi(k-1)$ having the same cycle structure as ρ. For example, if $r = 7$ and $\rho = [1,2,7][3,6][4,9]$, the $(k-1)!$ perms $\pi_k\rho\pi^-_k$ are just $[a_1,a_2,a_7][a_3,a_6][a_4,a_9]$ as $a_1 \ldots a_{k-1}$ runs through the images of all perms of $\Pi(k-1)$. Therefore the probability that $\sigma_{k,j}\pi_i\sigma^-_{k,j} \in \Pi(l-1)$ is

$$\binom{l-1}{r}\bigg/\binom{k-1}{r} = \frac{(l-1)(l-2)\ldots(l-r)}{(k-1)(k-2)\ldots(k-r)}\,.$$

Now let's compute the probability that $j_{k-1} = k-1$, …, $j_{q+1} = q+1$, $j_q < q$, and $j_l = l$, given a subscript q in the range $k > q > l$. We will assume that $\pi_{k-1}, \ldots, \pi_{q+1}, \pi_{q-1}, \ldots, \pi_2$ have been assigned some fixed values, while π_k and π_q run independently through all of their $(k-1)!\,(q-1)!$ possibilities. Under these circumstances we will prove that $\sigma_{k,j}\pi_i\sigma^-_{k,j}\sigma^-_{q,j_q}$ is uniformly distributed over $\Pi(q-1)$.

Let p be a positive integer less than q. Let $\alpha \in \Pi(q)$ take $q \mapsto p$ and have the same cycle structure as π_i. Also let β be an element of $\Pi(q-1)$. Then there is exactly one perm π_q that will make $\alpha\sigma_{q,p}^{-} = \beta$, namely

$$\pi_q = \beta^{-}\alpha\pi_{p+1}^{-}\dots\pi_{q-1}^{-}\,.$$

(This perm takes $q \mapsto q-1$ and fixes all points $> q$, so it meets the conditions necessary to be called π_q.) Moreover, when π_q has this value, the number of perms π_k such that $\sigma_{k,j}\pi_i\sigma_{k,j}^{-} = \alpha$ is independent of α, as we have observed in the previous case. Therefore the probability that we have both $j_q = p$ and $\sigma_{k,j}\pi_i\sigma_{k,j}^{-}\sigma_{q,p}^{-} = \beta$ is independent of β, and independent of p.

The uniform distribution of $\sigma_{k,j}\pi_i\sigma_{k,j}^{-}\sigma_{q,p}^{-}$ implies that the probability of satisfying all of the conditions $j_{k-1} = k-1, \dots, j_{q+1} = q+1$, $j_q < q$, and $j_l = l$ is equal to $1/l$ times the probability that $j_{k-1} = k-1$, $\dots$, $j_{q+1} = q+1$, and $j_q < q$, because the values $j_{q-1}\dots j_2$ are uniformly distributed. And we know from the previous analysis that this probability is

$$\frac{1}{l}\left(\frac{q(q-1)\dots(q-r+1)-(q-1)(q-2)\dots(q-r)}{(k-1)(k-2)\dots(k-r)}\right) = \frac{r}{l}\,\frac{(q-1)\dots(q-r+1)}{(k-1)\dots(k-r)}\,.$$

Finally, therefore, we can compute the probability that $j_l = l$, when k, j, i, and l are given as above and π_i has $i-r$ fixed points: It comes to

$$\frac{1}{(k-1)\dots(k-r)}\left((l-1)\dots(l-r) + \frac{r}{l}\sum_{l<q<k}(q-1)\dots(q-r+1)\right)$$

$$= \frac{1}{l} + \frac{(l-1)\dots(l-r)(l-r-1)}{(k-1)\dots(k-r)(l-r)}$$

$$< \frac{1}{l} + \frac{(l-1)\dots(l-r)}{(k-1)\dots(k-r)}\,.$$

Since $r \geq 2$, we obtain the desired upper bound

$$\Pr(j_l = l) < \frac{1}{l} + \frac{(l-1)(l-2)}{(k-1)(k-2)} < \frac{1}{l} + \frac{l^2}{k^2}\,.$$

This implies the desired lower bound $\Omega(n^5)$ on the total multiplication time. We can, for example, sum over $\Omega(n^4)$ values (k,j,i,l) with $1 < i \leq \frac{1}{4}n < l \leq \frac{1}{2}n < j \leq \frac{3}{4}n < k \leq n$; in each of these cases a multiplication will require $\Omega(n)$ steps with probability at least $1 - (1/l + l^2/k^2) > 1/2$ when $n > 72$.

Since the average running time is $\Omega(n^5)$, there must exist, for all n, a sequence of perms $\pi_2, \ldots, \pi_n$ that make the algorithm do $\Omega(n^5)$ operations. But it appears to be difficult to define such perms via an explicit construction. Nor is there an obvious way to prove the $\Omega(n^5)$ bound when the iterative implementation of step A2 is adopted in place of the recursive implementation, even in the totally random case.

8. More Meaningful Upper Bounds

The examples studied above show that it is misleading to characterize Algorithms A_n and B_n by merely saying that they will process m perms of $\Pi(n)$ with a worst-case running time of $O(n^5 + mn^2)$. In one sense this estimate is sharp, because we've seen that $\Omega(n^5)$ behavior may indeed occur; but our other examples, together with extensive computational experience, show that the procedures often run considerably faster in practice.

We can improve the estimate of Section 4 by introducing another parameter. Let g be the order of the group $\Gamma(n)$ that is generated. Then we have the following result:

Theorem. *A transversal system for a perm group of order g generated by m perms of $\Pi(n)$ can be found in at most $O\bigl(n^2(\log g)^3/\log n\bigr) + O(mn\log g)$ steps, using at most $O(n^2\log g/\log n) + O(n(\log g)^2)$ cells of memory.*

Proof. Let $s(k)$ and $t(k)$ be defined as before. Then $g = \prod_{k=1}^n s(k)$, and the number of membership tests is

$$m + \sum_{k=1}^{n}\bigl(s(k)t(k) - s(k) + 1\bigr).$$

Each membership test involves at most $O(\log g)$ multiplications by non-identity perms, because the number of indices k with $s(k) > 1$ cannot exceed $\theta(g)$, the total number of prime factors of g. This reasoning accounts for the term $O(mn\log g)$ in the theorem.

Moreover, each $t(k)$ is at most $\theta(g) = O(\log g)$, as we have argued before. Therefore we can complete the proof by showing that $\sum_{k=1}^n \bigl(s(k) - 1\bigr) = O(n\log g/\log n)$.

Given n and s, let us try to minimize the product $\prod_{k=1}^n s_k$ subject to the conditions

$$s = \sum_{k=1}^{n}(s_k - 1) \qquad \text{and} \qquad 1 \le s_k \le k\,.$$

If $s_{k-1} > s_k$, we can interchange $s_{k-1} \leftrightarrow s_k$ without violating the conditions; hence we may assume that $s_1 \le s_2 \le \cdots \le s_n$. Furthermore, if $1 < s_{k-1} \le s_k < k$, we can decrease the product by setting $(s_{k-1}, s_k) \leftarrow (s_{k-1} - 1, s_k + 1)$. Hence the product is smallest when we have $s_k = k$ for as many large k as possible:

$$s_n = n,\ \ s_{n-1} = n-1,\ \ \ldots,\ \ s_{q+1} = q+1,\ \ s_q = r,\ \ s_{q-1} = \cdots = s_1 = 1.$$

Here q and r are the unique integers such that

$$\binom{n}{2} - s - 1 = \binom{q}{2} - r \qquad \text{and} \qquad 1 \le r < q \le n.$$

(We assume that $0 \le s < \binom{n}{2}$.) The minimum product is

$$P(n, s) = r\frac{n!}{q!}\,.$$

The actual product in the algorithm is $g \ge P\bigl(n, \sum_{k=1}^{n}\bigl(s(k) - 1\bigr)\bigr)$, hence our proof will be complete if we can show that

$$s = O\Bigl(n\frac{\log P(n,s)}{\log n}\Bigr)\,.$$

But this is not difficult. If $s \ge n^2/4$ we have $q \le n/\sqrt{2}$, so $\log P(n,s) = \Theta(n \log n)$ and the result holds. At the other extreme, if $0 \le s < n$, we have $P(n,s) = s + 1$ and again the result is trivial. Otherwise we note that $n - q \ge \lfloor s/n \rfloor$, hence

$$P(n, s) \ge \frac{n!}{q!} > q^{\lfloor s/n \rfloor} > \Bigl(\frac{n}{2}\Bigr)^{s/n-1};$$

the relation $(s/n)\log n = O\bigl(nP(n,s)\bigr)$ follows immediately.

The space required to store the transversal perms $\sigma_{k,j}$ is

$$\sum_{k=1}^{n} k(s(k) - 1) = O(n^2 \log g / \log n).$$

The space required to store the strong generators can be reduced to $\sum kt(k)$ summed over those k with $s(k) > 1$, for if $s(k) = 1$ we have $T(k) = T(k-1)$. This sum has $O(\log g)$ terms, each of which is $O(n \log g)$. So the proof of the theorem is complete. $\square$

Inspection of this proof shows that the running time is actually bounded by a slightly smaller estimate than claimed, namely

$$O\bigl(n^2 l_n(g)^2 \log_n g\bigr) + O\bigl(n^2 l_n(g)^2\bigr) + O\bigl(mnl_n(g)\bigr)\,,$$

where $l_n(g) = \min\bigl(n, \theta(g)\bigr)$. The space bound is, similarly,

$$O(n^2 \log_n g) + O\bigl(nl_n(g)^2\bigr)\,.$$

And the examples in Sections 5 and 6 above show that even these improved bounds might be unduly pessimistic; sometimes a judicious relabeling of points will speed things up.

The storage occupied by strong generators is usually less than the storage required for perms of the transversal system, but it can be greater. For example, when n is even and the generators are respectively

$$\begin{gathered}[n-1,n]\\ [n-3,n-2]\,[n-1,n]\\ \vdots\\ [1,2]\ldots[n-3,n-2]\,[n-1,n]\end{gathered}$$

then $g = 2^{n/2}$ and the $nl_n(g)^2$ term dominates.

The values of $l_n(g)$ and $\log_n g$ are often substantially smaller than n, in groups of computational interest. For example, the Hall–Janko group J_2 has $g = 2^7 \cdot 3^3 \cdot 5^2 \cdot 7$ and $n = 100$ (see [6]); here $\theta(g) = 13$ and $\log_n g \approx 2.9$. The unitary group $U_6(2)$, which has order $g = 2^{15} \cdot 3^6 \cdot 5 \cdot 7 \cdot 11$, is represented as a perm group on $n = 672$ points in the Cayley library [10]; in this case $l_n(g) = 24$ and $\log_n g \approx 3.5$. Some representative large examples are Conway's perfect group ·0, for which $g = 2^{22} \cdot 3^9 \cdot 5^4 \cdot 7^2 \cdot 11 \cdot 13 \cdot 23$, $n = 196560$, and $\log_n g \approx 3.6$; and Fischer's simple group F'_{24}, for which $g = 2^{21} \cdot 3^{16} \cdot 5^2 \cdot 7^3 \cdot 11 \cdot 13 \cdot 17 \cdot 23 \cdot 29$, $n = 306936$, and $\log_n g \approx 4.4$. (See [3].)

9. Historical Remarks and Acknowledgments

The algorithm described above is a variant of a fundamental procedure sketched by Sims in 1967 [8], which he described more fully a few years later as part of a larger body of algorithms [9]. The principal difference between the method of [9] and the present method is that Sims essentially worked with sets of strong generators satisfying the condition $T(1) \subseteq T(2) \subseteq \cdots \subseteq T(n)$. Thus, for example, when $\sigma \in \Sigma(n)$ he would test the product $\sigma\tau$ for all strong generators τ; the present algorithm tests $\sigma\tau$ for such σ only with the perms τ of $T(n)$, namely the given generators π. His example, in which the group generated by $[1,2,4,5,7,3,6]$

and [2, 4][3, 5] required the verification of 54 products $\sigma\tau$, requires the testing of only 40 products in the present scheme. On the other hand, his method for representing the $\Sigma(k)$ as words in the generators was considerably more economical in its use of storage space, and space was an extremely critical resource at the time. Moreover, his way of maintaining strong generators blended well with the other routines in his system, so it is not clear that he would have regarded the method of the present paper as an improvement.

Polynomial bounds on the worst-case running time were not obvious from this original work. Furst, Hopcroft, and Luks showed in 1980 [5] that a transversal system and a set of strong generators could be found in $O(n^6)$ steps. (In their method the transversal system and strong generators were identical.) The author developed the present algorithm independently a year later, while preparing to write Volume 4 of *The Art of Computer Programming* and while advising an undergraduate student who was working on a research project with Persi Diaconis [4]. The present method became more widely known after the author discussed it informally at a conference in Oberwolfach on November 6, 1981; several people, notably Clement Lam, suggested clarifications of the rough notes that were distributed at that time. Eventually Professor Babai was kind enough to suggest that the notes of 1981 be published now, instead of waiting until Volume 4 has been completed. Those notes are reproduced with slight improvements in Sections 1–4 of the present paper. I am grateful to the referees and to Profs. Babai and Luks for several penetrating remarks that prompted the additional material in Sections 5–8.

Improved methods have been discovered in the meantime, notably by Jerrum [7], who has reduced the storage requirement to order n^2. Babai, Luks, and Seress [2] have developed a more complicated procedure whose worst-case running time is only $O(n^{4+\epsilon})$.

The word "perm," introduced experimentally in the author's Oberwolfach notes, does not seem to be winning any converts. (In fact, Pratt himself has forgotten that he once made this suggestion in conversation with the author.) However, the proposal to use the notation π^- for inverses, instead of the usual π^{-1}, has significantly greater merit, and the author hopes to see it widely adopted in future years. The shorter notation is easier to write on a blackboard and easier to type on a keyboard. Moreover, the longer notation α^{-1} is redundant, just as α^1 is redundant; in fact, α^{-1} stands for α^- raised to the first power! Thus there is no conflict between the two conventions, and a gradual changeover should be possible.

This research was supported in part by the National Science Foundation under grant CCR-86-10181, and by Office of Naval Research contract N00014-87-K-0502.

References

[1] László Babai, "On the length of subgroup chains in the symmetric group," *Communications in Algebra* **14** (1986), 1729–1736.

[2] László Babai, Eugene M. Luks, and Ákos Seress, "Fast management of permutation groups," *IEEE Symposium on Foundations of Computer Science* **29** (1988), 272–282. See also *SIAM Journal on Computing* **26** (1997), 1310–1342.

[3] J. H. Conway, "Three lectures on exceptional groups," in *Finite Simple Groups*, Proceedings of the Oxford Instructional Conference on Finite Simple Groups, 1969, edited by M. B. Powell and G. Higman (London: Academic Press, 1971), 215–247.

[4] Persi Diaconis, R. L. Graham, and William M. Kantor, "The mathematics of perfect shuffles," *Advances in Applied Mathematics* **4** (1983), 175–196.

[5] Merrick Furst, John Hopcroft, and Eugene Luks, "Polynomial-time algorithms for permutation groups," *IEEE Symposium on Foundations of Computer Science* **21** (1980), 36–41.

[6] Marshall Hall, Jr., and David Wales, "The simple group of order 604,800," *Journal of Algebra* **9** (1968), 417–450.

[7] Mark Jerrum, "A compact representation for permutation groups," *Journal of Algorithms* **7** (1986), 60–78.

[8] Charles C. Sims, "Computational methods in the study of permutation groups," in *Computational Problems in Abstract Algebra*, Proceedings of a conference held at Oxford University in 1967, edited by John Leech (Oxford: Pergamon, 1970), 169–183.

[9] Charles C. Sims, "Computation with permutation groups," in *ACM Symposium on Symbolic and Algebraic Manipulation* **2** (1971), 23–28.

[10] D. E. Taylor, "Pairs of generators for matrix groups," *The Cayley Bulletin* **3** (Department of Pure Mathematics, University of Sydney, 1987).

Addendum

No proof that the iterative implementation of step A2 may require $\Omega(n^5)$ steps has yet been found.

Chapter 21

An Algorithm for Brownian Zeros

[Originally published in Computing **33** *(1984), 89–94, under the title "An algorithm for Brownian zeroes."]*

An efficient binary technique is presented for determining, to prescribed accuracy, the location of the zeros of a stochastic function $B(x)$ that is a classical Brownian motion in one dimension.

Let $B(x)$ be a randomly generated function from the unit interval to the real numbers such that $B(0)$ has a given value a and such that, for all fixed $0 \le x < y \le 1$, the value of $B(y) - B(x)$ is normally distributed with mean zero and variance $y - x$. If $0 \le x < y < z \le 1$, the values of $B(y) - B(x)$ and $B(z) - B(y)$ are, in fact, supposed to be *independent* normal deviates with mean zero and respective variances $y - x$ and $z - y$. Since the sum of two standard normal deviates having variances v_1 and v_2 is normally distributed with variance $v_1 + v_2$, this definition makes sense. We shall call the random function $B(x)$ a *Brownian function*, since it is a classical one-dimensional Brownian motion as studied by Norbert Wiener and many others (see Wiener [6], Lévy [4], Itô–McKean [3], Mandelbrot [5]).

The purpose of this note is to deal with the generation of such functions $B(x)$ on a computer. More precisely, we want to consider the simulation of a random Brownian function when the goal is to locate the zeros of $B(x)$. It is necessary, of course, to explain quite carefully what this means, since it is obviously impossible to generate $B(x)$ for infinitely many values of x; "random function generation" is inherently more complex than random number generation. Moreover, it is impossible to list all of the zeros of a random $B(x)$, since there usually are uncountably many of them. Itô and McKean have proved, for example, that when $a = 0$ the zeros typically occur at a Cantor-like set of points whose Hausdorff–Besicovitch dimension is 1/2. (See [3, Problem 1.7.5 and Section 2.5].)

The truth is that we are interested in studying discrete approximations to a Brownian motion having a certain degree n of *resolution*, where n is a given integer. The output of our algorithm will be the set of integers j such that a particular randomly generated Brownian motion has $B(x) = 0$ for some x between j/n and $(j+1)/n$, where $0 \le j < n$. As n increases, we are essentially looking at the graph of $B(x)$ with a higher degree of magnification, so we might see more wiggles in the curve as it crosses the x axis; therefore we specify a particular degree of resolution when we ask for the zeros.

An obvious way to approach the problem just stated would be to compute all of the values $B(j/n)$ for $0 < j \le n$, by setting $B(j/n) = B((j-1)/n) + X_j/\sqrt{n}$, where the quantities $X_1, \ldots, X_n$ are independent normal deviates with mean zero and variance one. Then $B(x)$ has a zero between $(j-1)/n$ and j/n if $B((j-1)/n)B(j/n) \le 0$. However, this method is not accurate because it misses occasional zeros when $B((j-1)/n)$ and $B(j/n)$ have the same sign. The purpose of the present note is to point out that there is a simple algorithm that not only computes such zeros correctly, it also does the job by generating far fewer than n values of $B(x)$ in most cases.

The algorithm to be developed depends on two well-known ideas, embodied in Lemmas 1 and 2.

Lemma 1. *Let X_1, X_2, X_3, X_4 be independent random variables with the normal distribution, having mean zero and variance one. Then the joint distribution of the pairs $(X_1/\sqrt{2}, X_1/\sqrt{2} + X_2/\sqrt{2})$ is the same as the joint distribution of $(X_3/2 + X_4/2, X_3)$.*

Proof. The density function for $(X_1/\sqrt{2}, X_1/\sqrt{2}+X_2/\sqrt{2})$ at the point (x,y) is $\exp(-\frac{1}{2}(\sqrt{2}x)^2)/\sqrt{2\pi}$ times $\exp(-\frac{1}{2}(\sqrt{2}y - \sqrt{2}x)^2)/\sqrt{2\pi}$, since $X_1 = \sqrt{2}x$ and $X_2 = \sqrt{2}y - \sqrt{2}x$. Similarly, the density function for $(X_3/2 + X_4/2, X_3) = (x,y)$ is equal to $\exp(-\frac{1}{2}y^2)/\sqrt{2\pi}$ times $\exp(-\frac{1}{2}(2x-y)^2)/\sqrt{2\pi}$. In both cases, therefore, the density is equal to $\exp(-\frac{1}{2}(4x^2 - 4xy + 2y^2))/2\pi$. □

Lemma 2. *Let $a \cdot b > 0$. The probability that a Brownian motion with $B(0) = a$ and $B(1) = b$ has $B(x) = 0$ for at least one x in the range $0 \le x \le 1$ is e^{-2ab}.*

Proof. We shall use the well-known "reflection principle" to establish a measure-preserving correspondence between Brownian motions that have $B(0) = a$ and $B(1) = -b$ and Brownian motions that have $B(0) = a$ and $B(1) = b$ and at least one root of the equation $B(x) = 0$. Then the lemma will follow, since the ratio of Brownian motions with

$(B(0), B(1)) = (a, -b)$ to Brownian motions with $(B(0), B(1)) = (a, b)$, for fixed a, is

$$\exp(-\tfrac{1}{2}(a+b)^2)/\exp(-\tfrac{1}{2}(a-b)^2) = e^{-2ab}.$$

If $B(0) = a$ and $B(1) = -b$, the set $\{x \mid B(x) = 0\}$ is nonempty and closed, since $ab > 0$ and $B(x)$ is continuous; hence there is a smallest element x_0 such that $B(x_0) = 0$. The function $\overline{B}(x) = B(x)\cdot \operatorname{sign}(x_0 - x)$ satisfies $\overline{B}(0) = a$, $\overline{B}(1) = b$, and $\overline{B}(x_0) = 0$. Conversely if $\overline{B}(0) = a$ and $\overline{B}(1) = b$ and $\overline{B}(x) = 0$ has at least one root, let x_0 be the smallest root; then $B(x) = \overline{B}(x)\cdot \operatorname{sign}(x_0 - x)$ has $B(0) = a$ and $B(1) = -b$. This one-to-one correspondence is measure-preserving, since $\overline{B}$ can be obtained from B by negating normal deviates, and since the normal distribution is symmetric about zero. Notice furthermore that $B(x) = 0$ if and only if $\overline{B}(x) = 0$. □

The gist of Lemma 1 is that we can either compute $B(\frac{1}{2}) = B(0) + X_1/\sqrt{2}$ first and then $B(1) = B(\frac{1}{2}) + X_2/\sqrt{2}$, or we can compute $B(1) = B(0) + X_3$ first and then we can compute a consistent value of $B(\frac{1}{2})$ by computing $B(\frac{1}{2}) = \frac{1}{2}(B(0) + B(1)) + \frac{1}{2}X_4$. In general, if the values $B(y)$ and $B(z)$ have been generated but no values have been established for arguments between y and z, we can set

$$B(\tfrac{1}{2}(y+z)) = \tfrac{1}{2}(B(y) + B(z)) + \tfrac{1}{2}\sqrt{z-y}\,X,$$

where X is a standard normal deviate.

The gist of Lemma 2 is that we can decide whether or not the Brownian function we are generating will have a zero in an interval, and if it does we can locate the zeros by means of the reflective construction.

Therefore the zero-finding problem stated above can be solved by a recursive procedure. Let the resolution be $n = 2^m$, and let the value $B(0) = a$ be given. Assume that *uniform* is a procedure that returns a random real number uniformly distributed between 0 and 1, and that *normal* is a procedure that returns a random real number normally distributed with mean 0 and variance 1. The procedure call

$$\mathit{Bzero}(0, a, a + \mathit{normal}, 1, 1, 0, m)$$

will have the effect of generating a random $B(x)$ and locating its zeros, by printing out all values $j/2^m$ such that there is at least one zero between $j/2^m$ and $(j+1)/2^m$. For all such j, the procedure will also print the values of $B(j/2^m)$ and $B((j+1)/2^m)$.

In general, the procedure call

$$Bzero(y, a, b, r, s, k, m)$$

will have the following interpretation: All zeros between y and $y + 2^{-k}$ are to be located, with resolution 2^m, for a Brownian function $A(x)$ such that $A(y) = a$ and $A(y+2^{-k}) = b$. The parameters r and s are ± 1; they are set so that the function $A(x)$ is equal to $s \cdot B(x)$ for $y \le x \le y+2^{-k}$ if $r = +1$ or if there are no zeros, otherwise it is equal to the *reflection* of $s \cdot B(x)$ as in Lemma 2. The procedure also returns the value *false* if it found no zeros, *true* if it found at least one zero.

We shall specify the algorithm in an ALGOL-like programming language.

```
function Bzero(y, a, b, r, s : real; k, m : integer): Boolean;
var c : real;
begin if ab > 0 then
   if uniform < exp(−2^(k+1) ab) then
      begin b ← −b;  r ← −r;
      end;
   else return (false);
if k = m then print(y, s a, r s b);
else begin c ← (a + b)/2 + 2^(−1−k/2) normal;
   if Bzero(y, a, c, r, s, k + 1, m)
   then Bzero(y + 2^(−k−1), c, b, 1, r s, k + 1, m);
   else Bzero(y + 2^(−k−1), c, b, r, s, k + 1, m);
   end;
return (true);
end.
```

By changing the procedure so that it prints the values on each level instead of only when reaching level m, we obtain output for several resolutions simultaneously, with respect to a single random function $B(x)$ that is being generated.

A careful study of the procedure will reveal the slightly intricate maneuvering that is involved: When a uniform random number is drawn leading to the decision that an interval contains a zero, the function values are reflected in that interval so that a zero will indeed be found. Reflected values may in turn be reflected again, in the smaller subintervals. The algorithm tries to reflect the function by trying first to reflect in the first half-interval; then it either negates or keeps trying to reflect, in the second half-interval, depending on whether a zero was found.

In order to test the effectiveness of this algorithm, 100 experiments were done with $B(0) = 2/3$ and resolution 2^{20}. The probability that $B(x)$ has no zeros for $0 \le x \le 1$ when $B(0) = 2/3$ is

$$\int_{-2/3}^{2/3} e^{-x^2/2} dx/\sqrt{2\pi} = \operatorname{erf}(\sqrt{2}/3) \approx .4950,$$

and it turned out that exactly 50 of the 100 experiments fell into this category. The first experiment gave a function with zeros, however, and in fact this function had respectively (2, 2, 2, 2, 6, 12, 20, 24, 28, 40, 52, 64, 96, 122, 192, 254, 356, 518, 732, 1060) zeros at resolutions $(2^1, 2^2, \ldots, 2^{20})$. (Here we count two zeros between $j/2^k$ and $(j+1)/2^k$ if $B(j/2^k)$ and $B((j+1)/2^k)$ have the same sign.) A total of 1692 normal deviates were generated to find all of the 1060 zeros at resolution 2^{20}, and it turned out that exactly 1692 uniform deviates were also generated when making the test '$\mathit{uniform} < \exp(-2^{k+1}ab)$'. This is no coincidence, if we assume that the variables a, b, c are never exactly zero, since the fact that $ab < 0$ when c is generated implies that the number of calls to *uniform* will always be equal to, or one greater than, the number of calls to *normal*, depending on the sign of $B(0)B(1)$. Furthermore the total number of calls to *Bzero* is twice the number of calls to *normal*, minus one, so the number of normal deviates generated is a fairly good indicator of the running time of this algorithm. The total numbers of zeros found at respective resolutions $(2^1, \ldots, 2^{20})$, summed over all 100 experiments, came to (109, 155, 205, 297, 391, 541, 781, 1111, 1597, 2225, 3063, 4375, 6215, 8783, 12387, 17541, 24841, 35065, 49757, 70497), and the total number of normal deviates generated was 112722. Note that the number of zeros grows as the square root of the resolution, and the running time of the algorithm is bounded by $O(1)$ plus a constant times the number of zeros times the logarithm of the resolution. Thus the procedure is several orders of magnitude faster than a brute force algorithm would be.

Postscript

The method of generating $B(x)$ by successive bisections can be formulated as follows, given an infinite sequence $X_0, X_1, \ldots$ of normal deviates (see Lévy [4, Section 1]): Let $B(0) = a$ and $B(1) = a + X_0$; then for $j = 1, 2, \ldots,$ suppose that $2^k \le j < 2^{k+1}$ and define $\alpha_j = 2^{-k}j - 1$,

$$B(\alpha_j + 2^{-k-1}) = \tfrac{1}{2}(B(\alpha_j) + B(\alpha_j + 2^{-k})) + \tfrac{1}{2}2^{-k/2}X_j.$$

For example,

$$\begin{aligned} B(\tfrac{1}{2}) &= \tfrac{1}{2}(B(0)+B(1))+\tfrac{1}{2}X_1,\\ B(\tfrac{1}{4}) &= \tfrac{1}{2}(B(0)+B(\tfrac{1}{2}))+\tfrac{1}{2}X_2/\sqrt{2},\\ B(\tfrac{3}{4}) &= \tfrac{1}{2}(B(\tfrac{1}{2})+B(1))+\tfrac{1}{2}X_3/\sqrt{2}, \end{aligned}$$

etc. An inductive argument proves that we have

$$\begin{aligned} B(\alpha) = B(0) &+ \alpha X_0 + \|\alpha\| X_1 + 2^{-1/2}\|2\alpha\| X_{\lfloor 2(\alpha+1)\rfloor}\\ &+ 4^{-1/2}\|4\alpha\| X_{\lfloor 4(\alpha+1)\rfloor} + 8^{-1/2}\|8\alpha\| X_{\lfloor 8(\alpha+1)\rfloor} + \cdots, \end{aligned} \tag{$*$}$$

for all dyadic rationals α, where

$$\|x\| \;=\; \min_{\text{integer } n} |x-n| \;=\; \min(x-\lfloor x\rfloor, \lceil x\rceil - x).$$

Since this infinite series is correct for all dyadic rationals, it gives a binary representation of any Brownian function $B(x)$. (A formula equivalent to $(*)$, but expressed in rather different notation, appears at the bottom of page 406 in [1].)

An amusing corollary of the identity above follows from the fact that $B(\alpha) - B(0)$ has variance α: We obtain the formula

$$\alpha(1-\alpha) = \|\alpha\|^2 + \|2\alpha\|^2/2 + \|4\alpha\|^2/4 + \|8\alpha\|^2/8 + \cdots$$

for $0 \le \alpha \le 1$, which leads to an interesting doubly infinite series [2].

Incidentally, if the binary method is to be replaced by an n-ary method, the following generalization of Lemma 1 can be used: The joint distribution of

$$\Big(\frac{1}{\sqrt{n}}X_1,\ \frac{1}{\sqrt{n}}X_2,\ \ldots,\ \frac{1}{\sqrt{n}}X_{n-1},\ \frac{1}{\sqrt{n}}(X_1+\cdots+X_n)\Big)$$

is the same as the joint distribution of

$$\begin{aligned} \Big(\frac{1}{n}Y_1 + \frac{1}{\sqrt{n}}Y_2 + \frac{1-\sqrt{n}}{n(n-1)}S,\ \frac{1}{n}Y_1 + \frac{1}{\sqrt{n}}Y_3 + \frac{1-\sqrt{n}}{n(n-1)}S,\\ \ldots,\ \frac{1}{n}Y_1 + \frac{1}{\sqrt{n}}Y_n + \frac{1-\sqrt{n}}{n(n-1)}S,\ Y_1\Big), \end{aligned}$$

where $S = Y_2 + \cdots + Y_n$.

Open Problems

The author has been unable to deduce the expected number of normal deviates generated by the stated algorithm, as a function of a and m.

The problem of Brownian zeros is the one-dimensional analog of a much more difficult problem concerning random "fractal coastlines." Recent motion pictures have included scenes of computer-generated artificial planets whose "continents" are defined by the sea-level points of a Brownian sphere-to-line function (see [5, pages 217–224]). It would be interesting to develop an efficient algorithm for the generation of such images, by limiting the computation to the areas where it is necessary, somewhat as the algorithm above explores only intervals that actually contain zeros.

Acknowledgments

This paper was stimulated by B. Mandelbrot's book about fractals [5], and by Alan R. Siegel's suggestion to use the reflection principle in place of a complicated heuristic that the author had adopted when first trying to locate Brownian zeros. The author thanks the referee for pointing out reference [1].

This research was supported in part by National Science Foundation grant IST-7921977 and by Office of Naval Research contract N00014-76-C-0330.

References

[1] Z. Ciesielski, "Hölder conditions for realizations of Gaussian processes," *Transactions of the American Mathematical Society* **99** (1961), 403–413.

[2] R. L. Graham and D. E. Knuth, Problem E2982, *American Mathematical Monthly* **90** (1983), 54.

[3] Kiyosi Itô and Henry P. McKean, Jr., *Diffusion Processes and Their Sample Paths* (New York: Springer, 1965).

[4] P. Lévy, *Processus stochastiques et mouvement Brownien* (Paris: Gauthier-Villars, 1948).

[5] Benoit B. Mandelbrot, *Fractals: Form, Chance, and Dimension* (San Francisco: W. H. Freeman, 1977).

[6] Norbert Wiener, "Differential-space," *Journal of Mathematics and Physics* **2** (1923), 131–174.

Chapter 22

Semi-Optimal Bases for Linear Dependencies

[Originally published in Linear and Multilinear Algebra **17** *(1985), 1–4.]*

Let A be an $m \times n$ matrix of real or complex numbers, and let μ be a given constant ≥ 1. If A has rank m, it is possible to choose m columns of A such that, if B is the $m \times m$ matrix formed by these m columns, all entries of $B^{-1}A$ are less than or equal to μ in absolute value. Moreover, if $\mu > 1$, it is possible to find m such columns in a number of steps that is polynomial in m and n and inversely proportional to $\log \mu$.

Given n column vectors $a^1, \ldots, a^n$ of dimension m, it is well known that the $m \times n$ matrix $A = (a^1 \ldots a^n)$ has rank m if and only if it is possible to choose m columns $1 \leq j_1 < \cdots < j_m \leq n$ such that the square matrix $(a^{j_1} \ldots a^{j_m})$ formed by these columns is nonsingular. Furthermore, Cramer's rule implies that, if $B = (b^1 \ldots b^m)$ is any $m \times m$ matrix, the element in row i and column j of $\operatorname{adj}(B)A$ is

$$\det(b^1 \ldots b^{i-1} a^j b^{i+1} \ldots b^m)\,,$$

where $\operatorname{adj}(B)$ is the adjugate of B (the transpose of the matrix of B's cofactors).

Consequently, if $B = (a^{j_1} \ldots a^{j_m})$ is chosen so as to maximize the magnitude of the determinant over all $m \times m$ submatrices of A, and if A has rank m, all of the elements of $B^{-1}A = \operatorname{adj}(B)A/\det B$ will be at most 1 in absolute value.

We shall call such a choice of columns an *optimal basis* for A. The system of linear equations $a^1 x_1 + \cdots + a^n x_n = b$ is diagonalized when premultiplied by B^{-1}; the resulting formulas express the dependent variables $\{x_{j_1}, \ldots, x_{j_m}\}$ in terms of $n - m$ independent variables in such a way that errors in the values of the independent variables are not inflated.

The argument above proves the existence of optimal bases, but it does not prove anything about uniqueness. In fact, an optimal basis might be obtained even when B does not have the largest determinant; all we need, in order to get the elements of $B^{-1}A$ in the unit disk, is that B be *locally* maximum, in the sense that its determinant dominates all determinants obtained by replacing any one of its columns by another column of A. The optimal basis property cannot characterize $\max \det B$, because there are matrices like $\binom{1\;0\;1\;-1}{0\;1\;1\;\;\;1}$ in which the last two columns yield a larger determinant than the first.

It may, of course, be difficult to find an optimal basis, since there are $\binom{n}{m}$ sets of columns to choose from. Therefore it is desirable to weaken the condition slightly, and to search for a *semi-optimal basis* $B = (a^{j_1} \ldots a^{j_m})$ such that the magnitude of all elements of $B^{-1}A$ is at most μ, where μ is a given constant greater than 1.

A straightforward algorithm can be used to discover a semi-optimal basis in a reasonable amount of time, by means of an inductive method. Assuming that such a basis has already been found for the first $m-1$ rows of A, we can choose j_m such that $\left|\det(a^{j_1} \ldots a^{j_m})\right|$ is maximal, given the values of $j_1, \ldots, j_{m-1}$. It is not difficult to prove that the elements of $B^{-1}A$ will now all be $\le 2\mu$ in magnitude (see below); hence by Hadamard's inequality, the magnitude of each $m \times m$ submatrix of $B^{-1}A$ is at most $M = (2\mu\sqrt{m}\,)^m$. If the largest entry of $B^{-1}A$ occurs in row i and column j, and if it has magnitude $\mu' > \mu$, the idea is to replace j_i by j in the current basis; then continue with the new $B^{-1}A$ in the same fashion, until a semi-optimal basis is obtained. The replacement of j_i by j multiplies the determinant of the current basis by μ', hence the method must converge in at most $\log M/\log\mu$ steps.

Let e^m be the column vector with 1 in row m and zeros elsewhere. To complete the proof, we want to show that if $(a^{j_1} \ldots a^{j_{m-1}} e^m)^{-1}A$ has no elements exceeding magnitude μ in its first $m-1$ rows, and if we have $\left|\det(a^{j_1} \ldots a^{j_m})\right| \ge \left|\det(a^{j_1} \ldots a^{j_{m-1}} a^j)\right|$ for $1 \le j \le n$, then $(a^{j_1} \ldots a^{j_m})^{-1}A$ has no elements exceeding magnitude 2μ. The entries of the bottom row of $(a^{j_1} \ldots a^{j_m})^{-1}A$ are in the unit disk by hypothesis, so by symmetry it suffices to show that $\left|\det(a^{j_1} \ldots a^{j_{m-2}} a^j a^{j_m})\right| \le 2\mu\left|\det(a^{j_1} \ldots a^{j_m})\right|$ for $1 \le j \le n$. There are simple proofs, but it is instructive to obtain this result by setting $\nu = \mu$, $\lambda = 1$, $x^k = a^{j_k}$, $y = a^j$, and $z = e^m$ in the following more general inequality:

Lemma. *If*

$$\begin{aligned}\left|\det(x^1 \ldots x^{m-1} y)\right| &\le \lambda\left|\det(x^1 \ldots x^m)\right| \ne 0\,,\\ \left|\det(x^1 \ldots x^{m-2} yz)\right| &\le \mu\left|\det(x^1 \ldots x^{m-1} z)\right| \ne 0\,,\\ \left|\det(x^1 \ldots x^{m-2} x^m z)\right| &\le \nu\left|\det(x^1 \ldots x^{m-1} z)\right| \ne 0\,,\end{aligned}$$

then $\bigl|\det(x^1\ldots x^{m-2}yx^m)\bigr| \le (\lambda\nu+\mu)\bigl|\det(x^1\ldots x^m)\bigr|$.

This lemma is a consequence of the general identity

$$\begin{aligned}\det(x^1\ldots x^{m-2}vw)\det(x^1\ldots x^{m-2}yz)&\\ +\det(x^1\ldots x^{m-2}vy)\det(x^1\ldots x^{m-2}zw)&\\ +\det(x^1\ldots x^{m-2}vz)\det(x^1\ldots x^{m-2}wy)&=0\,,\end{aligned}\qquad(*)$$

which is interesting even in the case $m = 2$. For $m > 2$, there is an easy inductive proof of this identity that begins by zeroing all but one element of the first column of each determinant.

The lemma now follows because

$$\det(x^1\ldots x^{m-2}yx^m) = \biggl(-\frac{\det(x^1\ldots x^{m-2}x^{m-1}y)}{\det(x^1\ldots x^m)}\cdot\frac{\det(x^1\ldots x^{m-2}x^m z)}{\det(x^1\ldots x^{m-1}z)} + \frac{\det(x^1\ldots x^{m-2}yz)}{\det(x^1\ldots x^{m-1}z)}\biggr)\det(x^1\ldots x^m)\,.$$

Incidentally, it is interesting to trace the origin of identity $(*)$ using Muir's history [1]. The case $m = 2$ was discovered in 1748 by Vandermonde's teacher, Alexis Fontaine des Bertins, who wrote out 126 instances and then said "et cetera"! Gaspard Monge gave the case $m = 3$ in 1809; the generalization to all m was due to P. Desnanot in 1819, and it was independently rediscovered by Jacobi in 1841. Bézout found a generalization that connects six 3-dimensional vectors,

$$|uvw|\,|xyz| - |uvx|\,|wyz| + |uvy|\,|wxz| - |uvz|\,|wxy| = 0\,,$$

in 1779; an n-dimensional generalization in this vein was found by Reiss in 1829.

The author had hoped to use identity $(*)$ to prove that the elements of $B^{-1}A$ remain $\le 2\mu$ throughout the semi-optimal basis algorithm described above, but such a conjecture turned out to be false. For example, the algorithm begins by choosing the first three columns of the matrix

$$A = \begin{pmatrix} 1.00 & 0.00 & 0.90 & -0.99 & 0.99\\ 0.00 & 1.00 & -0.99 & 0.91 & 0.99\\ 0.00 & 0.00 & 1.00 & 0.99 & 0.80\end{pmatrix}$$

leading to

$$B^{-1}A = \begin{pmatrix} 1.00 & 0.00 & 0.00 & -1.8810 & 0.270\\ 0.00 & 1.00 & 0.00 & 1.8901 & 1.782\\ 0.00 & 0.00 & 1.00 & 0.9900 & 0.800\end{pmatrix}.$$

Column 4 should now replace column 2. But that makes the entry in column 5 of row 1 equal to $0.270 + 1.8810 \times 1.782/1.8901 > 2.04$, counterexampling the conjecture if $\mu = 1.02$.

This research was supported in part by the National Science Foundation under grant MCS-83-00984.

Bibliography

[1] Thomas Muir, *The Theory of Determinants in the Historical Order of Development* **1** (London: Macmillan, 1906).

Addendum

A simplified version of the algorithm in this paper is used by the METAFONT compiler to maintain dynamic systems of linear dependencies that are specified by the user. See Donald E. Knuth, *METAFONT: The Program*, Volume D of *Computers & Typesetting* (Reading, Massachusetts: Addison–Wesley, 1986), Part 28.

The problem of finding the largest subdeterminant of A was shown to be NP-complete by Christos H. Papadimitriou, "The largest subdeterminant of a matrix," *Bulletin de la Société Mathématique de Grèce* (nouvelle série) **25** (1984), 95–105. His construction does not, however, extend to a proof that it is NP-hard to find an optimal basis. That question remains open. Indeed, no matrices are known for which the algorithm sketched above doesn't find an optimal basis in polynomial time.

A faster algorithm to approximate the largest subdeterminant was devised by Leonid Khachiyan, "On the complexity of approximating extremal determinants in matrices," *Journal of Complexity* **11** (1995), 138–153.

Further information about identity $(*)$ appears in Donald E. Knuth, "Overlapping Pfaffians," *Electronic Journal of Combinatorics* **3**, 2 (1996), paper R5, 13 pages. [Reprinted as Chapter 7 of *Selected Papers on Discrete Mathematics*, CSLI Lecture Notes 106 (Stanford, California: Center for the Study of Language and Information, 2003), 105–121.]

Chapter 23

Evading the Drift in Floating-Point Addition

[Written with John F. Reiser. Originally published in Information Processing Letters **3** *(1975), 84–87, 164.]*

Consider a machine that does decimal floating-point arithmetic on normalized numbers with three significant digits. Suppose rounding is done for positive numbers by adding 5 in the fourth significant digit, then truncating the result to three significant digits. If $x = 400$ and $y = 3.5$ then we find $x \oplus y = 404$, $(x \oplus y) \ominus y = 401$, $((x \oplus y) \ominus y) \oplus y = 405$, and so on; repeated addition and subtraction of y will cause a steady increase in the values. This phenomenon is called *drift*, and it is caused by the bias towards $+\infty$ in the rounding rule.

This paper presents a proof that various "unbiased" forms of rounding, called *stable rounding*, will eliminate drift.

We shall consider floating-point numbers z with p significant digits in radix b, and we write $z = f(z) \cdot b^{e(z)}$ where $e(z)$ is the *exponent part* and $f(z)$ is the *fraction part.* If $z \neq 0$ we assume that $1 > |f(z)| \geq 1/b$, and that $f(z) \cdot b^p$ is an integer; conversely, 0 and all such z are floating-point numbers.

Let round$(\cdot)$ be a function from the real numbers to the floating-point numbers such that round(t) is the nearest floating-point number to t. When this nearest value is not unique, we shall assume for the moment that either of the two nearest neighbors can be used, although the example above shows that drift can occur when rounding is biased consistently. After examining the set of cases for which drift is possible, we will introduce further restrictions on the rounding function.

In practice there must be finite upper and lower bounds on floating-point exponents, and round(t) is undefined when $|t|$ is a sufficiently large or small positive number. But for our purposes we may assume that $e(z)$ can take on any integer value.

The floating-point sum of two floating-point numbers x and y is defined by the rule $x \oplus y = \operatorname{round}(x+y)$, and it is said to be *exact* if $x \oplus y = x + y$. The floating-point difference is defined similarly as $x \ominus y = \operatorname{round}(x-y)$.

Let us now consider the sequence of floating-point numbers x, w, x', w', x'', ..., where $w = x \oplus y$, $x' = w \ominus y$, $w' = x' \oplus y$, $x'' = w' \ominus y$, etc. If $x = x'$ then $w = w'$; if $w = w'$ then $x' = x''$; and so on. In order to examine the possibility of drift, we shall determine all floating-point values x and y for which $w \neq w'$ under the general rounding scheme described above.

The analysis consists of a somewhat tedious list of special cases, although enough simplifications are available to keep the total number of possibilities within reason. For each addition and subtraction we will consider the relative alignments of the operands and the result; and for each alignment we will consider the range of values of the rounding error, until reaching a case where it is clear that w can or cannot equal w'.

If any operation in the sequence x, w, x', w', x'', ..., is exact, so are all the subsequent ones. For example, if $w = x \oplus y = x + y$ then $x' = w \ominus y = \operatorname{round}((x+y)-y) = \operatorname{round}(x) = x$, since x is a floating-point number.

We may assume that x is positive, since our arguments will all be valid for negative x by changing appropriate signs; note that floating-point numbers as defined above are symmetrical about zero. With this understanding, we proceed to the case analysis.

Case I, $e(y) > e(x)$. Let $d = e(y) - e(x)$ and $c = e(w) - e(y)$; we have $d \geq 1$ and $c \leq 1$. For every minuscule letter z denoting a floating-point number, let the corresponding majuscule letter Z denote the real number $f(z) \cdot b^{e(z)-e(x)+p}$. Then X, Y, W are all integers, and $X + Y = W + R$ with $|R| \leq \frac{1}{2}b^{d+c}$. We may assume that $R \neq 0$, or else the addition will be exact; hence $d + c > 0$.

Case I·A, $c \leq 0$. Then $R = X + Y - W \equiv X$ (modulo b^{d+c}), since $Y \equiv 0$ (modulo b^d) and $W \equiv 0$ (modulo b^{d+c}). Therefore $W - Y = X - R$ has at most $(p+1) - (d+c) \leq p$ significant digits, and $w \ominus y = w - y$.

Case I·B, $c = 1$. Then $y > 0$ and $R \equiv X + Y_0 \cdot b^d$ (modulo b^{d+1}), where Y_0 is the pth digit of $f(y)$. It follows that $w \ominus y$ is exact, since $W - Y$ has at most $(p+1) - d \leq p$ significant digits.

Case II, $e(x) \geq e(y)$. Let $d = e(x) - e(y)$ and $c = e(w) - e(x)$; we have $d \geq 0$ and $c \leq 1$. Let the convention of Case I be redefined so that $Z = f(z) \cdot b^{e(z)-e(y)+p}$. Then X, Y, W are all integers such that

$X + Y = W + R$, with $|R| \le \frac{1}{2}b^{d+c}$; as before we may assume that $d + c > 0$.

Case II·A, $c \le -2$. Then $d + c > 0$ implies that $d \ge 3$, hence $X + Y \ge b^{p+d-1} - (b^p - 1) > b^{p+d-2}$, contradicting $c \le -2$.

Case II·B, $c = -1$. Then $d > 1$ and $Y < 0$; we have $|R| \le \frac{1}{2}b^{d-1}$ and $R \equiv Y$ (modulo b^{d-1}).

Case II·B1, $e(x') = e(x) = e(w) + 1$. Let $W - Y = X' + S$, where $|S| \le \frac{1}{2}b^d$. Then $|X - X'| = |R + S| \le \frac{1}{2}b^{d-1} + \frac{1}{2}b^d < b^d$ so $x' = x$.

Case II·B2, $e(x') = e(x) - 1 = e(w)$. Let $W - Y = X' + S$, where $S \equiv -Y$ (modulo b^{d-1}) and $|S| \le \frac{1}{2}b^{d-1}$.

Case II·B2a, $Y \not\equiv \frac{1}{2}b^{d-1}$ (modulo b^{d-1}). Then there are unique choices for S and R, hence $S = -R$ and $X - X' = R + S = 0$.

Case II·B2b, $Y \equiv \frac{1}{2}b^{d-1}$ (modulo b^{d-1}). Since $b^{p+d-1} \le X = X' + R + S \le (b^{p+d-1} - b^{d-1}) + \frac{1}{2}b^{d-1} + \frac{1}{2}b^{d-1} = b^{p+d-1}$ we have $X = b^{p+d-1}$, $X' = b^{p+d-1} - b^{d-1}$, and $R = S = \frac{1}{2}b^{d-1}$. It is possible to have $w' \ne w$, and this case is discussed further below.

Case II·C, $c = 0$. We have $d \ge 1$, $R \equiv Y$ (modulo b^d), $|R| \le \frac{1}{2}b^d$.

Case II·C1, $|R| < \frac{1}{2}b^d$.

Case II·C1a, $X > b^{p+d-1}$ or $R < 0$. Then we have[1] $\text{round}(W - Y) = \text{round}(X - R) = X$, so $x' = x$.

Case II·C1b, $X = b^{p+d-1}$ and $R > 0$. Then $Y > 0$, $e(x') = e(x) - 1$, and $W - Y = X' + S$ where $|S| \le \frac{1}{2}b^{d-1}$.

Case II·C1b1, $W > b^{p+d-1}$ or $S < \frac{1}{2}b^{d-1}$. Then $\text{round}(X' + Y) = \text{round}(W - S) = W$, so $w' = w$.

Case II·C1b2, $W = b^{p+d-1}$ and $S = \frac{1}{2}b^{d-1}$. Then $w \ne w'$ is possible; see below.

Case II·C2, $|R| = \frac{1}{2}b^d$. Then $w \ne w'$ is possible in a variety of subcases analyzed below.

Case II·D, $c = 1$. Then $y > 0$.

Case II·D1, $e(x') = e(w) = e(x) + 1$. Then $W - Y = X' + S$, where $|S| \le \frac{1}{2}b^{d+1}$. Furthermore $W - Y \ge b^{p+d} - \frac{1}{2}b^d$, and $b^{p+d} \le X' = X - R - S \le X + b^{d+1} \le (b^{p+d} - b^d) + b^{d+1}$. Hence $X' = b^{p+d}$ and $R + S \le -b^d$ and $S \ge -\frac{1}{2}b^d$. It follows that $S < \frac{1}{2}b^{d+1}$, and $W' = \text{round}(X' + Y) = \text{round}(W - S) = W$.

---

[1] By $\text{round}(W - Y)$ we really mean $b^{-e(y)+p}\,\text{round}(w - y)$, since the rounding function need not satisfy $\text{round}(bx) = b \cdot \text{round}(x)$ under the general conventions described here.

Case II·D2, $e(x') = e(w) - 1 = e(x)$. Then $W - Y = X' + S$, where $S \equiv -Y$ (modulo b^d) and $|S| \le \frac{1}{2}b^d$. We may assume that $d \ge 1$.

Case II·D2a, $e(w') = e(x')$. Then $X' + Y = W' + R'$, where $R' \equiv Y$ (modulo b^d) and $|R'| \le \frac{1}{2}b^d$. Since $b^{p+d} \le W = W' + S + R' \le (b^{p+d} - b^d) + \frac{1}{2}b^d + \frac{1}{2}b^d = b^{p+d}$, we must have $W = b^{p+d}$, $S = R' = \frac{1}{2}b^d$, $W' = b^{p+d} - b^d$. This case is discussed further below.

Case II·D2b, $e(w') = e(w)$. Then $X' + Y = W' + R'$, where $|R'| \le \frac{1}{2}b^{d+1}$, and $|W - W'| = |S + R'| < b^{d+1}$, so $w' = w$.

The case analysis is complete, and it shows that a general rounding rule as described above will produce $w = w'$ except in four cases. None of the cases can occur when the radix b is odd, hence we may assume in what follows that b is even. Three of the four remaining cases are "rare," but Case II·C2 is common enough to deserve special attention.

Case II·C2 occurs if and only if $e(w) = e(x) = e(y) + d$, where $d \ge 1$ and $Y = Y_1 b^d + R$, $R = \pm\frac{1}{2}b^d$, $W = X + Y_1 b^d$. (The example at the beginning of this note has $p = 3$, $b = 10$, $d = 2$, $Y_1 = 3$.) In order to prevent drift in such a case, we want to make $\mathrm{round}(W - \frac{1}{2}b^d) = \mathrm{round}(W + \frac{1}{2}b^d)$. Therefore let us further specify the round function so that, in cases of ambiguity, it always chooses the value whose fraction part times b^p is even, or so that it always chooses the odd value. These two rules will be called round-to-even or round-to-odd. *Stable rounding* is any rounding rule that satisfies

$$z = \mathrm{round}(z - \tfrac{1}{2}b^{e(z)-p}) \text{ if and only if } z = \mathrm{round}(z + \tfrac{1}{2}b^{e(z)-p})$$

for all floating-point numbers z with $|f(z)| \ne 1/b$. (When $f(z) = 1/b$ we always have $\mathrm{round}(z - \frac{1}{2}b^{e(z)-p}) = z - \frac{1}{2}b^{e(z)-p} \ne z$, but we might have $\mathrm{round}(z + \frac{1}{2}b^{e(z)-p}) = z$; similar remarks apply when $f(z) = -1/b$.) Note that stable rounding as defined here need not be strictly 'round-to-even' or 'round-to-odd'; it must only make a consistent choice of even or odd for each set of floating-point numbers having a given exponent and sign. In order to preserve the identity $x \ominus y = -(y \ominus x)$ we should further stipulate that $\mathrm{round}(-x) = -\mathrm{round}(x)$; but the following discussion does not require this stipulation.

With stable rounding we can prove that almost all of Case II·C2 satisfies $w = w'$:

Case II·C2a, $R = +\frac{1}{2}b^d$. Then $x' \le x$ and $w' \le w$ by monotonicity of the round function.

Case II·C2a1, $e(x') = e(x) - 1$. [Example: $x = 100$, $y = 10.5$, $w = 110$, $x' = 99.5$.] Then $w \ominus y$ is exact.

Case II·C2a2, $e(x') = e(x)$. Then $W - Y = X' + S$ where $S = \pm\frac{1}{2}b^d$. Since $X' = X - R - S$, we may assume that $S = R$ and $X' = X - b^d$.

Case II·C2a2a, $W > b^{p+d-1}$. Then since $W = \mathrm{round}(W + \frac{1}{2}b^d)$, we have $W' = \mathrm{round}(X' + Y) = \mathrm{round}(W - \frac{1}{2}b^d) = W$, by stable rounding.

Case II·C2a2b, $W = b^{p+d-1}$. Then $w \neq w'$ is possible, see below.

Case II·C2b, $R = -\frac{1}{2}b^d$. Then $x' \geq x$ and $w' \geq w$.

Case II·C2b1, $e(x') = e(x) + 1$. [Example: $x = 999$, $y = -11.5$, $w = 988$, $x' = 1000$.] Then $X' = \mathrm{round}(X + \frac{1}{2}b^d) = X + b^d$ and $W' = \mathrm{round}(W + \frac{1}{2}b^d) = W$ since $\mathrm{round}(W - \frac{1}{2}b^d) = W$.

Case II·C2b2, $e(x') = e(x)$. Then as in Case II·C2a2 we may assume that $X' = X + b^d$; and as in Case II·C2b1 this implies $W' = W$.

If stable rounding is applied to Case II·B2b, we always have $w' = w$, since $W' = \mathrm{round}(X' + Y) = \mathrm{round}(W - \frac{1}{2}b^{d-1}) = \mathrm{round}(W + \frac{1}{2}b^{d-1}) = W$.

Thus, stable rounding guarantees that $w' = w$ except in three "rare" cases (and their negatives) that can be completely characterized by the following formulas:

Case II·C1b2. $X = b^{2p-1}, Y = Y_1 b^{p-1} + \frac{1}{2}b^{p-1}$, where $0 < Y_1 < \frac{1}{2}b$, Y_1 is even, and round-to-odd is used for x'. [Example: $x = 100$, $y = 0.250$, $w = 100$, $x' = 99.7$, $w' = 99.9$.] Thus we have $W = b^{2p-1}$, $X' = (b^p - Y_1 - 1)b^{p-1}$, $W' = (b^p - 1)b^{p-1}$, $X'' = X'$.

Case II·C2a2b. [Example: $x = 169$, $y = -68.5$, $w = 100$, $x' = 168$, $w' = 99.5$.] $X = b^{p+d-1} + Y_1 b^d$, $Y = \frac{1}{2}b^d - Y_1 b^d$, $1 \leq d \leq p$, where $b^{p-d-1} < Y_1 \leq b^{p-d}$, Y_1 is odd, and round-to-even is used for w. Then $W = b^{p+d-1}$, $X' = X - b^d$, $W' = (b^p - \frac{1}{2}b)b^{d-1}$, $X'' = X'$.

Case II·D2a. [Example: $x = 936$, $y = 68.5$, $w = 1000$, $x' = 931$, $w' = 999$.] $X = (b^p - Y_1 + k)b^d$, $Y = Y_1 b^d + \frac{1}{2}b^d$, $1 \leq d < p$, where $0 \leq k < \frac{1}{2}b$, $b^{p-d-1} \leq Y_1 < b^{p-d}$, Y_1 is even, $Y_1 > k$, and round-to-odd is used for x'. Then $W = b^{p+d}$, $X' = (b^p - Y_1 - 1)b^d$, $W' = (b^p - 1)b^d$, $X'' = X'$.

In each of these cases, $x'' = x'$, so the drift is eliminated. Note however that there will be cases with $w \neq w'$ no matter what stable rounding rule is chosen. Thus we have proved the following results.

Theorem 1. *If stable rounding is used,* $(((x \oplus y) \ominus y) \oplus y) \ominus y = (x \oplus y) \ominus y$ *for all floating-point numbers* x *and* y *such that the operations are defined.* □

Theorem 2. *There exist* x *and* y *such that* $((x \oplus y) \ominus y) \oplus y \neq x \oplus y$, *regardless of what rounding scheme is used, when the radix is even.* □

A slightly weaker form of Theorem 1 was stated by Kahan [2, Chapter 3], with proof left to the reader, and it was his remarks that stimulated the present investigation. Kahan mentioned that stable rounding also eliminates drift in repeated multiplication and division by a nonzero constant.

David Matula [5] has shown that under suitable conditions the operation of conversion from floating decimal to floating binary and back again will be the identity transformation, using the classical function round($\cdot$) which handles ambiguous cases by rounding away from zero. Presumably the same result will hold with stable rounding also; and similar results can perhaps be obtained for exponentials followed by logarithms, and so on. Our case analysis is rather lengthy, and it would be very nice to discover a higher-level approach that would establish the identity $f(f(x)) = f(x)$, where $f(x) = \mathrm{round}(g^{-1}(\mathrm{round}(g(x))))$, for a large class of continuous monotonic functions $g(\cdot)$. The analysis above applies only to the case $g(x) = x + y$, when y is a floating-point number.

The history of stable rounding is rather interesting, though we are not sure we have traced it to its source. The earliest reference we have been able to find is in the first (1930) edition of Scarborough's classic treatise on numerical analysis [7, page 2]. Scarborough presents the round-to-even rule for floating-decimal calculations, without comment or reference to any source. His preface states that he was especially influenced by Runge's work, but we have not been able to find any remarks about rounding rules in Runge's books. The classic French treatise by Jules Tannery [8, page 277], first published in 1894, states that rounding is customarily done by adding unity whenever the next digit is 5 or more.

Tannery's argument for non-stable rounding is that precisely half of all real numbers have a 0, 1, 2, 3, or 4 in any given digit position, while the other half have 5, 6, 7, 8, or 9 in that position, hence the old rule tends to produce "compensating errors." On the other hand Scarborough added several paragraphs to his discussion of rounding in the second (1950) edition of his book; he stated, "It should be obvious to any thinking person that when a 5 is cut off, the preceding digit should be increased by 1 in only *half* the cases. ... Since even and odd digits occur with equal frequency, on the average, the rule that the odd digits be increased by 1 when a 5 is dropped is logically sound."

Actually Tannery's and Scarborough's contradictory arguments are both incorrect; for example, the logarithmic law of distribution of leading digits [3, Section 4.2.4] implies that the least significant digit of a floating-point number will be odd slightly less often than it will be even, assuming that b is even and $p > 1$. The real justification for stable rounding is that it eliminates drift.

Stable rounding is a trivial extension to any general rounding scheme, since it is easy (especially in hardware) to test at the appropriate time whether or not an intermediate result is midway between two consecutive floating-point numbers. Piper [6] indicates that the Bendix G-20 computer had round-to-odd. Scarborough [7] points out that round-to-even is preferable since for example it leads to less subsequent error when the result is divided by 2, a common operation.

There seems to be no good reason for designing floating-point hardware or firmware that does not perform stable rounding.

Acknowledgments

We wish to thank A. Householder and W. M. Kahan for helpful discussions.

The preparation of this paper was supported in part by a National Science Foundation Fellowship and by NSF grant GJ 36473X.

References

[1] T. J. Dekker, "A floating-point technique for extending the available precision," *Numerische Mathematik* **18** (1971), 224–242.

[2] W. Kahan, "Implementation of algorithms, part I," Lecture notes by W. S. Haugeland and D. Hough, Computer Science Technical Report No. 20 (Berkeley, California: University of California, 1973).

[3] Donald E. Knuth, *Seminumerical Algorithms*, Volume 2 of *The Art of Computer Programming* (Reading, Massachusetts: Addison–Wesley, 1969).

[4] Seppo Linnainmaa, "Analysis of some known methods of improving the accuracy of floating-point sums," *BIT* **14** (1974), 167–202.

[5] David W. Matula, "In-and-out conversions," *Communications of the ACM* **11** (1968), 47–50.

[6] C. A. Piper, "Round-off" (letter to the editor), *Communications of the ACM* **4**, 3 (March 1961), A13.

[7] James B. Scarborough, *Numerical Mathematical Analysis* (Baltimore: The Johns Hopkins Press, 1930).

[8] Jules Tannery, *Leçons d'Arithmétique*, 6th edition (Paris: Librairie Armand Colin, 1911).

[9] J. M. Yohe, "Roundings in floating-point arithmetic," *IEEE Transactions on Computers* **C-22** (1973), 577–586.

Addendum

The analogous question for multiplication and division is resolved in exercise 4.2.2–22 of the second edition of [3], published in 1981.

General results for monotone functions were proved by Harold G. Diamond, "Stability of rounded off inverses under iteration," *Mathematics of Computation* **32** (1978), 227–232.

Chapter 24

Deciphering a Linear Congruential Encryption

[Originally published in IEEE Transactions on Information Theory **IT-31** *(1985), 49–52.]*

The multiplier, the increment, and the seed value of a linear congruential random number generator on a binary computer can be deduced without great difficulty from the leading bits of the "random" numbers that are generated.

Introduction

Today's most popular random number generators are based on the "linear congruential recurrence"

$$x_{n+1} = (ax_n + c) \bmod 2^k, \tag{1}$$

where $a \bmod 4 = 1$ and $c \bmod 2 = 1$ and x_0 is arbitrary. This recurrence generates a sequence of k-bit fractions $x_0/2^k$, $x_1/2^k$, ... that is satisfactorily random for most purposes unless the multiplier a is chosen badly [2, pages 170–171]. Therefore it has been suggested that the leading bits of such a sequence might be useful for enciphering data. Let $k = h + l$, and consider the sequence of h-bit numbers $\lfloor x_0/2^l \rfloor$, $\lfloor x_1/2^l \rfloor$, ... that arises if we retain the h high-order bits and shift off the l (less random) low-order bits. For example, we might generate $x_{n+1} = (ax_n + c) \bmod 2^{32}$ and use the sequence of 16-bit numbers $\lfloor x_0/2^{16} \rfloor$, $\lfloor x_1/2^{16} \rfloor$, ... to encode a series of 16-bit data words. The purpose of this paper is to show that such a procedure is not very secure, because there is a way to determine the unknown constants a, c, and x_0 in about order 2^l steps, if we are given sufficiently many values of $\lfloor x_n/2^l \rfloor$.

More precisely, let h and l be positive integers with $h+l=k$, and let the sequence $\langle x_n \rangle$ be defined by (1). Let $\hat{x}_n$ stand for the h high-order bits of x_n, namely $\lfloor x_n/2^l \rfloor$, and let Problem A be the task of deducing a, c, and x_0 from the 2^l values of $\hat{x}_n$ for $0 \le n < 2^l$. We shall see that Problem A can be solved in $O(2^l)$ steps whenever $l > 1$. Let Problem B be the more general task of computing a, c, and x_0 from the first N values of $\hat{x}_n$; we shall develop a method that usually solves this problem in $O(k \cdot 2^l/N)^2$ steps when $h \ge 2$.

Solving Problem A may seem simple, since we merely need to determine $3l + 2h - 3$ bits of information (namely, the leading $h + l - 2$ bits of a, plus $h+l-1$ bits of c, plus l bits of x_0), from $h \cdot 2^l$ given bits. However, linear congruential sequences are good enough that there is no obvious way to break the code. The reader may wish to try the following example in order to get a feeling for the problem: Let $h = l = 3$, so that we are generating numbers by the rule $x_{n+1} = (ax_n + c) \bmod 64$ for some a, c, and x_0, and suppose the first eight values of $\lfloor x_n/8 \rfloor$ are

$$4\ 5\ 1\ 0\ 2\ 7\ 7\ 1. \tag{2}$$

Can you deduce the secret quantities a, c, and x_0 without simply trying out all 4096 possibilities (16 choices for a and 32 choices for c and 8 choices for x_0)?

Solution to Problem A

Let

$$x_{n,l} = \lfloor x_n/2^{l-1} \rfloor \bmod 2 \tag{3}$$

be the lth bit from the right of x_n. The most straightforward way to solve Problem A seems to be to determine the leftmost unknown bits $x_{n,l}$ first, thereby reducing the case (h, l) to the simpler case $(h+1, l-1)$, until we reach $l = 1$. The key fact we shall use is expressed in Lemma 1.

Lemma 1. *Let* $y_n^{(t)} = (x_{n+2^t} - x_n) \bmod 2^k$ *for* $0 \le t < k$. *Then* $y_{n+1}^{(t)} = ay_n^{(t)} \bmod 2^k$ *for all* n, *and each* $y_n^{(t)}$ *is an odd multiple of* 2^t.

Proof. The first statement follows from the observation that

$$y_{n+1}^{(t)} \equiv (ax_{n+2^t} + c) - (ax_n + c) \equiv ay_n^{(t)} \quad (\text{modulo } 2^k).$$

The second statement is obviously true if $a = 1$. When $a > 1$ we have $a = 1 + 2^f q_0$ for some $f \ge 2$ and some odd number q_0; it follows by induction that

$$a^{2^t} = 1 + 2^{f+t} q_t, \quad q_t \text{ odd}, \tag{4}$$

for all $t \geq 0$. Thus the second statement of the lemma follows from the well-known relation

$$x_{n+j} = \left(a^j x_n + \left(\frac{a^j - 1}{a - 1}\right)c\right) \bmod 2^k, \tag{5}$$

since $(a^{2^t} - 1)/(a - 1)$ is an odd multiple of 2^t. □

Lemma 2. *For $1 \leq l < k - 1$ there is a constant b_l such that*

$$x_{n,l} = \left(x_{n+2^{l-1},l+1} - x_{n,l+1} + b_l\right) \bmod 2. \tag{6}$$

Proof. According to Lemma 1, we have

$$y_n^{(l-1)} \equiv b_l 2^l + 2^{l-1} \quad (\text{modulo } 2^{l+1}) \tag{7}$$

for some b_l and all n. The result now follows by considering the addition $x_n + y_n^{(l-1)}$ in binary notation, since $x_{n,l}$ is the "carry" out of the lth place. □

Lemma 2 gives us a powerful way to determine the missing bits for Problem A: Once we have figured out the bits b_l, b_{l-1}, …, b_1, we can easily find all of the x_n. This will be plenty of information to deduce both a and c. Fortunately, we can determine the b's uniquely in most cases, without much work.

Let us assume that $h \geq 2$ and that we know the value of $\hat{x}_{2^l}$, as well as the values of $\hat{x}_n$ for $0 \leq n < 2^l$; this will be the case if we have reached case (h, l) from case $(h - 1, l + 1)$. Assume furthermore that $l > 1$. Then we shall see that $x_{0,l}$ can be determined uniquely; hence b_l will be known, and we'll be able to compute all of the lth bits $x_{n,l}$.

The condition $l > 1$ implies that $a^{2^{l-1}} \equiv 1$ (modulo 8); hence $y_{2^{l-1}}^{(l-1)} \equiv a^{2^{l-1}} y_0^{(l-1)} \equiv y_0^{(l-1)} \equiv b'_l 2^l + 2^{l-1}$ (modulo 2^{l+2}), where $b'_l = b_l$ or $b_l + 2$. Extending the proof of Lemma 2, we now have $\hat{x}_{2^{l-1}} \equiv \hat{x}_0 + b'_l + x_{0,l}$ and $\hat{x}_{2^l} \equiv \hat{x}_{2^{l-1}} + b'_l + 1 - x_{0,l}$ (modulo 4). The value of $x_{0,l}$ can therefore be read off from the relation

$$\hat{x}_0 - 2\hat{x}_{2^{l-1}} + \hat{x}_{2^l} \equiv 2x_{0,l} + 1 \quad (\text{modulo } 4). \tag{8}$$

If $h = 1$ or $l = 1$, or if we do not know $\hat{x}_{2^l}$, we simply try both possibilities $b_l = 0$ and $b_l = 1$. The following algorithm therefore solves Problem A for $l > 1$ in $O(2^l)$ steps:

D0. [Loop twice.] Do the following steps for both $b_l = 0$ and $b_l = 1$.

D1. [Reduce.] Use (6) to compute $x_{n,l}$ for $0 \le n < 2^{l-1}$. Also set $x_{2^{l-1},l} \leftarrow 1 - x_{0,l}$. Then reduce l by 1 and increase h by 1.

D2. [Done?] If $l > 1$, use equations (8) and (6) to compute b_l, then return to step D1.

D3. [Finish.] For $x_{0,1}x_{1,1}x_{2,1} = 010$ and 101 (two cases), compute a, where $a \equiv (x_2 - x_1)(x_1 - x_0)^{-1}$ (modulo 2^k). Then set $c \leftarrow (x_1 - ax_0) \bmod 2^k$. Check these values against the original data to Problem A. □

For example, the problem with $h = l = 3$ stated in the introduction can be resolved by essentially forming the following two tableaux of binary numbers corresponding to the two choices in D0:

$$\begin{array}{ll} x_0 = 1\,0\,0\,0\,0\,v & 1\,0\,0\,1\,1\,\bar{v} \\ x_1 = 1\,0\,1\,0\,1\,\bar{v} & 1\,0\,1\,1\,0\,v \\ x_2 = 0\,0\,1\,0\,1\,v & 0\,0\,1\,1\,0\,\bar{v} \\ x_3 = 0\,0\,0\,1\,*\,* & 0\,0\,0\,0\,*\,* \\ x_4 = 0\,1\,0\,1\,*\,* & 0\,1\,0\,0\,*\,* \\ x_5 = 1\,1\,1\,*\,*\,* & 1\,1\,1\,*\,*\,* \\ x_6 = 1\,1\,1\,*\,*\,* & 1\,1\,1\,*\,*\,* \\ x_7 = 0\,0\,1\,*\,*\,* & 0\,0\,1\,*\,*\,* \end{array}$$

The four choices for multipliers modulo 64 are $31/11 \equiv 61$, $33/9 \equiv 25$, $33/5 \equiv 45$, and $31/7 \equiv 41$; the possibilities for (a, c, x_0) are therefore $(61, 11, 32)$, $(25, 49, 33)$, $(45, 17, 39)$, and $(41, 23, 38)$. Only the last of these produces the original data, but by a strange coincidence the second solution produces the almost identical sequence 4 5 1 0 2 7 7 3!

Notice that only the least-significant bits of the input values $\hat{x}_0, \ldots, \hat{x}_{2^l-1}$ are inspected by the algorithm except in step D3. It is not difficult to verify that the two choices of b_l in step D0 simply complement the least significant l bits, so the algorithm can be streamlined.

This method solves Problem A only when $l > 1$. For completeness, let us also consider briefly the degenerate case $l = 1$. Problem A is not very interesting when $l = 1$, since only two values $\hat{x}_0$ and $\hat{x}_1$ are given; putting $x_0 = 2\hat{x}_0 + 1$ and $c = 1 + 2r$ and $a \equiv (2\hat{x}_1 - c)x_0^{-1}$ (modulo 2^k) gives 2^{h-1} solutions with $a \bmod 4 = 1$ whenever r has the appropriate parity, and there are also many solutions with $x_0 = 2\hat{x}_0$ and $x_1 = 2\hat{x}_1 + 1$. But if three values $\hat{x}_0$, $\hat{x}_1$, and $\hat{x}_2$ are given, the problem becomes more interesting: There are at most two solutions, found as in step D3 of the algorithm, namely

$$\begin{aligned} a &\equiv (2(\hat{x}_2 - \hat{x}_1) \pm 1)(2(\hat{x}_1 - \hat{x}_0) \mp 1)^{-1} \\ x_0 &= 2\hat{x}_0 + (1 \pm 1)/2 \\ c &\equiv 2\hat{x}_1 + (1 \mp 1)/2 - ax_0. \end{aligned}$$

If $a \bmod 4 = 3$ for one of these "solutions," then $a \bmod 4 = 3$ for the other, so neither is a true solution. Otherwise, one of these solutions has $a \bmod 8 = 1$ and the other has $a \bmod 8 = 5$.

Solution to Problem B

The algorithm we have given for Problem A is of no use for Problem B unless N exceeds 2^{l-1}, since the starting point of that whole attack was based on looking at $\hat{x}_{n+2^{l-1}} - \hat{x}_n$. Suppose, for example, that we have $h = l = 16$, but we know only the first $N = 30{,}000$ values of $\hat{x}_n$; the methods above are inapplicable, so we need a new idea. In this section we shall see that it is possible to crack the code for $h = l = 16$ in a reasonable time, even when N is as low as 258.

Before considering the new method in detail, we need to realize that Problem B usually has far more solutions than Problem A, since substantially less data has been given; yet most of the indeterminacy comes from solutions that are almost the same. The reason for this difficulty is obvious if we understand that if b is any constant, the sequence $\langle x'_n \rangle$ defined by

$$x'_n = (x_n + b) \bmod 2^k \tag{9}$$

satisfies the linear congruential recurrence

$$x'_{n+1} = (ax'_n + (c - (a-1)b)) \bmod 2^k. \tag{10}$$

If $-2^l \le b \le 2^l$, we will have $\hat{x}'_n = \hat{x}_n$ plus -1, 0, or 1; and if $-x_{\min} \le b < 2^l - x_{\max}$ we will have $\hat{x}'_n = \hat{x}_n$ for $0 \le n < N$, where $x_{\min}$ and $x_{\max}$ are the minimum and maximum values of $x_n \bmod 2^l$ for $0 \le n < N$. Thus, we have a family of solutions in which x_0 ranges over an entire interval; all of these solutions have the same multiplier and the same value of $(x_1 - x_0) \bmod 2^k$, and the increments c are easy to express as a function of x_0.

The indeterminacy of Problem B as it was stated leads us to change our definitions slightly. Let us say that a solution to Problem B is the determination of the multiplier a together with the first difference

$$y_0 = (x_1 - x_0) \bmod 2^k, \tag{11}$$

instead of asking for a, c, and x_0. For practical purposes, this knowledge is sufficient, since we will have

$$\begin{aligned}\hat{x}_n &= \left\lfloor \left(\left(x_0 + \frac{a^n - 1}{a - 1}\, y_0 \right) \bmod 2^k \right) \Big/ 2^l \right\rfloor \\ &= \left(\hat{x}_0 + \left\lfloor \frac{a^n - 1}{a - 1} \frac{y_0}{2^l} \right\rfloor + \epsilon_n \right) \bmod 2^h \end{aligned} \tag{12}$$

for all n, where ϵ_n is 0 or 1. (This equation follows from Lemma 1, since $y_j^{(0)} = a^j y_0 \bmod 2^k$.)

The following method determines the interval of permissible seed values x_0 such that the first N data points $\hat{x}_n$ are correct, given a solution (a, y_0) to Problem B in the new sense.

I1. [Initialize.] Set $x \leftarrow 2^l \hat{x}_0$, $u \leftarrow 0$, $v \leftarrow 2^l - 1$, $y \leftarrow y_0$, and $j \leftarrow 0$. (During this algorithm, we have the invariant relations $y = y_j^{(0)}$ and $x = x'_j$, assuming that $x'_0 = 2^l x_0$ and $x'_{j+1} = (ax'_j + c) \bmod 2^k$; the seed value x_0 will lead to the correct numbers $\hat{x}_n$ for $0 \le n \le j$ if and only if $2^l \hat{x}_0 + u \le x_0 \le 2^l \hat{x}_0 + v$.)

I2. [Done?] Increase j by 1, then terminate the algorithm if $j = N$.

I3. [Update.] Set $x \leftarrow (x + y) \bmod 2^k$ and $y \leftarrow ay \bmod 2^k$; also set $x^{(u)} \leftarrow \lfloor (x + u)/2^l \rfloor$ and $x^{(v)} \leftarrow \lfloor (x + v)/2^l \rfloor \bmod 2^h$. We must now have $\hat{x}_j = x^{(u)}$ or $x^{(v)}$, otherwise (a, y_0) was not a correct solution to Problem B. If $\hat{x}_j = x^{(u)} \ne x^{(v)}$, change v to $2^l - 1 - (x \bmod 2^l)$; otherwise if $\hat{x}_j = x^{(v)} \ne x^{(u)}$, change u to $2^l - (x \bmod 2^l)$. Then return to step I2. (We could actually terminate the algorithm now, if $u = v$.) □

Significant improvements to this algorithm are possible if $N = 2^l$. In that case the reader will find it interesting to construct an algorithm that deduces x_0 from a, y_0, and $(\hat{x}_0, \ldots, \hat{x}_{2^l - 1})$ in only $O(l)$ steps by an appropriate "binary search."

Let us now try to find a solution to Problem B when N is small. We shall assume that $h \ge 2$. Again we can apply the ideas of Lemma 1, but this time we will use smaller values of t. One of the key ideas needed is that the leading bits of $y_n^{(t)}$ are almost known, in the sense that

$$\lfloor y_n^{(t)}/2^l \rfloor + \kappa_n^{(t)} \equiv \hat{x}_{n+2^t} - \hat{x}_n \quad (\text{modulo } 2^h), \tag{13}$$

where $\kappa_n^{(t)}$ is 0 or 1. In fact, $\kappa_n^{(t)}$ is the carry out of the lth bit when x_n is added to $y_n^{(t)}$ in binary notation.

Let us set $t = \lfloor \lg(N-2) \rfloor$, the largest integer such that $2^t + 1 < N$. Knowing the values of $\hat{x}_0$, $\hat{x}_1$, $\hat{x}_{2^t}$, and $\hat{x}_{2^t+1}$, and recalling that $y_n^{(t)}$ is an odd multiple of 2^t, we find that there are 2^{l-t} possibilities for $y_0^{(t)}$ and 2^{l-t} possibilities for $y_1^{(t)}$. For example, if $h = l = 10$ and $t = 4$, we have $y_0^{(4)} \equiv 2^{10}(\hat{x}_{16} - \hat{x}_0 - \kappa_0^{(4)}) + 2^4 + 2^5 v_0$ (modulo 2^{20}), where $0 \le v_0 < 2^5$ and $0 \le \kappa_0^{(4)} \le 1$, and a similar formula holds for $y_1^{(4)}$. We will try all of these possibilities and see which (if any) are viable.

Given $y_0^{(t)}$ and $y_1^{(t)}$, we can determine $\alpha \equiv y_1^{(t)}/y_0^{(t)}$ (modulo 2^{k-t}); this number α is such that $y_n^{(t)} = \alpha^n y_0^{(t)}$, so if there are any integers n in the range $2 \le n < N - 2^t$, we may be able to reject the hypothesized values quickly by checking whether (13) holds for some $\kappa_n^{(t)} = 0$ or 1. If this optional test does not rule out the current $y_1^{(t)}$ and $y_0^{(t)}$ we proceed to *decrease* the value of t as follows. Lemma 1 implies that

$$y_n^{(t)} = (1 + a^{2^{t-1}})y_n^{(t-1)} \bmod 2^k, \tag{14}$$

and we have $a \equiv \alpha$ (modulo 2^{k-t}); hence $a^{2^{t-1}} \equiv \alpha^{2^{t-1}}$ (modulo 2^{k-1}). If $t > 1$, we have sufficient information to determine $y_0^{(t-1)}$ and $y_1^{(t-1)}$ (modulo 2^{k-1}) by (14), and the full values of $y_0^{(t-1)}$ and $y_1^{(t-1)}$ can be deduced from (13) when $h \ge 2$. We might find that (13) leads to a contradiction; otherwise we continue with $t \leftarrow t - 1$ and a new value of α as before. When $t = 1$ we have a similar situation, except that we must try two different values of a, and we need not compute $y_1^{(0)}$. If no contradictions appear, we have a tentative solution (a, y_0) to Problem B. A final test with the interval algorithm above will establish whether or not this is indeed a correct solution.

For example, let us consider again the problem in the introduction, but this time let us assume that only the first $N = 6$ values $(\hat{x}_0, \ldots, \hat{x}_5) = (4, 5, 1, 0, 2, 7)$ are given. With $t = 2$ there are four possibilities to try, namely $y_0^{(2)} = 44$ or 52, $y_1^{(2)} = 12$ or 20. Taking $(44, 12)$, we find $\alpha \equiv 9$ (modulo 16), so $a^2 \equiv 17$ (modulo 32) and $y_0^{(1)} \equiv 44/18 \equiv 6$ (modulo 32). By considering $\hat{x}_2$ and $\hat{x}_0$, we conclude that $y_0^{(1)} = 38$; similarly $y_1^{(1)} = 22$, and the next iteration has $\alpha \equiv 9$ (modulo 32). An impossible situation for $y_0^{(0)}$ occurs if $a = 9$, but $a = 41$ implies $y_0 = 7$, and this is a solution for $37 \le x_0 \le 39$. The next choice $(44, 20)$ leads to a contradiction when attempting to define $y_0^{(1)}$, and so does the third choice $(52, 12)$. The final choice $(52, 20)$ leads to a solution $a = 25$, $y_0 = 9$, which works for $32 \le x_0 \le 34$.

The total time to perform this algorithm is $O(2^{2l}k \log N/N^2)$ plus $O(N)$ per tentative solution (a, y_0). In practice, there are few tentative solutions (usually there is only one), unless h is small or N is very small. For example, the author made a series of experiments with $l = 16$, $a = 69069$, $c = 12345$, $x_0 = 54321$, $N = 259$, and varying h. For each of 65536 choices of $y_0^{(8)}$ and $y_1^{(8)}$, the algorithm tested $y_2^{(8)}$ for consistency. If this succeeded, the next step was to determine $y_0^{(7)}$, then $y_1^{(7)}$, then $y_2^{(7)}$, then $y_0^{(6)}$, and so on. When $h = 16$ it turned out that $y_2^{(8)}$ already gave a contradiction in 65533 cases, while two cases satisfied $y_2^{(8)}$ but not $y_0^{(7)}$; the remaining case led to the correct multiplier 69069, and it turned out that $53899 \le x_0 \le 54356$. When $h = 8$ the problem was a little harder, but 65087 choices were rejected because of $y_2^{(8)}$, 443 because of $y_0^{(7)}$, and 5 because of $y_1^{(7)}$. When $h = 4$, it was necessary to go still deeper; 579 incorrect multipliers passed $y_1^{(7)}$ safely, but only two of them got as far as $y_2^{(6)}$, and none reached $y_0^{(5)}$. When $h = 3$, one of the incorrect multipliers got as far as $y_1^{(2)}$ before getting stuck at $y_2^{(2)}$. Only when $h = 2$ were there any incorrect multipliers reaching the stage $y_0^{(0)} = y_0$. (This happened because no failure can occur at $y_0^{(t-1)}$ or $y_1^{(t-1)}$ when $h = 2$. It would have been much better to test $y_3^{(t)}$, $y_4^{(t)}$, … to weed out bad multipliers with less work.) Even so, none of the 405 bad multipliers passing y_0 survived as far as ten iterations of Algorithm I. Apparently, spurious multipliers can be weeded out rather quickly in nearly all cases.

If we assume that the original sequence is indeed a valid congruential sequence, there is no need to do the $O(N)$ validation steps unless more than one tentative solution turns up. Thus we have a method that will usually deduce the solution (a, y_0) in only $O(l^2)$ steps when N is of order 2^l.

The author experimented with a few other approaches that did not lead to substantial improvements over the method above. However, the ideas will be sketched here because they may be useful in some future investigations. Instead of blindly guessing the values of $y_0^{(t)}$ and $y_1^{(t)}$, we can gather statistics that indicate their approximate magnitude. Let f be given by (4), and let $f + u + t = l + 1$; then $y_{n+2^u}^{(t)} \equiv y_n^{(t)} + 2^{l+1}$ (modulo 2^{l+2}). We can use this relation for $n = 0$, 2^4, $2 \cdot 2^4$, $3 \cdot 2^4$, … to determine the carry digits $\kappa_n^{(t)}$. (It is easy to determine whether $f = 2$, 3, …, $l+1-t$ or $f > l+1-t$, by a similar method.) The expected value of $\kappa_n^{(t)}$ is exactly $(y_n^{(t)}/2^t) \bmod 2^{l-t}$, so we can obtain good estimates of the unknown portions of $y_0^{(t)}$ and $y_1^{(t)}$ by taking sufficiently large samples. When we are rather confident about the values of $y_0^{(t)}$ and $y_1^{(t)}$, we need not use a method that detects inconsistencies rapidly. We might prefer instead to deduce the value of y_0 as soon as possible; then we can decrease t by more than one unit at a time, using the identity

$$y_n^{(t)} = \left(\frac{a^{2^t} - 1}{a^{2^j} - 1}\right) y_n^{(j)} \bmod 2^k, \quad \text{for } 0 \le j \le t, \tag{15}$$

which generalizes (14).

Although we have seen that linear congruential sequences do not lead to secure encryptions by themselves, slight modifications would defeat the methods presented here. In particular, there appears to be no way to decipher the sequence of high-order bits generated by linear congruential sequences that have been shuffled by the method of Bays and Durham [1], if we modify that method to use hidden bits (not part of $\hat{x}_n$) to control the shuffling. Furthermore, there is no obvious way to deduce the multiplier of a sequence defined by $x_{n+1} = ax_n \bmod p$ from the leading bits of x_n when p is a prime number like $2^{32} - 99$. On the other hand, McCarthy [3] and Reeds [4]–[6] have shown that other ways of using linear congruential sequences can be successfully deciphered.

This work was supported in part by the National Science Foundation under grant MCS-77-23638 and by the Office of Naval Research under contract N00014-76C-0330.

References

[1] Carter Bays and S. D. Durham, "Improving a poor random number generator," *ACM Transactions on Mathematical Software* **2** (1976), 59–64.

[2] Donald E. Knuth, *Seminumerical Algorithms*, Volume 2 of *The Art of Computer Programming*, second printing (Reading, Massachusetts: Addison–Wesley, November 1971).

[3] John McCarthy, "Proposed criterion for a cipher to be probable-word-proof," *Communications of the ACM* **18** (1975), 131–132.

[4] James Reeds, "'Cracking' a random number generator," *Cryptologia* **1** (1977), 20–26.

[5] James Reeds, "Cracking a multiplicative congruential encryption algorithm," in *Information Linkage Between Applied Mathematics and Industry*, edited by Peter C. C. Wang (New York: Academic Press, 1979), 467–472.

[6] James Reeds, "Solution of challenge cipher," *Cryptologia* **3** (1979), 83–95.

Addendum

For further development of these methods, see Alan M. Frieze, Johan Hastad, Ravi Kannan, Jeffrey C. Lagarias, and Adi Shamir, "Reconstructing truncated integer variables satisfying linear congruences," *SIAM Journal on Computing* **17** (1988), 262–280; Joan Boyar, "Inferring sequences produced by a linear congruential generator missing low-order bits," *Journal of Cryptology* **1** (1989), 177–184; Joan Boyar, "Inferring sequences produced by pseudo-random number generators," *Journal of the Association for Computing Machinery* **36** (1989), 129–141; Hugo Krawczyk, "How to predict congruential generators," *Journal of Algorithms* **13** (1992), 527–545; Antoine Joux and Jacques Stern, "Lattice reduction: A toolbox for the cryptanalyst," *Journal of Cryptology* **11** (1998), 161–185; Simon R. Blackburn, Domingo Gomez-Perez, Jaime Gutierrez, and Igor E. Shparlinski, "Reconstructing noisy polynomial evaluation in residue rings," *Journal of Algorithms* **61** (2006), 47–59.

Chapter 25

Computation of Tangent, Euler, and Bernoulli Numbers

[Written with Thomas J. Buckholtz. Originally published in Mathematics of Computation **21** *(1967), 663–688.]*

Some elementary methods are described by which tangent numbers, Euler numbers, and Bernoulli numbers can be calculated much more easily and rapidly on electronic computers than by means of the traditional recurrence relations that have been used for over a century. These methods have been used to extend the existing tables. Some theorems about the periodicity of the tangent numbers, which were suggested by the tables, are also proved.

1. Introduction

The tangent numbers T_n, Euler numbers E_n, and Bernoulli numbers B_n are defined to be the coefficients in the following power series:

$$\tan z = T_0/0! + T_1 z/1! + T_2 z^2/2! + \cdots = \sum_{n\geq 0} T_n z^n/n!, \qquad (1)$$

$$\sec z = E_0/0! + E_1 z/1! + E_2 z^2/2! + \cdots = \sum_{n\geq 0} E_n z^n/n!, \qquad (2)$$

$$z/(e^z - 1) = B_0/0! + B_1 z/1! + B_2 z^2/2! + \cdots = \sum_{n\geq 0} B_n z^n/n!. \qquad (3)$$

Much of the older mathematical literature uses a slightly different notation for these numbers, to take account of the zero coefficients. Thus we find many papers where $\tan z$ is written $T_1 z + T_2 z^3/3! + T_3 z^5/5! + \cdots$, $\sec z$ is written $E_0 + E_1 z^2/2! + E_2 z^4/4! + \cdots$, and $z/(e^z - 1)$ is written $1 - z/2 + B_1 z^2/2! - B_2 z^4/4! + B_3 z^6/6! - \cdots$. Some other authors have used essentially the notation defined above but with different signs; in particular our E_{2n} is often accompanied by the sign $(-1)^n$.

In Section 2 we present simple methods for computing T_n, E_n, and B_n, which are readily adapted to electronic computers. More details of the computer program are explained in Section 3. Tables of T_n and E_n for $n \le 120$, and B_n for $n \le 250$, are appended to this paper, thereby extending the hitherto published values of T_n for $n < 60$ [6], E_n for $n \le 100$ [2, 3], and B_n for $n \le 220$ [7, 4].

Using the methods of this paper it is not difficult to extend the tables much further, and the authors have submitted a copy of the values of T_n ($n \le 835$), E_n ($n \le 808$), B_n ($n \le 836$) to the Unpublished Mathematical Tables repository of this journal.

Section 4 shows how the formulas of Section 2 lead to some simple proofs of arithmetical properties of these numbers.

2. Formulas for Computation

The traditional method of calculating T_n and E_n is to use recurrence relations, such as the following: Let $\cos z = \sum_{n \ge 0} C_n z^n/n!$; then the coefficient of $z^n/n!$ in $(\tan z)(\cos z)$ is

$$\sum_k \binom{n}{k} T_k C_{n-k},$$

and in $(\sec z)(\cos z)$ it is

$$\sum_k \binom{n}{k} E_k C_{n-k}.$$

Hence, making use of the fact that $T_{2n} = E_{2n+1} = 0$, we have the recurrence relations

$$\binom{2n+1}{1} T_1 - \binom{2n+1}{3} T_3 + \cdots + (-1)^n \binom{2n+1}{2n+1} T_{2n+1} = 1, \quad n \ge 0; \qquad (4)$$

$$\binom{2n}{0} E_0 - \binom{2n}{2} E_2 + \cdots + (-1)^n \binom{2n}{2n} E_{2n} = 0, \quad n > 0. \qquad (5)$$

The disadvantage of these formulas is that the binomial coefficients as well as the numbers T_n and E_n become very large when n is large, so a time-consuming multiplication of multiple-precision numbers is implied. As Lehmer [4] has observed, we may simplify the calculations if we remember the values of

$$\binom{2n+1}{k} T_k, \quad \binom{2n}{k} E_k$$

so that when n increases by 1 we need only multiply

$$\binom{2n+1}{k} T_k$$

by

$$\frac{(2n+2)(2n+3)}{(2n+2-k)(2n+3-k)}$$

to get the next value. But the method to be described here is even simpler, and has other advantages.

The tangent numbers may be evaluated by noting that $D(\tan^n z)$ is $n(\tan^{n-1} z)(1+\tan^2 z)$; hence the nth derivative of $\tan z$ is a polynomial in $\tan z$. Indeed, we have $D^n(\tan z) = P_n(\tan z)$, where the polynomials $P_n(x)$ are defined by

$$P_0(x) = x, \quad P_{n+1}(x) = (1+x^2)P_n'(x). \tag{6}$$

Thus if we write

$$D^n(\tan z) = T_{n,0} + T_{n,1}\tan z + T_{n,2}\tan^2 z + \cdots,$$

the coefficients $T_{n,k}$ satisfy the recurrence equation

$$T_{0,k} = \delta_{1k}; \quad T_{n+1,k} = (k-1)T_{n,k-1} + (k+1)T_{n,k+1}. \tag{7}$$

Since $T_n = D^n(\tan z)\,|_{z=0} = T_{n,0}$, and since $T_{n,k}$ is zero except for at most $(n+3)/2$ values of k, formula (7) shows that the calculation of all $T_{n+1,k}$ from the values of $T_{n,k}$ essentially requires only $(n+2)/2$ multiplications of a small number k by a large number $T_{n,k}$, plus $n/2$ additions of large numbers. We are interested only in $T_{n,0}$ for odd values of n, so we might try to use the relation

$$T_{n+2,k} = (k-2)(k-1)T_{n,k-2} + 2k^2 T_{n,k} + (k+1)(k+2)T_{n,k+2};$$

but a count of the operations involved shows that this change provides little if any improvement over (7), hence the simpler form (7) is preferable.

Similarly, we have

$$D(\sec z \tan^n z) = \sec z(n\tan^{n-1} z + (n+1)\tan^{n+1} z).$$

Hence if we write

$$D^n(\sec z) = (\sec z)(E_{n,0} + E_{n,1}\tan z + E_{n,2}\tan^2 z + \cdots) \tag{8}$$

we have the recurrence

$$E_{0,k} = \delta_{0k}; \quad E_{n+1,k} = kE_{n,k-1} + (k+1)E_{n,k+1}. \tag{9}$$

Since $E_n = E_{n,0}$, this relation yields an efficient method for calculating the Euler numbers. A somewhat similar recurrence was used by Joffe [3] to calculate Euler numbers; his method requires essentially the same amount of computation, but as explained in the next section there is a way to modify (9) to obtain a considerable advantage.

The identities $\tan(\pi/4 + z/2) = \tan z + \sec z$ and

$$D^n(\tan(\pi/4 + z/2)) = 2^{-n}P_n(\tan(\pi/4 + z/2))$$

imply that the sums of the numbers $T_{n,k}$ have a very simple form:

$$2^{-n}P_n(1) = 2^{-n}\sum_{k\geq 0} T_{n,k} = \begin{cases} E_n, & n \text{ even;} \\ T_n, & n \text{ odd.} \end{cases} \tag{10}$$

This relation can be used to advantage when both E_n and T_n are being calculated.

The definition of $\tan z$ implies that

$$\begin{aligned} \tan z &= \frac{\sin z}{\cos z} = \frac{(e^{iz} - e^{-iz})}{i(e^{iz} + e^{-iz})} = \frac{1}{z}\Big(\frac{2iz}{e^{2iz}+1} - iz\Big) \\ &= \frac{1}{z}\Big(\frac{2iz}{e^{2iz}-1} - \frac{4iz}{e^{4iz}-1} - iz\Big) = \frac{1}{z}\Big(-iz + \sum_{n\geq 0}((2iz)^n - (4iz)^n)\frac{B_n}{n!}\Big); \end{aligned}$$

and by equating coefficients we obtain the well-known identity

$$B_n = -i^{-n}nT_{n-1}/(2^n(2^n-1)), \quad n > 1. \tag{11}$$

Hence the Bernoulli numbers may be obtained from the tangent numbers by a calculation that is especially simple on a binary computer.

The celebrated von Staudt–Clausen theorem [8, 1] states that

$$B_{2n} = C_{2n} - \sum_{\substack{p \text{ prime} \\ p-1 \text{ divides } 2n}} \frac{1}{p}, \tag{12}$$

where C_{2n} is an integer. Table 3 below expresses B_n in this form. We will show that, using (12), the calculation of (11) can be carried out without any multiple-precision division.

3. Details of the Computation

By the recurrence (7) we may discard the value of $T_{n,k}$ once $T_{n+1,k+1}$ has been calculated, so only about n of the values $T_{n,k}$ need to be retained in the computer memory at any one time. A further technique can be employed when the memory size has been exceeded; for example, suppose we start with the computation of $T_{n,k}$ for $n \leq 4$:

	$k=0$	$k=1$	$k=2$	$k=3$	$k=4$	$k=5$	$k=6$
$n=0$	0	1					
$n=1$	1	0	1				
$n=2$	0	2	0	2			
$n=3$	2	0	8	0	6		
$n=4$	0	16	0	40	0	24	

Suppose further that very little memory space is available, so that we cannot completely evaluate all of the entries for $n = 5$; we might obtain

$n=5$	16	0	136	0	240	0	*

where '*' denotes an unknown value. The calculation may still proceed, keeping track of unknown values:

$n=6$	0	272	0	1232	0	*	
$n=7$	272	0	3968	0	*		
$n=8$	0	7936	0	*			
$n=9$	7936	0	*				etc.

In this way we are able to compute the values of about twice as many tangent numbers as were produced before overflow occurred, avoiding much of the calculation of the $T_{n,k}$.

Since the numbers T_n become very large (T_{835} has 1916 digits, and T_n is asymptotically $2^{n+2}n!/\pi^{n+1}$ when n is odd), care needs to be taken for storage allocation of the numbers $T_{n,k}$ if we are to make efficient use of memory space. The program we prepared makes use of two rather small areas of memory (say A and B), each of which is capable of holding any one of the numbers $T_{n,k}$, plus a large number of consecutive locations used for all the remaining values. By sweeping cyclically through this large memory area, it is possible to store and retrieve the values in a simple manner.

For the sake of illustration let us suppose the word size of our computer is very small, so that only one decimal digit may be stored per word; and suppose there are just 14 words of memory used for the table of $T_{n,k}$. After the calculation of the values for $n = 4$, the memory might have the following configuration:

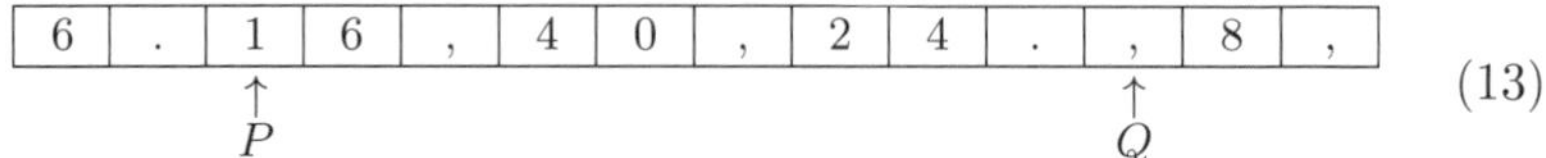

(13)

Here P and Q represent variables in the program that point to the current places of interest in the memory; P points to the number that will be accessed next, and Q points to the place where the next value is to be written. Only locations from P to $Q - 1$ contain information that will be used subsequently by the program. The symbols '.' and ',' represent special negative codes that delimit the numbers in an obvious fashion. As we begin the calculation for $n = 5$, we set area A to zero and a variable k to 1. The basic cycle is then:

a) Set area B to k times the next value indicated by P, and move P to the right.
b) Store the value of $A+B$ into the locations indicated by Q, and move Q to the right.
c) Transfer the contents of B to area A.
d) Increase k by 2.

In the case of (13) we would change the memory configuration to

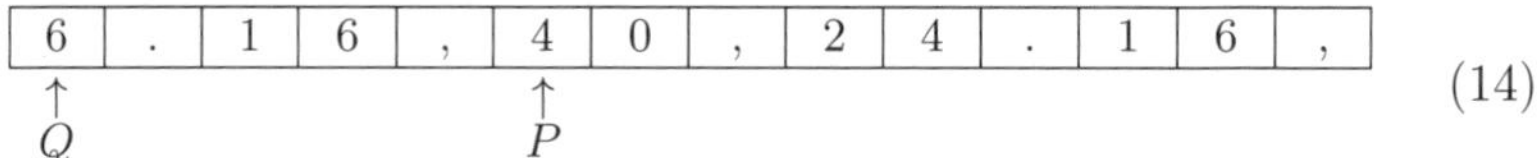

(14)

with $A = 16$ and $k = 3$. Notice that the pointer P has moved to the right past the value 16; the value $16 = 0 + 1 \cdot 16$ has been stored; and the pointer Q has moved to the right, then to the far left (treating the memory as a circular store). The next two iterations of steps (a)–(d) give

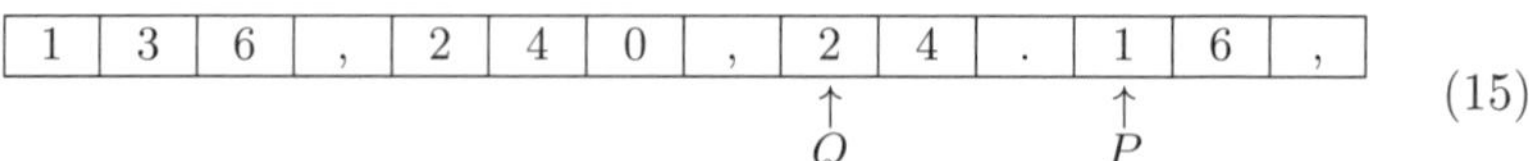

(15)

with $A = 120$ and $k = 7$. At this point the terminating '.' has been sensed, and the program attempts to store the value from area A. But this would make pointer Q pass P, so the "memory overflow" condition is sensed, and the memory configuration becomes

(16)

where '$*$' is another internal code symbol. The computation for $n = 6$ is similar, but it uses a different initialization since n is even; after $n = 6$ has been processed we will have

2	3	2	,	$*$	4	0	,	$*$	2	7	2	,	1
					$\uparrow$ Q				$\uparrow$ P				

(17)

and so on.

The discussion above has been simplified slightly for purposes of exposition. In the actual program, it is preferable to keep the numbers stored with least significant digit first, so that for example (16) would really be

6	3	1	,	0	4	2	,	$*$	2	1	6	1	,
									$\uparrow$ Q		$\uparrow$ P		

(18)

in order to simplify the multiple-precision operations. A few other changes in the sequence of operations were made in order to use memory a little more efficiently; for example, the value $T_{n,0}$ need never be retained.

A similar method may be used for E_n. This arrangement of the computation gives a substantial advantage over Joffe's method [3] because of the '$*$', and it has advantages over (10) for the same reason.

It remains to consider the calculation of the Bernoulli number B_{2n} from T_{2n-1}. Consider formula (12). If p is an odd prime, $2^{p-1} \equiv 1$ (modulo p); hence if $p - 1$ divides $2n$, then $2^{2n} - 1$ is divisible by p. So we first compute the integer

$$N = (-1)^{n-1} 2n T_{2n-1} + \sum_{\substack{p \text{ prime} \\ p-1 \text{ divides } 2n}} \frac{2^{2n}(2^{2n}-1)}{p} \tag{19}$$

by referring to an auxiliary table of primes that may be calculated at the beginning of the program. Then it is merely a question of computing

$$C_{2n} = N/(2^{2n}(2^{2n}-1)) = N/2^{4n} + N/2^{6n} + N/2^{8n} + \cdots. \tag{20}$$

The calculation of $N/2^k$ is of course merely a "shift right" operation in a binary computer, so all the terms of the infinite series on the right side of (20) are readily computed. This series converges very rapidly, and we know C_{2n} is an integer, so we need only carry out the calculation indicated in (20) until it converges one word-size (35 bits) to the right of the decimal point. It is simple to check at the same time that C_{2n} is indeed very close to an integer, in order to verify the computations.

4. Periodicity of the Sequences

Examination of the tables produced by the computer program shows that the unit's digits of the nonzero tangent numbers repeat endlessly in the pattern 2, 6, 2, 6, 2, 6, starting with T_3; furthermore the two least significant digits ultimately form a repeating period of length 10, namely 16, 72, 36, 92, 56, 12, 76, 32, 96, 52, 16, 72, ... The three least significant digits have a period of length 50, and for four digits the period length is 250. These empirical observations suggest that theoretical investigation of period length might prove fruitful.

Theorem 1. *Let p be an odd prime, and let λ be the period length of the sequence $\langle T_n \bmod p\rangle$. Then*

$$\lambda = \begin{cases} p-1, & p \equiv 1 \text{ (modulo 4)}; \\ 2(p-1), & p \equiv 3 \text{ (modulo 4)}; \end{cases} \tag{21}$$

and

$$T_{n+\lambda} \equiv T_n \quad \text{(modulo } p) \quad \text{for all } n \geq 0. \tag{22}$$

Proof. It is clear from the recurrence relation (7) that the sequence $\langle T_n \bmod p\rangle$ is determined by the recurrence equation

$$y_{n+1} = Ay_n \tag{23}$$

where the vector y_n and the matrix A are defined by

$$A = \begin{bmatrix} 0 & 2 & & & & & \\ 1 & 0 & 3 & & & & \\ & 2 & 0 & 4 & & & \\ & & 3 & 0 & & & \\ & & & & \ddots & & \\ & & & & & 0 & p{-}1 \\ & & & & & p{-}2 & 0 \end{bmatrix}, \quad y_n = \begin{bmatrix} T_{n,1} \\ T_{n,2} \\ T_{n,3} \\ T_{n,4} \\ \vdots \\ T_{n,p-2} \\ T_{n,p-1} \end{bmatrix}. \tag{24}$$

The reason is that $T_{n,k}$ can contribute nothing to any subsequent value of $T_{n \bmod p}$ when $k \geq p$.

We will show below that the minimum polynomial equation satisfied by A is

$$A^{p-1} - (-1)^{(p-1)/2} I \equiv 0 \quad \text{(modulo } p); \tag{25}$$

hence (22) is valid for the value of λ given by (21). It remains to show that λ is the true period length of the sequence, not merely a multiple of the period.

Accordingly, suppose $T_{n+\lambda'} \equiv T_n$ (modulo p) for some positive $\lambda' \le \lambda$ and all large n. In view of (22) this congruence must hold for all $n \ge 0$. Let $y = y_{\lambda'} - y_0$; then $f(A^n y) \equiv 0$ for all $n \ge 0$, where f denotes the projection onto the first component of the vector $A^n y$. But this relation implies $n!\,\alpha_n \equiv 0$ (modulo p) for all components α_n of y; hence $y \equiv 0$, that is, $y_0 \equiv y_{\lambda'} = A^{\lambda'} y_0$. It follows that $y_n \equiv A^{\lambda'} y_n$ for all $n \ge 0$, and since the vectors $y_0, \ldots, y_{p-2}$ are obviously linearly independent we must have $A^{\lambda'} \equiv I$ (modulo p). Therefore, $\lambda' \ge \lambda$, and the proof is complete.

It remains to verify (25), which seems to be a nontrivial identity. The minimum polynomial of A must clearly be of degree $p-1$, since $y_0, \ldots, y_{p-2}$ are linearly independent; therefore, it suffices to calculate the characteristic polynomial of A. Let

$$D_n = \det \begin{bmatrix} x & -(n-1) & & & & \\ -n & x & -(n-2) & & & \\ & -(n-1) & x & & & \\ & & & \ddots & & \\ & & & & x & -1 \\ & & & & -2 & x \end{bmatrix}; \qquad (26)$$

then $D_n = xD_{n-1} - (n-1)nD_{n-2}$ so we have

$$\begin{aligned} D_1 &= x, \\ D_2 &= x^2 - 1\cdot 2, \\ D_3 &= x^3 - (1\cdot 2 + 2\cdot 3)x, \\ D_4 &= x^4 - (1\cdot 2 + 2\cdot 3 + 3\cdot 4)x^2 + 1\cdot 2\cdot 3\cdot 4, \\ D_5 &= x^5 - (1\cdot 2 + 2\cdot 3 + 3\cdot 4 + 4\cdot 5)x^3 + (1\cdot 2\cdot 3\cdot 4 + 1\cdot 2\cdot 4\cdot 5 + 2\cdot 3\cdot 4\cdot 5)x, \end{aligned}$$

and in general

$$D_n = x^n - s_{n,1}x^{n-2} + s_{n,2}x^{n-4} - s_{n,3}x^{n-6} + \cdots, \qquad (27)$$

where

$$s_{n,k} = \sum a_1(a_1+1)a_2(a_2+1)\ldots a_k(a_k+1) \qquad (28)$$

is summed over all values $1 \le a_1 \ll a_2 \ll \cdots \ll a_k < n$. (Here the notation $u \ll v$, for integers u and v, denotes the relation $v \ge u+2$.) Thus, $s_{n,k}$ is the sum of all products of k of the pairs $1\cdot 2$, $2\cdot 3$, …, $(n-1)\cdot n$, with no "overlapping" pairs allowed in the same term.

To evaluate $s_{p-1,k} \bmod p$, it is convenient to allow also the pairs $(p-1)\cdot p$ and $p\cdot 1$, since these contribute nothing to the sum. Thus, for example, we have

$$\begin{aligned} s_{6,2} \equiv 1{\cdot}2{\cdot}3{\cdot}4 + 1{\cdot}2{\cdot}4{\cdot}5 + 1{\cdot}2{\cdot}5{\cdot}6 + 1{\cdot}2{\cdot}6{\cdot}7 + 2{\cdot}3{\cdot}4{\cdot}5 \\ + 2{\cdot}3{\cdot}5{\cdot}6 + 2{\cdot}3{\cdot}6{\cdot}7 + 2{\cdot}3{\cdot}7{\cdot}1 + 3{\cdot}4{\cdot}5{\cdot}6 + 3{\cdot}4{\cdot}6{\cdot}7 \\ + 3{\cdot}4{\cdot}7{\cdot}1 + 4{\cdot}5{\cdot}6{\cdot}7 + 4{\cdot}5{\cdot}7{\cdot}1 + 5{\cdot}6{\cdot}7{\cdot}1 \end{aligned}$$

(modulo 7). Let us say that two of the terms $a_1(a_1+1)\ldots a_k(a_k+1)$ and $a'_1(a'_1+1)\ldots a'_k(a'_k+1)$ are "equivalent" if, for some r and t and for all j, we have $a_j \equiv a'_{1+(j+r) \bmod p} + t$; thus, in the example above the terms

$$1{\cdot}2{\cdot}4{\cdot}5,\ 2{\cdot}3{\cdot}5{\cdot}6,\ 3{\cdot}4{\cdot}6{\cdot}7,\ 4{\cdot}5{\cdot}7{\cdot}1,\ 5{\cdot}6{\cdot}1{\cdot}2,\ 6{\cdot}7{\cdot}2{\cdot}3, \text{ and } 7{\cdot}1{\cdot}3{\cdot}4$$

are mutually equivalent. It is impossible for a term to be equivalent to itself when $0 < t < p$, since this would imply $a_1 + \cdots + a_k \equiv a_1 + \cdots + a_k + kt$, making $t \equiv 0$. Therefore, each equivalence class consists of precisely p terms. When $k < (p-1)/2$ the sum over an equivalence class has the form

$$\sum_{0 \le t < p} (a_1+t)(a_1+t+1)\ldots(a_k+t)(a_k+t+1)$$

where the summand is a polynomial of degree $\le p-2$ in t. Any such summation may be expressed modulo p as a sum of terms of the form

$$c \sum_{0 \le t < p} \binom{t}{j} = c\binom{p}{j+1} \equiv 0 \quad (\text{modulo } p)$$

for some constant c, since $0 \le j < p-1$; so $s_{p-1,k} \equiv 0$. It follows that

$$D_{p-1} \equiv x^{p-1} + (-1)^{(p-1)/2}(p-1)! \quad (\text{modulo } p), \tag{29}$$

and an application of Wilson's theorem completes the proof of (25). □

Theorem 2. *Let p be an odd prime, and let λ be the period length of the sequence $\langle E_n \bmod p\rangle$. Then*

$$\lambda = \begin{cases} p-1, & p \equiv 1 \text{ (modulo 4)}; \\ 2(p-1), & p \equiv 3 \text{ (modulo 4)}; \end{cases} \tag{30}$$

and

$$E_{n+\lambda} \equiv E_n \quad (\text{modulo } p) \quad \text{for all } n \ge 1. \tag{31}$$

Proof. Make the following changes in the proof of Theorem 1:

$$A = \begin{bmatrix} 0 & 1 & & & & & \\ 1 & 0 & 2 & & & & \\ & 2 & 0 & 3 & & & \\ & & 3 & 0 & & & \\ & & & & \ddots & & \\ & & & & & 0 & p-1 \\ & & & & & p-1 & 0 \end{bmatrix}, \quad y_n = \begin{bmatrix} E_{n,0} \\ E_{n,1} \\ E_{n,2} \\ E_{n,3} \\ \vdots \\ E_{n,p-2} \\ E_{n,p-1} \end{bmatrix}. \tag{32}$$

Then the minimum polynomial equation satisfied by A is

$$A^p - (-1)^{(p-1)/2} A \equiv 0 \quad (\text{modulo } p). \tag{33}$$

The proof is a straightforward modification of the previous proof. □

The congruences (22) and (31) were obtained long ago by Kummer (see for example [5, page 270]), but he did not show that the true period length could not be a proper divisor of the number λ given by (21) and (30). More general congruences due to Kummer make it possible to establish further results about the period length:

Theorem 3. *Let p be an odd prime, and let λ be given by* (30). *Then*

$$T_{n+\lambda p^{k-1}} \equiv T_n \qquad (\text{modulo } p^k), \quad n \geq k; \tag{34}$$

$$E_{n+\lambda p^{k-1}} \equiv E_n \qquad (\text{modulo } p^k), \quad n \geq k. \tag{35}$$

Proof. Assume that $n \geq k$ and define the sequence $\langle u_m \rangle$ by the rule

$$u_m = (-1)^{(p-1)m/2} T_{n+(p-1)m}, \quad m \geq 0. \tag{36}$$

Kummer's congruence for the tangent numbers may be written

$$\Delta^k u_m \equiv 0 \quad (\text{modulo } p^k), \quad m \geq 0, \quad k \geq 1, \tag{37}$$

where $\Delta^k u_m$ denotes

$$\binom{k}{0} u_{m+k} - \binom{k}{1} u_{m+k-1} + \binom{k}{2} u_{m+k-2} - \cdots + (-1)^k \binom{k}{k} u_m.$$

We will prove that (37) implies

$$u_{m+p^{r-1}} \equiv u_m \quad (\text{modulo } p^r), \quad m \geq 0, \quad r \geq 1, \tag{38}$$

and this will establish (34). Equation (35) follows in the same way if we let

$$u_m = (-1)^{(p-1)m/2} E_{n+(p-1)m}, \quad m \geq 0.$$

Assume that (37) is valid for some sequence of real numbers (not necessarily integers) u_0, u_1, $\ldots$; thus, $\Delta^k u_m$ is an integer multiple of p^k when $k \geq 1$, but not necessarily when $k = 0$. We will prove that the sequence u_m/p, u_{m+p}/p, u_{m+2p}/p, $\ldots$, for fixed m, also satisfies Eq. (37), and this will suffice to prove (38) by induction on r.

Let E be the operator $Eu_m = u_{m+1}$. Equation (37) may be written $(E-1)^k u_m \equiv 0$ (modulo p^k), and our goal as stated in the preceding paragraph is to show that $(E^p - 1)^k (u_m/p) \equiv 0$ (modulo p^k); that is, $(E^p - 1)^k u_m \equiv 0$ (modulo p^{k+1}). Let

$$f(E) = E^{p-2} + 2E^{p-3} + \cdots + (p-2)E + (p-1);$$

then $E^p - 1 = (E-1)(p + f(E)(E-1))$, hence

$$(E^p - 1)^k u_m = \sum_{0 \leq j \leq k} \binom{k}{j} p^j (E-1)^{2k-j} f(E)^{k-j} u_m$$

and each term in the sum on the right is an integer multiple of p^{2k}. Hence, we have proved in fact that $(E^p - 1)^k u_m \equiv 0$ (modulo p^{2k}), which is more than enough to complete the proof of the theorem. □

Notice that Eqs. (34) and (35) do not necessarily give the true period length of the sequence mod p^k when $k > 1$; although (34) is "best possible" when $p = 5$ and $k = 2, 3, 4$, the tangent numbers have the same period length modulo 9 as they do modulo 3.

The tangent number T_{2n+1} is divisible by 2^n, so the period length of T_n mod 2^r is 1 for all r. Equation (35) is valid for $\lambda = 2$ when $p = 2$, since Kummer's congruence (37) holds for $u_m = E_{n+2m}$. In particular, we may combine the results proved above to show that, for any modulus m, the sequences $\langle T_n \bmod m\rangle$ and $\langle E_n \bmod m\rangle$ are periodic, and that the period length divides $2\phi(m)$, where ϕ denotes Euler's totient function.

TABLE 1. The first 60 nonzero tangent numbers

n	T_n
1	1.
3	2.
5	16.
7	272.
9	7936.
11	353792.
13	22368256.
⋮	
119	3257 2969544137 3711110813 9491520587 0894578681 8558730200 7333881055 9724342116 8172307776 6222847786 5964664757 3601851664 4828218413 9690510871 7176120451 6527175740 8580920993 7947000832.

[It is no longer desirable to devote 16 pages of a book to these numerical values, which can now be computed by standard software in a few microseconds]

TABLE 2. The first 61 nonzero Euler numbers

n	E_n
0	1.
2	1.
4	5.
6	61.
8	1385.
⋮	
120	248839 1574782987 1631690245 5408489408 2372867090 7090814055 5499968530 1842243985 7255460434 6369071792 7997103011 5914025391 0784871444 2940830046 2747699810 6540373770 6481607384 7531472025.

TABLE 3. The first 250 Bernoulli numbers

$B_0 = 1$, $B_1 = -1/2$, $B_{2n+1} = 0$ for $n \geq 1$, and the values of B_{2n} for $1 \leq n \leq 125$ appear below in the form $C_{2n} - \{p_1, p_2, \ldots, p_k\}$. This notation stands for $C_{2n} - 1/p_1 - \cdots - 1/p_k$; thus $B_4 = 1 - \{2, 3, 5\} = 1 - 1/2 - 1/3 - 1/5 = -1/30$. The Bernoulli numbers are expressed in this form here because the integers C_{2n} have not been tabulated before.

n	B_n
2	$1 - \{2, 3\}$.
4	$1 - \{2, 3, 5\}$.
6	$1 - \{2, 3, 7\}$.
8	$1 - \{2, 3, 5\}$.
10	$1 - \{2, 3, 11\}$.
12	$1 - \{2, 3, 5, 7, 13\}$.
14	$2 - \{2, 3\}$.
16	$-6 - \{2, 3, 5, 17\}$.
⋮	
250	1843 5261467838 9394126646 2015977022 3239649247 7000034429 4569758971 5176258684 2335348377 7655844047 4285697821 8329220039 6978835905 6020603056 3580448568 1973565915 1510513686 8316837867 5226653094 2856333382 8622890759 5799693397 1198209110 9285643939 6181295360 9407215690 8622535217 4286407738 3938476752 5254881572 $-\{2, 3, 11, 251\}$.

This research was supported in part by National Science Foundation grant GP 3909.

References

[1] Thomas Clausen, "Theorem," *Astronomische Nachrichten* **17** (1840), 351–352.

[2] S. A. Joffe, "Calculation of the first thirty-two Eulerian numbers from central differences of zero," *The Quarterly Journal of Pure and Applied Mathematics* **47** (1916), 103–126.

[3] S. A. Joffe, "Calculation of eighteen more, fifty in all, Eulerian numbers from central differences of zero," *The Quarterly Journal of Pure and Applied Mathematics* **48** (1919), 193–271.

[4] D. H. Lehmer, "An extension of the table of Bernoulli numbers," *Duke Mathematical Journal* **2** (1936), 460–464.

[5] Niels Nielsen, *Traité élémentaire des Nombres de Bernoulli* (Paris: Gauthier-Villars, 1923).

[6] J. Peters and J. Stein, *Zehnstellige Logarithmen der Zahlen von 1 bis 100000 nebst einem Anhang mathematischer Tafeln* (Berlin: Reichsamt für Landesaufnahme, 1922).

[7] С. З. Серебренниковъ, "Таблица первыхъ девяноста чиселъ Бернулли," [S. Z. Serebrennikoff, "A table of the first ninety Bernoulli numbers,"] *Mémoires de l'Academie Impériale des Sciences de St.-Petersbourg*, Classe des Sciences Physiques et Mathématiques (8) **16**, 10 (1905), 1–8.

[8] K. G. C. von Staudt, "Beweis eines Lehrsatzes, die BERNOULLIschen Zahlen betreffend," *Journal für die reine und angewandte Mathematik* **21** (1840), 372–374.

Addendum

If I were writing this paper today, I would mention the convergent sum

$$T_n + E_n = \frac{2^{n+2}n!}{\pi^{n+1}}\left(1 + \left(-\frac{1}{3}\right)^{n+1} + \left(\frac{1}{5}\right)^{n+1} + \left(-\frac{1}{7}\right)^{n+1} + \cdots\right)$$

instead of merely stating the first term in T_n's asymptotics. And I'd cite D. H. Lehmer's paper "Lacunary recurrence formulas for the numbers of Bernoulli and Euler," *Annals of Mathematics* (2) **36** (1935), 637–649.

However, I wouldn't actually be writing this paper at all today — because I've learned that in 1877 L. Seidel had already discovered a much better way to compute these numbers, by generating the triangle

$$\begin{matrix} & & 1 & & \\ & 0 & & 1 & \\ & 1 & 1 & 0 & \\ 0 & 1 & & 2 & 2 \\ 5 & 5 & 4 & 2 & 0 \end{matrix}$$

in which sums are formed alternately left-to-right, then right-to-left. Seidel's triangle ["Ueber eine einfache Entstehungsweise der Bernoulli'schen Zahlen und einiger verwandten Reihen," *Sitzungsberichte der mathematisch-physikalischen Classe der königlich bayerischen Akademie der Wissenschaften zu München* **7** (1877), 157–187] has been rediscovered many times, for example by A. J. Kempner in 1933, R. C. Entringer in 1966, M. D. Atkinson in 1986, and V. I. Arnold in 1991.

Chapter 26

Euler's Constant to 1271 Places

[Originally published in Mathematics of Computation **16** *(1962), 275–281.]*

The value of Euler's or Mascheroni's constant

$$\gamma = \lim_{n\to\infty}(1 + \tfrac{1}{2} + \cdots + \tfrac{1}{n} - \ln n)$$

has now been determined to 1271 decimal places, thus extending the previously known value of 328 places. A calculation of partial quotients and best rational approximations to γ was also made.

1. Historical Background

Euler's constant was, naturally enough, first evaluated by Leonhard Euler, who obtained the value 0.577218 in 1734 [1]. By 1781 he had calculated it more accurately as 0.5772156649015325 [2]. The calculations were carried out more precisely by several later mathematicians, among them Gauss, who obtained

$$\gamma = 0.577215664901532860606 53.$$

Various British mathematicians continued the effort [3, 4]; an excellent account of the work done on evaluation of γ before 1870 is given by Glaisher [5]. Finally, the famous mathematician-astronomer J. C. Adams [6] laboriously determined γ to 263 places. Adams thereby extended the work of Shanks, who had obtained 110 places (101 of which were correct).

Adams's result stood until 1952, when Wrench [8] calculated 328 decimal places. Although much work has been done trying to decide whether γ is rational, the evaluation has not been carried out any more precisely. With the use of high-speed computers, the constants e and π have been evaluated to many thousands of decimal places [9, 14]; a complete bibliography for π appears in [14]. The evaluation of γ to many places is considerably more difficult.

2. Evaluation of γ

The technique used here to calculate γ is essentially that used by Adams and earlier mathematicians. A complete derivation of the method is given by Knopp [7, §64B4]. We use Euler's summation formula in the form

$$\sum_{i=1}^{n} f(i) = \int_1^n f(x)\,dx + \frac{1}{2}\bigl(f(n)+f(1)\bigr) + \sum_{j=1}^{k} \frac{B_{2j}}{(2j)!}\bigl(f^{(2j-1)}(n) - f^{(2j-1)}(1)\bigr) + R_k, \tag{1}$$

where B_m denotes a Bernoulli number, defined symbolically by

$$e^{Bx} = \frac{x}{e^x - 1}. \tag{2}$$

With this notation, $B_1 = -\frac{1}{2}$, $B_2 = \frac{1}{6}$, $B_3 = 0$, $B_4 = -\frac{1}{30}$, etc. The remainder R_k in this formula is given by

$$R_k = \frac{1}{(2k+1)!} \int_1^n P_{2k+1}(x) f^{(2k+1)}(x)\,dx; \tag{3}$$

here $P_{2k+1}(x)$ is a periodic Bernoulli polynomial, symbolically

$$P_{2k+1}(x) = (\{x\}+B)^{2k+1} = (-1)^{k-1}(2k+1)! \sum_{r=1}^{\infty} \frac{2\sin 2r\pi x}{(2r\pi)^{2k+1}} \tag{4}$$

where $\{x\}$ is the fractional part of x.

Now we put $f(x) = 1/x$, obtaining from (1)

$$1 + \frac{1}{2} + \cdots + \frac{1}{n} = \ln n + \frac{1}{2} + \frac{1}{2n} + \frac{B_2}{2}\left(1 - \frac{1}{n^2}\right) + \cdots + \frac{B_{2k}}{2k}\left(1 - \frac{1}{n^{2k}}\right) - \int_1^n \frac{P_{2k+1}(x)}{x^{2k+2}}\,dx. \tag{5}$$

Taking the limit in (5) as $n \to \infty$, we find

$$\gamma = \frac{1}{2} + \frac{B_2}{2} + \cdots + \frac{B_{2k}}{2k} - \int_1^\infty \frac{P_{2k+1}(x)}{x^{2k+2}}\,dx. \tag{6}$$

Subtracting (5) from (6) gives

$$\gamma = 1 + \frac{1}{2} + \cdots + \frac{1}{n} - \ln n - \frac{1}{2n} + \frac{B_2}{2n^2} + \cdots + \frac{B_{2k}}{2kn^{2k}} - \int_n^\infty \frac{P_{2k+1}(x)}{x^{2k+2}}\,dx. \quad (7)$$

If the remainder is discarded and we consider (7) as an infinite series in k, it diverges as $k \to \infty$. Yet it still yields a good method for calculating γ, since

$$|P_{2k+1}(x)| \le \frac{2(2k+1)!}{(2\pi)^{2k+1}} \sum_{r=1}^{\infty} \frac{1}{r^{2k+1}} \quad (8)$$

and by applying Stirling's formula to (8) we obtain

$$\left| \int_n^\infty \frac{P_{2k+1}(x)}{x^{2k+2}}\,dx \right| \le \frac{4}{n}\sqrt{\frac{k}{\pi}}\left(\frac{k}{n\pi e}\right)^{2k}. \quad (9)$$

Put $k = 250$ and $n = 10000$ to obtain a remainder

$$\left| \int_{10000}^\infty \frac{P_{501}(x)}{x^{502}}\,dx \right| < 10^{-1269}, \quad (10)$$

so these values may be used in (7) to determine γ to at least 1269 places. This particular choice of k and n was made for convenience on a decimal computer, in an attempt to obtain the greatest precision in a reasonable amount of time.

3. Details of the Computation

The sum $1 + \frac{1}{2} + \cdots + \frac{1}{10000}$ was evaluated as

$$S_{10000} = \frac{3}{2} + \frac{7}{12} + \cdots + \frac{19999}{99990000} = 9.787606036\ldots; \quad (11)$$

combining terms in this way reduced the number of necessary divisions. The natural logarithm of 10000 was then determined by the formula

$$\ln 10000 = -252\ln(1-.028) + 200\ln(1+.0125) + 92\ln(1-.004672). \quad (12)$$

Such an expansion was designed for fast convergence and for convenience on a decimal computer. It is a simple matter to obtain such an expansion by hand calculation: We seek integers (x, y, z) such that $2^x 3^y 5^z \approx 1$

and $y \geq 0$. If three linearly independent solutions are obtained, one can calculate $\ln 2$, $\ln 3$, and $\ln 5$, and, in particular, $\ln 10$. If $2^{x_1}3^{y_1}5^{z_1} > 1$ and $2^{x_2}3^{y_2}5^{z_2} < 1$, suitable positive integral combinations of (x_1, y_1, z_1) and (x_2, y_2, z_2) will give closer approximations. The method is to find small values of (x, y, z) so that $x + y\log_2 3 + z\log_2 5 \approx 0$, then combine these to get better and better approximations. The expansion (12) corresponds to the solutions $(-1, 5, -3)$, $(-4, 4, -1)$, and $(6, 5, -6)$. For a binary computer the extra requirement $z \geq 0$ makes the task more difficult, but solutions can be used such as

$$\ln 10000 = 160 \ln 2^{-32}3^7 5^9 - 864 \ln 2^{-11}3^4 5^2 + 292 \ln 2^{-15}3^8 5. \qquad (12a)$$

Finally, Bernoulli numbers $B'_{2k} = 10^{-8k}B_{2k}$ were evaluated using the recursion relation

$$\binom{2k+1}{2k}B'_{2k} + 10^{-8}\binom{2k+1}{2k-2}B'_{2k-2} + \cdots + 10^{8-8k}\binom{2k+1}{2}B'_2 = (2k-1)/(2\cdot 10^{8k}). \qquad (13)$$

From the fact that

$$\left|\frac{B_{2k}}{B_{2k-2}}\right| \approx \frac{2k(2k-1)}{4\pi^2} \qquad (14)$$

it can be seen that the recursion (13) does not cause truncation errors to propagate. Furthermore, 1300 decimal places were used in all calculations.

When using (13) to calculate B'_{2k}, first all the positive terms were added together, then all the negative terms added together, and finally the two were combined. This gave extra speed to the calculations. Care was also taken to avoid multiplying by zero. The evaluation of B'_{2k} becomes more difficult as k increases, because of the number of terms and the size of the binomial coefficients. Since the B_n alternate in sign, the actual error in the calculation of γ is less than $B'_{502}/502 \approx 0.25 \times 10^{-1271}$, so the value obtained here should be correct to 1271 decimals. The fact that the final answer agrees with Adams's value and that numerous checks were made on all the arithmetical routines provides a good basis for guaranteeing the stated accuracy of the results. Dr. Wrench has independently verified the approximations to 1039 decimal places.

The present calculations were performed on a Burroughs 220 computer. The evaluation of S_{10000} required approximately one hour, and each of the logarithms required about six minutes. Evaluation of the 250 Bernoulli numbers was the most troublesome part of the calculations, and the total time for their calculation was approximately eight hours. A table of the Bernoulli numbers B' to 1270 decimal places has been deposited in the Unpublished Mathematical Tables file of the journal *Mathematics of Computation.*

4. Determination of Partial Quotients

To find best rational approximations to γ, we represent it as a continued fraction

$$\gamma = a_1 + \cfrac{1}{a_2 + \cfrac{1}{a_3 + \cfrac{1}{a_4 + \cdots}}}. \tag{15}$$

Put $P_1 = Q_0 = 1$, $Q_1 = P_0 = 0$, and for $i \geq 1$ let

$$P_{i+1} = a_i P_i + P_{i-1}, \qquad Q_{i+1} = a_i Q_i + Q_{i-1}. \tag{16}$$

In matrix notation,

$$\begin{pmatrix} P_{i+1} & P_i \\ Q_{i+1} & Q_i \end{pmatrix} = \begin{pmatrix} a_1 & 1 \\ 1 & 0 \end{pmatrix} \begin{pmatrix} a_2 & 1 \\ 1 & 0 \end{pmatrix} \cdots \begin{pmatrix} a_i & 1 \\ 1 & 0 \end{pmatrix}. \tag{17}$$

Then $\lim_{i\to\infty} P_i/Q_i = \gamma$. The fractions P_i/Q_i represent the best approximations to γ in the sense that

$$|Q_i\gamma - P_i| < |q\gamma - p|, \qquad \text{if } q < Q_i \text{ and } i \geq 3. \tag{18}$$

We have then $a_i \neq 0$ for all $i > 1$ if and only if γ is irrational; and the sequence of partial quotients a_i will be periodic if and only if γ is quadratic, that is, if and only if

$$\gamma = r + \sqrt{s}, \quad r \text{ and } s \text{ rational.}$$

For proofs of these well-known results see Cassels [10].

The algorithm used to determine the partial quotients a_i, using limited decimal precision, is as follows:

Set $\gamma_1 = \gamma$, and

$$a_i = [\gamma_i], \quad \gamma_{i+1} = \{\gamma_i\}^{-1}, \quad i \geq 1, \tag{19}$$

where $[x]$ and $\{x\}$ denote the integer and fraction parts of x. Starting with decimal numbers r_1 and s_1 such that

$$r_1 \leq \gamma_1 \leq s_1,$$

we successively find numbers r_i and s_i such that

$$r_i \le \gamma_i \le s_i. \tag{20}$$

If $[r_i] \ne [s_i]$, then the algorithm terminates. If $[r_i] = [s_i]$, then $[r_i] = a_i$ and

$$\{r_i\} \le \{\gamma_i\} \le \{s_i\}.$$

Hence

$$\{s_i\}^{-1} \le \gamma_{i+1} \le \{r_i\}^{-1}.$$

Choose decimal numbers r_{i+1} and s_{i+1} so that $r_{i+1} \le \{s_i\}^{-1}$ by truncation, $s_{i+1} \ge \{r_i\}^{-1}$ by rounding up. Then the algorithm continues, until $[r_i] \ne [s_i]$.

The method used for calculating $\{s_i\}^{-1}$ when $\{s_i\}$ has several hundred decimal places was adapted from that of Pope and Stein [13]. Approximately six seconds were required to obtain each quotient. If t partial quotients are desired, the total time is proportional to t^3.

Table 1 gives the value of γ to 1271 decimal places. Table 2 gives the first 372 partial quotients of γ. Only 372 are given, although the value in Table 1 would probably have yielded over 1000 partial quotients. Table 3 gives for the reader's convenience the first few "best rational approximations" to γ. Here the ratio 228/395 gives a remarkably good value, correct to six decimal places.

From Table 2 one can compute

$$Q_{373} \approx 3.021 \times 10^{195}, \tag{21}$$

and we can conclude that if γ is rational its denominator must be larger than Q_{373}. Another consequence is that only about 385 decimal places of Table 1 were needed to obtain the 372 partial quotients. The referee has pointed out that Lehman [11] had already calculated the first 315 partial quotients for γ on the basis of Wrench's 328-place value [8]. These are in perfect agreement with the values obtained here.

The partial quotients of γ, as calculated in Table 2, appear to be "random" in some sense. Almost all real numbers have partial quotients satisfying

$$\lim_{n\to\infty} \sqrt[n]{a_2 a_3 \ldots a_{n+1}} = K \tag{22}$$

where $K \approx 2.685$ is Khintchine's constant [12]. In this case,

$$\sqrt[371]{a_2 a_3 \ldots a_{372}} \approx 2.79$$

is a reasonable approximation to K.

TABLE 1. The value of Euler's constant

.57721 56649 01532 86060 65120 90082 40243 10421 59335 93992
35988 05767 23488 48677 26777 66467 09369 47063 29174 67495
14631 44724 98070 82480 96050 40144 86542 83622 41739 97644
92353 62535 00333 74293 73377 37673 94279 25952 58247 09491
60087 35203 94816 56708 53233 15177 66115 28621 19950 15079
84793 74508 57057 40029 92135 47861 46694 02960 43254 21519
05877 55352 67331 39925 40129 67420 51375 41395 49111 68510
28079 84234 87758 72050 38431 09399 73613 72553 06088 93312
67600 17247 95378 36759 27135 15772 26102 73492 91394 07984
30103 41777 17780 88154 95706 61075 01016 19166 33401 52278
93586 79654 97252 03621 28792 26555 95366 96281 76388 79272
68013 24310 10476 50596 37039 47394 95763 89065 72967 92960
10090 15125 19595 09222 43501 40934 98712 28247 94974 71956
46976 31850 66761 29063 81105 18241 97444 86783 63808 61749
45516 98927 92301 87739 10729 45781 55431 60050 02182 84409
60537 72434 20328 54783 67015 17739 43987 00302 37033 95183
28690 00155 81939 88042 70741 15422 27819 71652 30110 73565
83396 73487 17650 49194 18123 00040 65469 31429 99297 77956
93031 00503 08630 34185 69803 23108 36916 40025 89297 08909
85486 82577 73642 88253 95492 58736 29596 13329 85747 39302
37343 88470 70370 28441 29201 66417 85024 87333 79080 56275
49984 34590 76164 31671 03146 71072 23700 21810 74504 44186
64759 13480 36690 25532 45862 54422 25345 18138 79124 34573
50136 12977 82278 28814 89459 09863 84600 62931 69471 88714
95875 25492 36649 35204 73243 64109 72682 76160 87759 50880
95126 20840 45444 77992 3(0)

TABLE 2. Partial quotients $a_1, a_2, \ldots, a_{372}$

000 001 001 002 001 002 001 004 003 013 005 001 001 008 001 002 004 001 001
040 001 011 003 007 001 007 001 001 005 001 049 004 001 065 001 004 007 011 001
399 002 001 003 002 001 002 001 005 003 002 001 010 001 001 001 001 002 001 001
003 001 004 001 001 002 005 001 003 006 002 001 002 001 001 001 002 001 003 016
008 001 001 002 016 006 001 002 002 001 007 002 001 001 001 003 001 002 001 002
013 005 001 001 001 006 001 002 001 001 011 002 005 006 001 001 001 006 001 002
002 001 005 006 002 001 001 007 013 004 001 002 004 001 004 001 001 023 001 009
005 002 001 001 001 008 003 002 004 002 033 005 001 002 001 003 002 004 002 001
005 012 001 017 006 002 032 005 003 001 006 001 003 001 002 001 018 001 002 017
001 006 001 021 001 006 001 071 018 001 006 058 002 001 013 055 001 103 001 014
001 005 008 001 002 010 002 001 001 003 003 002 001 182 001 004 003 002 004 001
002 001 001 001 006 001 001 001 006 001 003 002 069 002 001 006 002 002 012 001
001 001 008 001 002 003 002 001 052 001 025 004 002 018 001 040 001 018 001 002
014 001 002 002 010 001 001 002 006 071 007 001 010 002 001 001 001 002 001 003
002 004 001 006 003 001 001 029 001 029 001 001 003 004 007 001 001 010 002 002
030 001 021 003 012 001 039 008 007 001 002 001 002 002 001 001 002 003 001 013
001 002 003 001 001 001 001 008 007 001 001 001 004 002 005 012 001 015 005 001
007 001 005 001 001 001 006 005 001 041 001 005 001 009 013 001 001 005 021 025
008 005 001 014 001 001 001 006 003 001 100 001 265

TABLE 3. Best rational approximations

1 / 2	.50
3 / 5	.60
4 / 7	.571
11 / 19	.579
15 / 26	.5769
71 / 123	.57724
228 / 395	.5772152
3035 / 5258	.57721567
15403 / 26685	.5772156642
18438 / 31943	.5772156654
33841 / 58628	.57721566487

The author wishes to acknowledge his gratitude to the Burroughs Corporation and to the Case Institute of Technology for the use of their Burroughs 220 computers.

References*

[1] Leonh. Euler, "De progressionibus harmonicis observationes," *Commentarii academiæ scientiarum imperialis Petropolitanæ* **7** (1734), 150–161. Reprinted in his *Opera Omnia*, series 1, volume 14, 87–100.

[2] Leonhard Euler, "De summis serierum numeros Bernoullianos involventium," *Novi commentarii academiæ scientiarum imperialis Petropolitanæ* **14** (1769), 129–167. Reprinted in his *Opera Omnia*, series 1, volume 15, 91–130. See especially §24. The calculation is given in detail in his subsequent publication, "De numero memorabili in summatione progressionis harmonicæ naturalis occurrente," *Acta academiæ scientiarum imperialis Petropolitanæ* **5 II** (1781), 45–75, §1–§15; reprinted in his *Opera Omnia*, series 1, volume 15, 569–583.

[3] William Shanks, "On the Calculation of the Numerical Value of Euler's Constant, which Professor Price, of Oxford, calls E," *Proceedings of the Royal Society of London* **15** (1867), 429–432; **16** (1867), 154; **16** (1868), 299–300; **18** (1869), 49; **20** (1871), 29–34.

[4] J. W. L. Glaisher, "On the calculation of Euler's constant," *Proceedings of the Royal Society of London* **19** (1870), 514–524.

[5] J. W. L. Glaisher, "On the history of Euler's constant," *The Messenger of Mathematics*, new series, **1** (1872), 25–30.

* These references are intentionally listed in historical (not alphabetical) order.

[6] J. C. Adams, "Note on the value of Euler's constant; likewise on the Values of the Napierian Logarithms of 2, 3, 5, 7, and 10, and of the Modulus of common Logarithms, all carried to 260 places of Decimals," *Proceedings of the Royal Society of London* **27** (1878), 88–94; **42** (1887), 22–25.

[7] Konrad Knopp, *Theory and Application of Infinite Series* (London: Blackie and Son, 1951).

[8] J. W. Wrench, Jr., "A new calculation of Euler's constant," *Mathematical Tables and Other Aids to Calculation* **6** (1952), 255.

[9] D. J. Wheeler, "The Calculation of 60,000 Digits of e by the Illiac," Digital Computer Laboratory Internal Report No. 43 (Urbana, Illinois: University of Illinois, 1953).

[10] J. W. S. Cassels, *An Introduction to Diophantine Approximation* (Cambridge: Cambridge University Press, 1957), 1–11.

[11] R. Sherman Lehman, "A study of regular continued fractions," Ballistic Research Laboratory Report No. 1066 (Aberdeen, Maryland: Aberdeen Proving Ground, February 1959), 57 pages.

[12] John W. Wrench, Jr., "Further evaluation of Khintchine's constant," *Mathematics of Computation* **14** (1960), 370–371.

[13] David A. Pope and Marvin L. Stein, "Multiple precision arithmetic," *Communications of the ACM* **3** (1960), 652–654.

[14] Daniel Shanks and John W. Wrench, Jr., "Calculation of π to 100,000 decimals," *Mathematics of Computation* **16** (1962), 76–99.

Addendum

Instead of using real numbers γ_i to compute the partial quotients as in (19), I should have used rational approximations and simply proceeded as in Euclid's algorithm. This improvement was pointed out by John W. Wrench, Jr., and Daniel Shanks, "Questions concerning Khintchine's constant and the efficient computation of regular continued fractions," *Mathematics of Computation* **20** (1966), 444–448.

A better way to compute γ, using the formula

$$\gamma + \ln n = \sum_{k=1}^{\infty} \frac{(-1)^{k-1} n^k}{k!\, k} - \int_n^{\infty} \frac{e^{-x}}{x}\, dx$$

(which avoids the bottleneck of my calculations, because it doesn't need Bernoulli numbers) was used by Dura W. Sweeney to calculate 3566 decimal digits. ["On the computation of Euler's constant," *Mathematics of*

Computation **17** (1963), 170–178.] His method was extended by W. A. Beyer and M. S. Waterman, "Error analysis of a computation of Euler's constant," *Mathematics of Computation* **28** (1974), 599–604, who believed that they had calculated γ to 7114 digits; but it turned out that only the first 4879 of those digits were correct. Richard Brent reached 20800 correct digits shortly afterwards, by using the formula above in a more efficient way; he also computed 20000 partial quotients of both γ and e^γ. ["Computation of the regular continued fraction for Euler's constant," *Mathematics of Computation* **31** (1977), 771–777.]

The first million digits of γ were obtained by Thomas Papanikolaou in 1997. By early 2009, nearly 30 billion digits had been computed, by Alexander J. Yee and Raymond Chan. (Further details can nowadays be found readily on the Internet.)

Chapter 27

Evaluation of Polynomials by Computer

[Originally published in Communications of the ACM **5** *(1962), 595–599.]*

In many applications it becomes necessary to evaluate

$$P = y^n + a_1 y^{n-1} + a_2 y^{n-2} + \cdots + a_n,$$

where the a's are constants known to the programmer. For example, one may wish to compute

$$P = y^4 + 3y^3 + 5y^2 + 7y + 9.$$

A beginner using ALGOL would tend to write

$$P := y \uparrow 4 + 3 \times y \uparrow 3 + 5 \times y \uparrow 2 + 7 \times y + 9.$$

But experienced programmers seeing this would smile and jot down the "professional" way to compute it:

$$P := (((y + 3) \times y + 5) \times y + 7) \times y + 9.$$

This ingenious way of rewriting the expression (sometimes called Horner's rule) needs only 3 multiplications and 4 additions, while the previous way of writing it requires 8 multiplications (or even 9, depending on how $y \uparrow 4$ is evaluated) and 4 additions. Recently Motzkin [3] showed that for certain polynomials there is yet a better way to do the evaluation, and his discovery inspired the study that led to this paper.

The polynomial above can, for example, be calculated using the sequence

$$z := y \times (y + 1); \quad P := (z + y - 1) \times (z + 4) + 13.$$

Here 2 multiplications and 5 additions are used, so a multiplication operation has been traded for an addition operation. Using floating-point hardware, a multiplication will take perhaps 5 times as long as an addition. On the other hand in fixed-point arithmetic (for example in the common case of a sine or arctangent subroutine), a multiplication will often take up to 20 or more times as long. However, if the calculations are done by floating-point subroutines on a machine that has no floating-point hardware, addition and multiplication are often roughly comparable, so nothing has been gained.

We will show, however, that for degree 5 or more the *total* number of operations, counting additions and multiplications with equal weight, can even be reduced. Specifically, we will show that an arbitrary polynomial of degree n requires at most the following numbers of multiplications and additions:

n	$\times$	$+$
4	2	5
5	3	5
6	3	7
7	4	8
8	5, usually 4	9

Preliminary Considerations

The method given in [3] shows how a sixth degree polynomial can be computed with 3 multiplications and 7 additions. Horner's rule takes $n-1$ multiplications and n additions; so the new approach saves two multiplications at the expense of one extra addition. Unfortunately, however, that method requires the solving of a quadratic equation, and the quadratic might not have any real roots. There are, in fact, many sixth degree polynomials that cannot be handled by the algorithm of [3], and a major modification of that method would apparently be required to make it cover all cases.

One would of course expect to be able to handle *special* polynomials in a more efficient manner than a general polynomial; after all, any method whatsoever will be valid for certain conditions on the coefficients of the polynomial. So we're led to wonder whether *every* sixth degree polynomial can be done with 3 multiplications. While trying to prove this impossible, the author discovered that it was indeed possible.

First, we observe that the general fourth degree equation can always be evaluated by a scheme of the form

$$z := y \times (y + a); \quad P := (z + y + b) \times (z + c) + d.$$

Treating the polynomial as

$$P = y^4 + Ay^3 + By^2 + Cy + D,$$

one way to precompute a, b, c, and d is to let

$$\begin{aligned} a &= (A-1)/2; \\ b &= B(a+1) - C - a(a+1)^2; \\ c &= B - b - a(a+1); \\ d &= D - bc. \end{aligned}$$

Sixth Degree Polynomials

Moving to degree six, we wish to evaluate

$$P = y^6 + Ay^5 + By^4 + Cy^3 + Dy^2 + Ey + F.$$

This can always be done by computations of the form

$$\begin{aligned} z &:= y \times (y+a); \\ w &:= (z+b) \times (y+c); \\ P &:= (w+z+d) \times (w+e) + f. \end{aligned}$$

The calculation of a, b, c, d, e, and f is much more trouble this time, but it is worthwhile if the polynomial is to be evaluated frequently as in a library subroutine. A relatively simple computer program can be written to calculate the new constants; a good compiler might also even do such calculation while compiling.

We start by solving the following set of equations, into which new variables p, q, r, and s are introduced:

$$2p + 1 = A; \quad (1)$$

$$p(p+1) + 2q + a = B; \quad (2)$$

$$p(2q+a) + q + r + s = C; \quad (3)$$

$$p(r+s) + r + q(q+a) = D; \quad (4)$$

$$ar + q(r+s) = E. \quad (5)$$

Equation (1) determines p, and then equation (2) determines a as a linear function of q. Equation (3) determines $r + s$ as another linear function of q, and using this fact in (4) we have r as a quadratic in q.

Then equation (5) becomes a cubic equation in q. More precisely, we can obtain the modified equations

$$a = B' - 2q, \tag{2'}$$
$$r + s = C' - q, \tag{3'}$$
$$r = q^2 + D'q + D'', \tag{4'}$$
$$2q^3 + E'q^2 + E''q + E''' = 0, \tag{5'}$$

where

$$\begin{aligned} B' &= B - p(p+1), \\ C' &= C - pB', \\ D' &= p - B', \\ D'' &= D - pC', \\ E' &= 2D' - B' + 1, \\ E'' &= 2D'' - B'D' - C', \\ E''' &= E - B'D''. \end{aligned}$$

Equation (5′), being a cubic equation, always has a real root q. Once q is known we can find a, r, s by the equations above and then finish by

$$c = p - a; \tag{6}$$
$$b = q - ac; \tag{7}$$
$$d = s - bc; \tag{8}$$
$$e = r - bc; \tag{9}$$
$$f = F - rs. \tag{10}$$

Example. Suppose we are given the polynomial

$$P = y^6 + 13y^5 + 49y^4 + 33y^3 - 61y^2 - 37y + 3.$$

We find $p = 6$ and obtain the cubic equation

$$2q^3 - 8q^2 + 2q + 12 = 0.$$

This cubic has the roots 2, 3, and -1, which lead to $(r, s) = (-5, -6)$, $(-1, -11)$, and $(-5, -3)$, respectively. There are therefore three equally good ways to compute the given polynomial:

$$z := y \times (y + 3);$$
$$w := (z - 7) \times (y + 3);$$
$$P := (w + z + 15) \times (w + 16) - 27;$$

$$z := y \times (y + 1);$$
$$w := (z - 2) \times (y + 5);$$
$$P := (w + z - 1) \times (w + 9) - 8;$$

$$z := y \times (y + 9);$$
$$w := (z + 26) \times (y - 3);$$
$$P := (w + z + 75) \times (w + 73) - 12.$$

The first solution is actually a little better, since the computation of $y + 3$ needs to be performed only once.

Arbitrary Polynomials of Even Degree

We now develop a different algorithm for arbitrary degrees. This general method will not always be the best possible; but it gives best possible results for a large class of polynomials (more than 50%) of degree 6, 7, or 8, and probably of higher orders as well. Let the given polynomial be

$$P = y^n + a_1 y^{n-1} + a_2 y^{n-2} + \cdots + a_n, \quad n \geq 5.$$

First suppose $n = 2m$ is even. Then P can be written as

$$z^n + z^{n-1} + b_2 z^{n-2} + \cdots + b_n,$$

where[1] $z = y + t$ and $t = (a_1 - 1)/n$.

The next step is the crucial one — to try to solve the $(m-1)$st degree equation

$$\alpha^{m-1} + b_3 \alpha^{m-2} + b_5 \alpha^{m-3} + \cdots + b_{n-1} = 0.$$

Call this equation a *reduction* equation.

---

[1] Another, very similar, algorithm can be used with $t = (a_1 + 1)/n$. This modification sometimes leads to fewer multiplications, for example when $P = y^6 + y^5 + y$. It might also be numerically more accurate, if a_1 is nearly equal to 1, although the accuracy depends also on other conditions.

If this polynomial has a real root α, we can apply a squaring rule,

$$P = (z^{n-2} + z^{n-3} + c_2 z^{n-4} + \cdots + c_{n-2})(z^2 - \alpha_n) + \beta_n,$$

where

$$\begin{aligned} c_2 &= b_2 + \alpha, & c_3 &= b_3 + \alpha,\\ c_4 &= b_4 + c_2\alpha, & c_5 &= b_5 + c_3\alpha,\\ c_6 &= b_6 + c_4\alpha, & c_7 &= b_7 + c_5\alpha,\\ &\vdots & &\vdots\\ c_{n-2} &= b_{n-2} + c_{n-4}\alpha, & c_{n-1} &= b_{n-1} + c_{n-3}\alpha,\\ \beta_n &= b_n + c_{n-2}\alpha, & \alpha_n &= \alpha. \end{aligned}$$

As a check we can verify that $c_{n-1} = 0$.

If the reduction equation has no real root this is a shame, and we apply Horner's rule twice to reduce the degree,

$$P = ((z^{n-2} + z^{n-3} + c_2 z^{n-4} + \cdots + c_{n-2})z + \alpha_n)z + \beta_n,$$

where $c_i = b_i$, $\alpha_n = b_{n-1}$, and $\beta_n = b_n$. The same process is now iterated on the reduced polynomial.

Eventually we will have reduced n to 4, and we will be left with

$$(z^2 + z + c)(z^2 - \alpha_4) + \beta_4,$$

which is easy to evaluate.

The final algorithm is then

$$\begin{aligned} z &:= y + t;\\ w &:= z \uparrow 2;\\ P1 &:= w + z + c;\\ P2 &:= P1 \times (w - \alpha_4) + \beta_4 \text{ or } (P1 \times z + \alpha_4) \times z + \beta_4;\\ P3 &:= P2 \times (w - \alpha_6) + \beta_6 \text{ or } (P2 \times z + \alpha_6) \times z + \beta_6; \end{aligned}$$

and so on. The choices in the latter steps depend on whether the reduction equation was solvable or not. (The reduction equations for $P2$, $P4$, $P6$, etc., are of odd degree. Thus they are always solvable, and the "choice" implied by the word "or" above is really nonexistent.) This scheme gives us approximately $n/2$ multiplications and $n+1$ additions. Actually there are $n - r - 1$ multiplications, where r is the number of reduction equations we were able to solve.

The algorithm just described is equivalent to Motzkin's for $n = 6$, but this presentation reveals the inner mechanism of his method.

Example. Suppose

$$P = y^8 + y^7 + 3y^6 + 2y^5 + 3y^4 - y^3 + 3y^2 - 2y + 1.$$

Here $t = 0$, so $z = y$ and the reader can see immediately that the example is rigged. We try first to solve the first reduction equation, $\alpha^3 + 2\alpha^2 - \alpha - 2 = 0$, and take the root $\alpha = -2$; so

$$P = (y^6 + y^5 + y^4 + y^2 - y + 1)(y^2 + 2) - 1.$$

The next reduction equation is $\alpha^2 - 1 = 0$ and we take $\alpha = -1$; thus

$$P = ((y^4 + y^3 - y + 1)(y^2 + 1) + 0)(y^2 + 2) - 1.$$

The final reduction equation is $\alpha - 1 = 0$; so we have

$$P = (((y^2 + y + 1)(y^2 - 1) + 2)(y^2 + 1) + 0)(y^2 + 2) - 1.$$

Arbitrary Polynomials of Odd Degree

If n is odd, there are several alternatives. We could simply reduce the degree by one with one application of Horner's rule and then apply the preceding method. Or we can reduce the degree by 2, using the squaring rule — a simple and straightforward modification of the squaring rule for even exponents. For example, if $n = 7$ we solve the cubic reduction equation $\alpha^3 + b_2\alpha^2 + b_4\alpha + b_6 = 0$. With this approach the problem can be reduced to that of evaluating a fifth degree equation. The fifth degree equation is reduced to degree 4 by using Horner's rule. This procedure shows that an arbitrary seventh degree polynomial can be computed using 4 multiplications and 8 additions, and the derivation of the constants is relatively simple. It seems that a good method to use would be to reduce n by 2 if $n = 4k-1$, and to reduce by 1 if $n = 4k+1$, thus avoiding the solution of at least one equation of even degree.

Another method is often useful when $n = 2m - 1$ is odd. With $P = y^n + a_1y^{n-1} + \cdots + a_n$ as before, we need not use the transformation $z = y+t$, but we can work with the original equation. Form the reduction equation $\alpha^{m-1} + a_2\alpha^{m-2} + a_4\alpha^{m-3} + \cdots + a_{n-1} = 0$. If α solves this equation the squaring rule can be used. By reducing n in steps of 2 it will eventually be reduced to a cubic polynomial evaluated in the form $(y + a_1)(y^2 - \alpha) + \beta$. More precisely, the algorithm will be

$$\begin{aligned} w &:= y \uparrow 2; \\ P1 &:= y + a_1; \\ P2 &:= P1 \times (w - \alpha_3) + \beta_3 \text{ or } ((P1 \times y) + \alpha_3) \times y + \beta_3; \\ P3 &:= P2 \times (w - \alpha_5) + \beta_5 \text{ or } ((P2 \times y) + \alpha_5) \times y + \beta_5; \end{aligned}$$

etc. There are n additions instead of $n+1$, and $n-1-r$ multiplications.

For example, let $n = 5$ and

$$P = y^5 + a_1 y^4 + a_2 y^3 + a_3 y^2 + a_4 y + a_5.$$

Suppose we can solve $\alpha^2 + a_2\alpha + a_4 = 0$. Then we have

$$P = (y^3 + a_1 y^2 + c_2 y + c_3)(y^2 - \alpha_5) + \beta_5$$

where

$$c_2 = a_2 + \alpha, \quad c_3 = a_3 + a_1\alpha, \quad \alpha_5 = \alpha, \quad \beta_5 = a_5 + c_3\alpha.$$

The final reduction equation is $\alpha + c_2 = 0$, which is easily solved. Thus we have finally,

$$P = (((y + a_1)(y^2 + c_2) + \beta_3)(y^2 - \alpha_5)) + \beta_5,$$

where $\beta_3 = c_3 - a_1 c_2$. The evaluation has been accomplished with 3 multiplications and 5 additions, provided that the quadratic equation could be solved.

Remarks

It can be shown that by solving a cubic equation, any fifth degree polynomial can be evaluated with 3 multiplications and 5 additions, and that this is the best possible; the algorithm is:

$$\begin{aligned} z &:= y + t; \\ w &:= z \uparrow 2; \\ P &:= ((w + a) \times w + b) \times (z + c) + d. \end{aligned}$$

Here a, b, c, d, and t are rather difficult to find, but their evaluation is no worse than the determination of the constants in the sixth degree algorithm given earlier.

Ostrowski [2] has shown that Horner's rule minimizes the total number of operations, counting additions on an equal par with multiplications, for $n \leq 4$. We have shown, on the other hand, that Horner's rule is never minimal for $n > 4$.

Another often important consideration has not been mentioned yet, namely the question of error analysis. Whichever way is chosen to evaluate these polynomials, it should be investigated to see if the proper amount of accuracy is obtained, for the necessary range of y.

The Special Case y^n

Certain polynomials have special properties that allow them to be evaluated in fewer steps. The eighth degree polynomial of an earlier example required only 4 multiplications and 7 additions. And of course a polynomial where all the coefficients of odd degree vanish requires in general fewer operations.

We consider here only the most special case of all, $P = y^n$. In this case a good compiler should definitely convert this somehow into a minimum number of multiplications. But it is a very difficult combinatorial problem to decide how to best compute y^n.

One method that is fairly well known (see, for example, Floyd [1, pages 50–51]), and which one might call the Binary Method, can be described as follows:

a) Write n as a number in the binary system; for example, if n is 13 in decimal, then n is 1101 in binary.

b) Replace each "1" by SX and each "0" by S; for example, 1101 becomes $SXSXSSX$.

c) Cancel the SX at the left end. The resulting string can be interpreted as instructions: $S =$ "square" and $X =$ "multiply by y."

In our example we calculate y^{13} with the sequence $SXSSX$:

1. Take y;
2. Square (giving y^2);
3. Multiply by y (giving y^3);
4. Square (giving y^6);
5. Square (giving y^{12});
6. Multiply by y (giving y^{13}).

We have used 5 multiplications, which in this case can be shown to be the minimum.

Another algorithm, believed to be new, might be called the Factor Method. Here we generate sequences $S(n,m)$ as follows:

a) If n is not prime, $S(n,m) = S(n/p,m)S(p,nm/p)$, where p is the smallest prime factor of n.

b) If n is prime, $S(n,m) = S(n-1,m)X_m$.

c) $S(1,m) =$ null string.

Interpretation: X_m means "multiply by y^m, which is a previously calculated result." $S(n,m)$ is the sequence for the calculation of y^{mn} assuming that y^m has been calculated. The goal is to find $S(n,1)$.

For example,

$$\begin{aligned} S(13,1) &= S(12,1)X_1 \\ &= S(6,1)S(2,6)X_1 \\ &= S(3,1)S(2,3)X_6X_1 \\ &= S(2,1)X_1X_3X_6X_1 \\ &= X_1X_1X_3X_6X_1. \end{aligned}$$

The Factor Method turns out to be identical to the Binary Method when $n = 13$.

Another way to describe this algorithm is:

a) Factor n into primes, $n = p_1p_2\ldots p_r$.

b) If n is prime, calculate $(y^{n-1})y$.

c) If n is composite, calculate $(\ldots((y^{p_1})^{p_2})\ldots)^{p_r}$.

To compare the two methods, the Binary Method uses $r+s-1$ multiplications, where r is the greatest integer less than or equal to $\log_2 n$ and s is the number of 1s in the binary representation of n. The Factor Method uses M_n multiplications, where

$$M_n = \begin{cases} M_{n-1}+1, & \text{if } n \text{ is prime;} \\ M_r + M_s, & \text{if } n = rs \text{ is composite.} \end{cases}$$

The Factor Method is slightly better than the Binary Method overall: For $n \le 150$, there are

93 cases where the two methods are equal,
32 cases where the Factor Method is one better,
16 cases where the Binary Method is one better,
8 cases where the Factor Method is two better,
1 case ($n = 129$) where the Binary Method is two better.

The smallest cases where the Factor Method excels are $n = 15$, 27, 30, 31*, 39, 45. The smallest cases where the Binary Method excels are $n = 33$, 49, 65, 66, 67, 69.

But there are cases where we can do better than *both* methods. For $n \leq 70$ there are five known cases where this is true. (They are $n = 23$, 43, 46, 47, 59.) A third method, which is the best technique known to the author, might be called the Tree Method since it develops a tree.

The Tree Method is a good exercise in list processing techniques. Start at level 0 with a single node labeled 1. To get to the next level when a level is completed, process the nodes on the preceding level from left to right. For each node, try adding each of the values above this node, successively from the top. If any new values are obtained, they become nodes branching out to the next level. Duplicate values obtained are discarded.

For example, we might develop three levels as in Figure 1 on the next page. Discarding values that are duplicates, we get Figure 2. Making the trial additions from top to bottom leads to a much better tree than making them from bottom to top. However, the bottom-to-top method is much more convenient to program, and it can be used if the new branches from a given node are attached from right to left. The tree begins looking as in Figure 3.

We have concentrated solely on the total number of multiplications as the criterion of excellence in this discussion. Clearly if n is an *unknown* variable, the Binary Method is preferable. In fact that method would be quite suitable to incorporate in the hardware of a binary computer, as an exponentiation operator.

If y is a floating-point number, there is of course a point of diminishing returns when n gets large, since we will eventually be better off by taking logarithms and exponentials. Since the Tree Method uses r

---

* EDITOR'S NOTE [by Robert W. Bemer]. Since much usage may be made of these methods in the floating-point mode, one could consider also the relative timing of multiplication and division. In some machines these have even been equal. Thus, for y^{31} other possibilities exist, using "D" to indicate a division:

Binary	(8)	$SXSXSXSX \equiv$	$X_1X_1X_3X_1X_7X_1X_{15}X_1$
Factor	(7)		$X_1X_2X_2X_6X_{12}X_6X_1$
Division	(6)		$X_1X_2X_4X_8X_{16}D_1$

The reader is invited to try y^{127} and see if some algorithm could be derived for mixed operations.

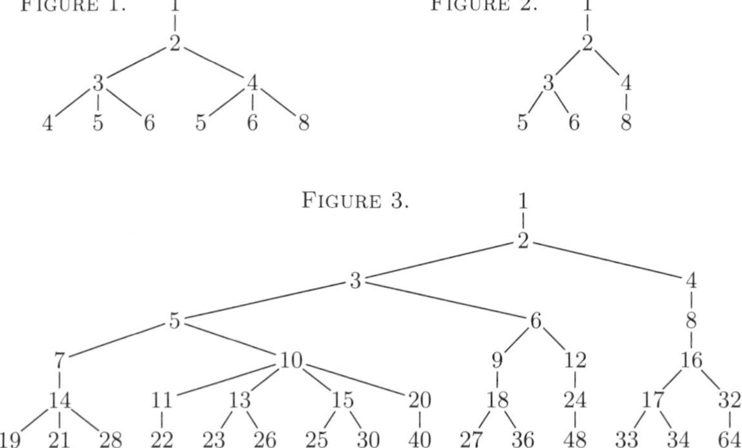

FIGURE 1. FIGURE 2. FIGURE 3.

multiplications at level r, it is possible to stop generating the tree at a certain level; then we have the set of all the "interesting" values of n. The tree in Figure 3, for example, shows all n for which it is known that y^n needs at most 6 multiplications. The Factor Method, on the other hand, is valuable when n is large and an application requires frequent calculation of y^n.

The user's manual for a certain commercial system (which will not be mentioned here by name) has a subroutine for "float to fix exponentiation" of $y \uparrow n$ that uses $n - 1$ multiplications, since it says "there is small probability of this routine being used with a very large n." The stated reasoning may seem to be valid at first; but it didn't take long before a user had to calculate y^{70} a very large number of times, and so the user rewrote the subroutine. This remark is included here to offer some justification for having a good power method.

References

[1] Robert W. Floyd, "An algorithm for coding efficient arithmetic operations," *Communications of the ACM* **4** (1961), 42–51.

[2] A. M. Ostrowski, "On two problems in abstract algebra connected with Horner's rule," *Studies in Mathematics and Mechanics Presented to Richard von Mises* (New York: Academic Press, 1954), 40–48.

[3] John Todd, *A Survey of Numerical Analysis* (New York: McGraw–Hill, 1962), 3–4.

Addendum

This paper, prepared when I was a graduate student in mathematics and supporting myself by writing compilers, is a pleasant reminder of the halcyon days when the literature of computer techniques was still in its infancy. I soon learned that the problem of optimum exponentiation was known to mathematicians as the problem of minimum *addition chains*, and that the Binary Method could be traced back more than 1600 years to ancient India; indeed, a similar method had been known in Egypt, prior to 2000 B.C.! The Factor Method, likewise, was "well known"; it had been published already by E. de Jonquières in 1894. The whole subject of polynomial evaluation was eventually destined to occupy more than 90 pages in my book *Seminumerical Algorithms* (Volume 2 of *The Art of Computer Programming*, 1997), Sections 4.6.3 and 4.6.4.

Robert W. Bemer, the editor to whom I had submitted this paper, was the IBM Corporation's Director of Systems Programming at the time. Many years later he showed me an internal memorandum that he had addressed to "All Members of Systems Programming" on 27 December 1962:

> `I should like to call your attention to an article by Mr. D. E. Knuth on Page 595 of the December issue of the Communications of the ACM. This is a remarkably useful technique and deserves study. In particular, Quality and Reliability will require that proposals for all subroutines which require polynomial approximations must make use of this method wherever required by the degree of the approximation.`

Chapter 28

Minimizing Drum Latency Time

[Originally published in Journal of the Association for Computing Machinery **8** *(1961), 119–150. "The author's first journal paper, written at the close of his senior year at Case Institute of Technology."]*

The need for latency minimization in many programs for certain computers is quite pronounced, as quantitative results indicate. Optimizing of computer instructions includes many things besides merely assigning locations optimally. There are rigorous mathematical methods for formulating the latency problem as a series of integer programming problems, and a mechanical technique has also been developed for iterating on chosen locations in an attempt to improve the choices. The most practical way discovered for minimizing latency, however, has been to let people do the job manually with a semi-systematic procedure and to use computers for the routine phases and for checking the human output.

I. Comparison of Present Methods

Introduction

On many digital computers in existence today the running time of a given problem depends on the locations of instructions and data stored in memory, and for programs that are to be run long hours on such a computer it pays to optimize the placement of these locations. This technique of minimizing latency has been mechanized to the extent that rough approximations to optimal running speed can be obtained by letting the computer itself handle the placement of instructions, and this machine treatment has proved to be quite economical for a large variety of problems. But for routines that are to find extensive use, nothing but the best optimization will do, and for this purpose no substitute has been found for the human "artist" who runs off for a while with a supply of pencils, paper, and erasers, and a latency chart, and comes back with a hand-optimized program.

Rather than speak continually in generalities, let us consider a particular machine and a particular piece of coding from the start. By discussing a specific computer we will be able to get definite facts on which to base our judgments. The computer on which our discussion will be based is the IBM 650, a magnetic drum machine, which was chosen not only because the author is most familiar with it but also because it is the most commonly known computer now in existence. Each machine has its own special problems for latency minimization, but the methods to be described here are very general and will be applicable to most other computers.

Example Program

The coding chosen for illustration is the following routine that performs floating-point addition:[1]

```
FAD   STD EXIT
      SLT  0008          separate
      STU ARG1             and store
      STL EXP1             the input
      RAU ACC              arguments
      SRT  0002
      STL EXP2
      STU ARG2
      RSM  8002          test
      AML EXP1             difference
      SLT  0001            of exponents
      NZU BIG              and prepare
      SRT  0005            shift command
      LDD SRT   STORE
BIG   LDD SRD   STORE    big move
STORE SDA SHIFT
      BMI SWICH
      RAU ARG2
      LDD ARG1  SHIFT
SWICH RAU ARG1           switch arguments
      LDD EXP2             and exponents
      STD EXP1             in special case
      LDD ARG2  SHIFT
```

[1] This routine evolved from the work of George Petznick, William Lynch, George Haynam, and the author. *[A brief explanation for 21st-century readers appears in the "historical postscript" at the close of this chapter.]*

```
24  SRT   SRT  0000 ADD
25  SRD   SRD  0000 ADD     shift and
26  ADD   AUP  8001           add
27        NZE       ZERO
28        SCT  0000         normalize
29        STL EXP2            result
30        SRT  0002
31        ALO  8002         round
32        SLO  8002
33        SCT  0004
34        STU ARG1
35        ALO EXP2          adjust
36        SLT  0008           exponent
37        RAM  8002
38        SLO ADD
39        SML EXP1          check if it
40        BMI       LOW       is in bounds
41        NZU HIGH
42        SRT  0008
43        SML ARG1          attach exponent
44        SIA ACC             to argument and
45        RAL  8001 EXIT      store result
46  ZERO  STL ACC   EXIT
```

Inputs to this routine are the number in the lower accumulator and the number in location ACC (which is machine location 0000), both in standard 650 floating-point notation.[2] It has been chosen because it is a typical routine for which hand optimization would be desirable, because lines 02–08 and 27–46 may be incorporated with floating-point multiplication and division routines to make a "package," and because the ending instructions on lines 27–46 are interesting in their own right: They normalize *before* rounding, and they check for precisely the proper range of the exponents; hence they give slightly better accuracy than any other routines in the literature, and they also enable the use of unnormalized numbers as arguments.[3] The coding as written is in an

[2] That is, an excess-50 exponent at the right of the eight-digit mantissa. *[That is, a positive real number of the form* $f \times 10^e$ *with* $0 \le f < 1$ *is represented by the ten-digit integer* $100A + 50 + e$, *where* A *is an eight-digit approximation to the number* $10^8 f$. *For example, the representation of* 1.0 *is* 1000000051. *A negative number such as* -1.0 *is represented in a similar way by the negative integer* -1000000051.*]*

[3] Strictly speaking, inputs that are badly unnormalized, such as 0000000099, might lead to an overflow at line 28; such numbers should be prohibited.

assembly language that is compatible with SOAP II, Case SOAP III, Wis-SOAP II, TASS, SuperSoap, etc., so any reader who understands 650 programming should be able to comprehend it. (However, a reader who wishes to understand the given coding completely is hereby warned that a few tricks *have* been employed. The worst of these is in line 38, which uses the fact that `AUP` is operation code 10 in machine language; furthermore, the 0004 of line 33 was adjusted so that the exponent comes out properly even though at most two shifts occur. These technicalities are not really crucial points in the following discussion.)

Definition of Optimization

Various opinions are in circulation about what to consider under the classification of optimizing. A narrow interpretation would be to say that it includes only the determination of best drum locations to use for the data addresses and instruction addresses of a given sequence of coding. Optimizing in its broadest sense, on the other hand, would include taking a program and completely rewriting it, perhaps changing the basic method of attack considerably, and coming up with the form that runs fastest or takes the least memory space.

The author's intention when he uses the word optimization falls in between these two extremes. Optimizing, as used here, means the polishing of a piece of coding in machine language, to minimize its running time, such that the basic structure or idea (i.e., flow chart) behind the coding remains unchanged, but certain alterations in the order or form of the instructions are allowed. In addition to making use of the published timing information, these include:

(1) The permutation of commutative instructions; that is, the reordering of instructions that may be interchanged without altering their effect. For example, the *shift-left* in line 02 could be interchanged with the *store-distributor* in line 01, since the distributor does not change on shifts. Hand optimization would include the permutation of those two instructions, if it is worthwhile from a timing standpoint.

(2) Choosing different constants and/or temporary storage where a choice of several locations is possible. For example, if there were two `AUP` commands, either of them could be used as a constant in line 38; and the temporary storages `ARG2`, `ACC`, or `SHIFT` would work as well as `EXP2` in lines 29 and 35, or as well as `ARG1` in lines 34 and 43.

(3) Giving certain branches of the program priority in the optimization because they are likely to be used most often. This includes a decision as to how many shifts to anticipate in lines 24 and 28.

(4) Using a little accumulated knowledge about the machine, such as what is in the distributor at a given time, to make minor changes "by ear" when opportunity presents itself. For example, we can change the pair

```
STL TEMP
RAL  8002
```

to the pair

```
STL TEMP
RAL  8001
```

which is always two word-times faster; or, if more convenient and the signs of the lower and upper accumulators are equal, we could write

```
SUP  8003
STL TEMP
```

instead. This technique is used mostly in conjunction with changes of type (1). A clever trick of this type was used by G. R. Trimble [9, page 39, locations 0063, 0020, 0032] when he saved a drum revolution by changing

```
SLT  0004
ALO ARG1
```

into the three instructions

```
LDD ARG1
SLT  0004
ALO  8001
```

because of the location of `ARG1` on the drum.

(5) Using facts about the program that follow from an understanding of what is happening. For example, we may know at some stage whether or not a complement cycle will occur, whether certain interlocks will be set while certain lines of coding are executed, in what range the addresses of masked or indexed instructions will fall, where the timing ring is set on `LDI` or `STI` commands, how long a table lookup can be expected to take, and so on. In our example there is an illustration of this in line 33, where we know that exactly one or two shifts will always occur.

The distinction between "hand optimization" and "clairvoyant recoding" will be a little clearer if we consider some borderline cases.

Lines 07, 08, 09 could be changed to

```
STU ARG2
STL EXP2
RSM  8001
```

or to

```
STU ARG2
RSM  8002
STD EXP2
```

if the running time would work out better, because all three sequences of instructions have the same effect on memory and the accumulator. Such changes would be considered admissible under points (1) and (4).

Another change could be made by noticing that although EXP1 is tested in line 39 it might just as well be EXP2 if suitable changes are made in the preceding program.[4] Such a change, however, would mean a more drastic change in the basic structure of the flow chart, so it *would not* be considered as merely hand optimization. Of course, the programmer who knows about such a possibility ought to investigate it in order to see if it works out better.

All five of the permissible alterations listed above really deserve to be used when they can be, for they are not extremely sophisticated. The time saved by hand optimizing is generally not merely a few word times but rather increments of drum revolutions (intervals of 4.8 milliseconds), so every little trick helps. A program that does not make use of those five techniques when they are beneficial should not be considered to have minimum latency.

Here is another example of the alteration techniques permissible in hand optimizing: Suppose you want to increase the numbers in COUNT and STEP each by one. You might code this in the obvious way from a flow chart:

```
RAL COUNT
ALO ONE
STL COUNT
RAL STEP
ALO ONE
STL STEP
```

But if you know that both COUNT and STEP will have the same sign (extra knowledge: technique (5)) you can use techniques (1) and (4),

---

[4] Namely, we can take out lines 21 and 22, and insert LDD EXP1, STD EXP2 between lines 18 and 19. It would also be better in this connection to change lines 09, 10, and 17, respectively to RAM 8002, SML EXP1, and BMI SWICH, so that the case of equal exponents goes through faster.

permutation and hardware knowledge, to save a drum revolution:

```
RAL COUNT
AUP STEP
ALO ONE
AUP  8001
STL COUNT
STU STEP
```

Various Optimization Schemes

The author has attempted to optimize the given floating-addition routine by various methods, to see exactly what kind of running times they actually give. Seven different schemes were used (see Tables 1–3):

(1) HAND OPTIMIZATION, in the sense above, using the five alteration tricks where they seemed to provide an advantage. The steps in obtaining this result will be described in the sequel. The additional condition that only machine locations 0100 through 0153 were to be used was also imposed, because such restrictions often occur when hand optimizing is desired.

(2) SOAP OPTIMIZATION. This familiar technique is used effectively by the computer itself, because it involves little thought, just good memory: Locations are picked optimally, or nearly so, when they are first encountered. This scheme runs the risk of finding locations placed poorly when they are to be used again. (Discussion of some fairly obvious methods for helping SOAP to optimize your programs may be found in [6].)

(3) SOAP OPTIMIZATION RESTRICTED to the sequential locations 0100–0153. This condition, which was also imposed on the hand optimized version, is used here for comparison; locations 0100–0199 were available under optimization types 2, 4, 6, and 7. As stated before, the sequential-location restriction occurs frequently in actual problems where hand optimizing is desired — for instance, in subroutines that are to be relocated, or in five- or seven-per-card loading routines, or in trace routines that must fit into a band or two, or in programs that need to be optimized by hand because they do a lot of manipulation with large blocks of consecutive working storage.

(4) THE CLASSIC "FIVE AND TEN SYSTEM." This method was used extensively in the pre-SOAP days: It is a first-come-first-served operation just like the SOAP system, but it is done by hand, by a person who doesn't bother to learn the rules of the optimizing charts. Just two simple rules are used instead [1]:

a) To optimize non-shift commands: data address = location plus 5; instruction address = data address plus 5.

b) To optimize shift commands: instruction address = location plus 10.

(5) SEQUENTIAL ASSIGNMENT. This method is the easiest when programming without the aid of an assembly system, but it turns out to be quite a timewaster: locations are assigned consecutively as they are used. (A variant of this, "Runcible assignment," adds 11 to the drum level each time while staying in the same band, thus picking 0100, 0111, 0122, 0133, 0144, 0105, 0116, ..., 0128, 0139, 0150, etc., in order. The latter method was not tried on this particular program, but it has the feature of not requiring an availability table.)

(6) SHOAP OPTIMIZATION. SHOAP is an acronym for "Symbolic Horribly Optimizing Assembly Program."[5] This is the same as SOAP, except that the *worst* locations are picked instead of the best ones.

(7) HAND SHOAP OPTIMIZATION. This process is obvious by analogy with SHOAP and hand optimization.

The actual assemblies are illustrated on the following pages so that a reader can verify the results. It is left as an exercise to get a better time for the hand optimization (hereafter called "HAND SOAP"), or a worse time for the HAND SHOAP version, subject to the same conditions as imposed above; such assemblies may or may not exist.

TABLE 1. Hand optimization

```
                         HAND SOAP
FAD    STD EXIT          0106 +24 0109 0113
       SLT  0008         0113 +35 0008 0122
       STU ARG1          0122 +21 0136 0139
       STL EXP1          0139 +20 0143 0146
       RAU ACC           0146 +60 0000 0105
       SRT  0002         0105 +30 0002 0114
       STL EXP2          0114 +20 0120 0124
       STU ARG2          0124 +21 0128 0131
       RSM  8002         0131 +68 8002 0140
       AML EXP1          0140 +17 0143 0149
       SLT  0001         0149 +35 0001 0111
       NZU BIG           0111 +44 0130 0117
       SRT  0005         0117 +30 0005 0129
       LDD SRT   STORE   0129 +69 0132 0137
BIG    LDD SRD   STORE   0130 +69 0134 0137
STORE  SDA SHIFT         0137 +22 0141 0100
       BMI SWICH         0100 +46 0115 0104
       RAU ARG2          0104 +60 0128 0133
       LDD ARG1  SHIFT   0133 +69 0136 0141
SWICH  RAU ARG1          0115 +60 0136 0101
       LDD EXP2          0101 +69 0120 0123
       STD EXP1          0123 +24 0143 0102
       LDD ARG2  SHIFT   0102 +69 0128 0141
```

```
SRT    SRT  0000 ADD     0132 +30 0000 0152
SRD    SRD  0000 ADD     0134 +31 0000 0152
ADD    AUP  8001         0152 +10 8001 0121
       NZE       ZERO    0121 +45 0125 0144
       SCT  0000         0125 +36 0000 0142
       STL ACC           0142 +20 0000 0103
       SRT  0002         0103 +30 0002 0110
       ALO  8002         0110 +15 8002 0119
       SLO  8002         0119 +16 8002 0127
       SCT  0004         0127 +36 0004 0135
       STU SHIFT         0135 +21 0141 0147
       ALO ACC           0147 +15 0000 0108
       SLT  0008         0108 +35 0008 0118
       RAM  8002         0118 +67 8002 0138
       SML EXP1          0138 +18 0143 0148
       SLO ADD           0148 +16 0152 0107
       BMI       LOW     0107 +46 0112 0150
       NZU HIGH          0112 +44 0151 0116
       SRT  0008         0116 +30 0008 0126
       SML SHIFT         0126 +18 0141 0145
       SIA ACC           0145 +23 0000 0153
       RAL  8001 EXIT    0153 +65 8001 0109
ZERO   STL ACC   EXIT    0144 +20 0000 0109
```

[5] SHOAP II and SHOAP III assembly programs have been used for demonstrations at Case Institute since 1957. A deck can be made by altering at most eleven instructions in the programs for SOAP II or Case SOAP III.

TABLE 2. Mechanical "optimizing" schemes

				SOAP	(TIGHT) SOAP	FIVE AND TEN	SEQUENTIAL	SHOAP
FAD	STD	EXIT		0103 +24 0106 0109	0103 +24 0106 0109	0100 +24 0105 0110	0100 +24 0101 0102	0104 +24 0106 0108
	SLT	0008		0109 +35 0008 0127	0109 +35 0008 0127	0110 +35 0008 0120	0102 +35 0008 0103	0108 +35 0008 0114
	STU	ARG1		0127 +21 0132 0135	0127 +21 0132 0135	0120 +21 0125 0130	0103 +21 0104 0105	0114 +21 0117 0119
	STL	EXP1		0135 +20 0139 0142	0135 +20 0139 0142	0130 +20 0135 0140	0105 +20 0106 0107	0119 +20 0122 0124
	RAU	ACC		0142 +60 0000 0105	0142 +60 0000 0105	0140 +60 0000 0155	0107 +60 0000 0108	0124 +60 0000 0154
	SRT	0002		0105 +30 0002 0111	0105 +30 0002 0111	0155 +30 0002 0115	0108 +30 0002 0109	0154 +30 0002 0110
	STL	EXP2		0111 +20 0115 0118	0111 +20 0115 0118	0115 +20 0170 0175	0109 +20 0110 0111	0110 +20 0164 0116
	STU	ARG2		0118 +21 0122 0125	0118 +21 0122 0125	0175 +21 0180 0185	0111 +21 0112 0113	0116 +21 0169 0121
	RSM	8002		0125 +68 8002 0133	0125 +68 8002 0133	0185 +68 8002 0145	0113 +68 8002 0114	0121 +68 8002 0128
	AML	EXP1		0133 +17 0139 0143	0133 +17 0139 0143	0145 +17 0135 0190	0114 +17 0106 0115	0128 +17 0122 0126
	SLT	0001		0143 +35 0001 0149	0143 +35 0001 0149	0190 +35 0001 0150	0115 +35 0001 0116	0126 +35 0001 0132
	NZU	BIG		0149 +44 0153 0104	0149 +44 0153 0104	0150 +44 0106 0111	0116 +44 0117 0118	0132 +44 0134 0135
	SRT	0005		0104 +30 0005 0117	0104 +30 0005 0117	0111 +30 0005 0121	0118 +30 0005 0119	0135 +30 0005 0140
	LDD	SRT	STORE	0117 +69 0120 0123	0117 +69 0120 0123	0121 +69 0126 0131	0119 +69 0120 0121	0140 +69 0142 0144
BIG	LDD	SRD	STORE	0153 +69 0156 0123	0153 +69 0107 0123	0106 +69 0161 0131	0117 +69 0122 0121	0134 +69 0136 0144
STORE	SDA	SHIFT		0123 +22 0177 0130	0123 +22 0128 0131	0131 +22 0136 0141	0121 +22 0123 0124	0144 +22 0146 0148
	BMI	SWICH		0130 +46 0183 0134	0131 +46 0134 0136	0141 +46 0146 0101	0124 +46 0125 0126	0148 +46 0100 0101
	RAU	ARG2		0134 +60 0122 0128	0136 +60 0122 0129	0101 +60 0180 0186	0126 +60 0112 0127	0101 +60 0169 0172
	LDD	ARG1	SHIFT	0128 +69 0132 0177	0129 +69 0132 0128	0186 +69 0125 0136	0127 +69 0104 0123	0172 +69 0117 0146
SWICH	RAU	ARG1		0183 +60 0132 0137	0134 +60 0132 0137	0146 +60 0125 0181	0125 +60 0104 0128	0100 +60 0117 0120
	LDD	EXP2		0137 +69 0115 0168	0137 +69 0115 0119	0181 +69 0170 0176	0128 +69 0110 0129	0120 +69 0164 0166
	STD	EXP1		0168 +24 0139 0192	0119 +24 0139 0144	0176 +24 0135 0191	0129 +24 0106 0130	0166 +24 0122 0174
	LDD	ARG2	SHIFT	0192 +69 0122 0177	0144 +69 0122 0128	0191 +69 0180 0136	0130 +69 0112 0123	0174 +69 0169 0146
SRT	SRT	0000	ADD	0120 +30 0000 0193	0120 +30 0000 0145	0126 +30 0000 0137	0120 +30 0000 0131	0142 +30 0000 0198
SRD	SRD	0000	ADD	0156 +31 0000 0193	0107 +31 0000 0145	0161 +31 0000 0137	0122 +31 0000 0131	0136 +31 0000 0198
ADD	AUP	8001		0193 +10 8001 0101	0145 +10 8001 0108	0137 +10 8001 0147	0131 +10 8001 0132	0198 +10 8001 0103
	NZE		ZERO	0101 +45 0154 0155	0108 +45 0112 0113	0147 +45 0102 0107	0132 +45 0133 0134	0103 +45 0105 0156
	SCT	0000		0154 +36 0000 0178	0112 +36 0000 0138	0102 +36 0000 0112	0133 +36 0000 0135	0105 +36 0000 0160
	STL	EXP2		0178 +20 0115 0119	0138 +20 0115 0121	0112 +20 0170 0127	0135 +20 0110 0136	0160 +20 0164 0115
	SRT	0002		0119 +30 0002 0175	0121 +30 0002 0130	0127 +30 0002 0187	0136 +30 0002 0137	0115 +30 0002 0170
	ALO	8002		0175 +15 8002 0184	0130 +15 8002 0140	0187 +15 8002 0197	0137 +15 8002 0138	0170 +15 8002 0176
	SLO	8002		0184 +16 8002 0144	0140 +16 8002 0100	0197 +16 8002 0157	0138 +16 8002 0139	0176 +16 8002 0182
	SCT	0004		0144 +36 0004 0107	0100 +36 0004 0114	0157 +36 0004 0117	0139 +36 0004 0140	0182 +36 0004 0138
	STU	ARG1		0107 +21 0132 0185	0114 +21 0132 0141	0117 +21 0125 0132	0140 +21 0104 0141	0138 +21 0117 0118
	ALO	EXP2		0185 +15 0115 0169	0141 +15 0115 0124	0132 +15 0170 0177	0141 +15 0110 0142	0118 +15 0164 0168
	SLT	0008		0169 +35 0008 0187	0124 +35 0008 0146	0177 +35 0008 0138	0142 +35 0008 0143	0168 +35 0008 0123
	RAM	8002		0187 +67 8002 0145	0146 +67 8002 0110	0138 +67 8002 0148	0143 +67 8002 0144	0123 +67 8002 0130
	SLO	ADD		0145 +16 0193 0147	0110 +16 0145 0150	0148 +16 0137 0142	0144 +16 0131 0145	0130 +16 0198 0102
	SML	EXP1		0147 +18 0139 0194	0150 +18 0139 0147	0142 +18 0135 0192	0145 +18 0106 0146	0102 +18 0122 0125
	BMI		LOW	0194 +46 0197 0148	0147 +46 0101 0151	0192 +46 0198 0152	0146 +46 0147 0148	0125 +46 0127 0178
	NZU	HIGH		0197 +44 0151 0102	0101 +44 0116 0126	0198 +44 0103 0108	0147 +44 0149 0150	0127 +44 0180 0131
	SRT	0008		0102 +30 0008 0121	0126 +30 0008 0148	0108 +30 0008 0118	0150 +30 0008 0151	0131 +30 0008 0186
	SML	ARG1		0121 +18 0132 0138	0148 +18 0132 0102	0118 +18 0125 0182	0151 +18 0104 0152	0186 +18 0117 0167
	SIA	ACC		0138 +23 0000 0157	0102 +23 0000 0152	0182 +23 0000 0156	0152 +23 0000 0153	0167 +23 0000 0152
	RAL	8001	EXIT	0157 +65 8001 0106	0152 +65 8001 0106	0156 +65 8001 0105	0153 +65 8001 0101	0152 +65 8001 0106
ZERO	STL	ACC	EXIT	0155 +20 0000 0106	0113 +20 0000 0106	0107 +20 0000 0105	0134 +20 0000 0101	0156 +20 0000 0106

TABLE 3. Hand de-optimization

```
                          HAND SHOAP
FAD    STD EXIT           0106 +24 0108 0110
       SLT  0008          0110 +35 0008 0116
       STU ARG1           0116 +21 0118 0120
       STL EXP1           0120 +20 0124 0112
       RAU ACC            0112 +60 0000 0104
       SRT  0002          0104 +30 0002 0160
       STU ARG2           0160 +21 0113 0115
       STL EXP2           0115 +20 0168 0119
       RSM EXP2           0119 +68 0168 0122
       AML EXP1           0122 +17 0124 0126
       SLT  0001          0126 +35 0001 0132
       NZU BIG            0132 +44 0129 0135
       SRT  0005          0135 +30 0005 0140
       LDD SRT    STORE   0140 +69 0142 0144
BIG    LDD SRD    STORE   0129 +69 0125 0144
STORE  SDA SHIFT          0144 +22 0146 0148
       BMI SWICH          0148 +46 0133 0123
       RAU ARG2           0123 +60 0113 0166
       LDD ARG1   SHIFT   0166 +69 0118 0146
SWICH  RAL EXP2           0133 +65 0168 0121
       STL EXP1           0121 +20 0124 0117
       RAU ARG1           0117 +60 0118 0114
       LDD ARG2   SHIFT   0114 +69 0113 0146
```

```
SRT    SRT  0000 ADD      0142 +30 0000 0130
SRD    SRD  0000 ADD      0125 +31 0000 0130
ADD    AUP  8001          0130 +10 8001 0134
       NZE       ZERO     0134 +45 0137 0109
       SCT  0000          0137 +36 0000 0192
       STL SHIFT          0192 +20 0146 0198
       SRT  0002          0198 +30 0002 0103
       ALO  8002          0103 +15 8002 0105
       SLO  8002          0105 +16 8002 0107
       SCT  0004          0107 +36 0004 0101
       ALO SHIFT          0101 +15 0146 0145
       STU SHIFT          0145 +21 0146 0147
       SLT  0008          0147 +35 0008 0141
       RAM  8002          0141 +67 8002 0139
       SML EXP1           0139 +18 0124 0128
       SLO ADD            0128 +16 0130 0184
       BMI       LOW      0184 +46 0136 0149
       NZU HIGH           0136 +44 0199 0138
       SRT  0008          0138 +30 0008 0194
       SML SHIFT          0194 +18 0146 0100
       SIA ACC            0100 +23 0000 0102
       RAL  8001 EXIT     0102 +65 8001 0108
ZERO   STL ACC   EXIT     0109 +20 0000 0108
```

Comparison of Speeds

The running times of these routines, as actually verified by computer tests, are summarized in Table 4. The weighted averages included there are computed under the assumption, taken from studies made by D. W. Sweeney, that the exponents of the two arguments are equal one-half of the time, and differ by at most seven almost all the other times. The first column contains the execution time when 1.0 is added to 10.0, where the number 10.0 was input in the lower accumulator and 1.0 was input in `ACC`. This column, then, has the execution time when the positive branch is taken in line 17 of the program. The second column is the execution time when the inputs are reversed, 10.0 being in `ACC` and 1.0 in the lower accumulator, thus causing the negative branch at line 17. The case of equal exponents would have the time shown in the "$1.0 + 10.0$" column, so the execution time in that column should occur about three times as often as the other.

TABLE 4. Timing results in word-time units

Method	$1.0 + 10.0$	$10.0 + 1.0$	Weighted Average
0. Theoretical minimum	302	352	315
1. HAND SOAP	403	503	428
2. SOAP	653	703	666
3. Tight SOAP	803	853	816
4. Five and ten	955	1005	968
5. Sequential	2301	2401	2326
6. SHOAP	2802	2952	2840
7. HAND SHOAP	2902	3102	2952

Comparison of SHOAP running times with HAND SHOAP times might be surprising at first, since they are comparable while the SOAP and HAND SOAP times are not. But this brings out a significant point: It is often more correct to think of hand optimization as providing a *difference* in running speeds rather than a *ratio*. Notice that the difference between SHOAP and HAND SHOAP is roughly the same as the difference between SOAP and HAND SOAP. These differences, for various pieces of coding, are approximately proportional to the length of coding, but actually become somewhat greater per line as the program lengthens; for as a location is used more and more, SOAP optimization generally becomes worse and worse. That is why a program like the Bell Interpretive System [10] would have been very much slower in operation if it had been merely SOAPed. (The reader is referred to the listing of that system as one of the most beautiful pieces of hand optimizing in existence.)

When SOAP was restricted to choosing locations from a small sequential set (mode 3 operation) the difference between it and HAND SOAP became highly significant. This fact points up another possible danger in reliance on SOAP optimization.

Notice that hand optimization beats sequential programming by a factor of 5.43 to 1. The classic comparison between these two schemes, the square-root subroutine optimized both ways [9, page 16], gave a factor of only 2.11 to 1. Here again the concept of difference rather than ratio gives a truer representation of the benefits of hand optimization. The square-root subroutine is not a typical case, because that particular program includes so large a proportion of divisions and multiplications as to make the arithmetic interlock times destroy much of the hand-optimization effects. Its shortness also contributes to its poor showing.

The first entry in Table 4 shows the theoretical lower limit to the times for the routine, calculated under the assumption that every location is located perfectly. If an immediate-access core-storage unit were available, it is interesting that the lower limit *could* actually be achieved with SOAP optimization (type 2) alone, by using just six core locations: `EXIT`, `ARG1`, `EXP1`, `ARG2`, `EXP2`, and `SHIFT`.

II. A Mathematical Solution

In this section we shall reduce the problem of optimizing in a narrow sense to a series of equations that are equivalent to the original problem. The great generality allowed in programs by the 650 coding structure makes this reduction difficult to describe unless we build up some notational conventions.

The Input Pseudo-Code

We will require that the instructions to be optimized have been prepared in a special language, a somewhat restricted dialect of SOAP. The actual transformation into this language is a mechanical process that need not be carried out in practice — it is used here merely for convenience in presentation.

We are given a list of v instructions (and perhaps data), namely Q_1, Q_2, …, Q_v. Each instruction Q_j is an ordered quadruple

$$(L_j, O_j, D_j, I_j),$$

where O_j is some 650 operation code and L_j, D_j, I_j must be either absolute addresses or special symbols of the form

$$\mathrm{SS}\alpha + n, \qquad \alpha = 1, 2, \ldots, \ell.$$

(Here n is an integer that is usually zero; in such a case we write just SSα instead of SS$\alpha + 0$.) By this convention we insist that blank addresses, for example, be filled in with symbols, and that the regional addresses known in advance be changed to their absolute equivalents. Regions that are to be placed optimally are assigned to a symbol SSα and regional addresses of this type can be written SS$\alpha + n$ with $n \neq 0$.

Symbolic addresses (in the SOAP sense) are all to be changed into absolute addresses when they are predefined, as is `ACC` in our example, and into special symbols SS1, SS2, etc., when they are to be optimized. Under these conventions, our example program would therefore begin as follows:

```
(SS1,  24,  SS2,    SS3)
(SS3,  35,   0008,  SS4)
(SS4,  21,  SS5,    SS6)
(SS6,  20,  SS7,    SS8)
(SS8,  60,   0000,  SS9)
            etc.
```

For each operation O_j we require that the exact optimizing information be known. In particular, we want special knowledge about the operands to be supplied in some way whenever it is required by the `DIV`, `MPY`, `SCT`, `DVR`, and `TLU` commands. Information about the setting of the timing ring is sometimes necessary, and we also need to know in some cases whether or not a "complement cycle" occurs.[6]

[6] The occurrence of complement cycles is defined in footnote 20.

Absolute addresses L are classified as to whether they fall into the set $\mathcal{D}$ of drum locations, $\mathcal{A}$ of locations 8000–8007, or $\mathcal{R}$ of other locations, and we write for example, $L \in \mathcal{A}$ to indicate that L is an 800X location. Immediate access core storage locations will be omitted from this discussion for simplicity, and so will tagged addresses. Methods for accommodating them will be discussed later.

To each α we associate the set $N(\alpha)$ of all n that occur with α in the form SS$\alpha + n$. We also have the sets $K(\alpha) \subseteq \mathcal{D}$ which consist of the "available" locations into which the symbol SSα may be placed.[7] We allow $K(\alpha)$ to be different for different α's.

Operating Sequences

Finally, we are supplied with a set of sequences S_k of finite length, for $k = 1, \ldots, w$, whose elements are numbers between 1 and v, possibly followed by the letter D or I. These are called "operation sequences," because they indicate in exactly what order the instructions are to be executed at running time.

For example, S_1 might be a sequence like

$$1,\ 2,\ 3\mathrm{D},\ 1,\ 4,\ 3\mathrm{I},\ 15.$$

This sequence indicates that instruction Q_3 is a "branch" instruction, and at running time S_1 is taken to mean the following: "Execute Q_1; then Q_2; then Q_3 taking the D-address branch; then Q_1; then Q_4; then Q_3 taking the I-address branch; and finally Q_{15}; in that order."

In general we write S_k as

$$a_{k1},\ a_{k2},\ \ldots,\ a_{km_k}$$

and we write $[j] = [j\mathrm{D}] = [j\mathrm{I}] = j$. Each element a_{kr} is a number j if O_j is a non-branch instruction, and must be of the form jD or jI if O_j is a branch command. Each sequence S_k is further restricted by requiring that

1) If a_{kr} is of the form j or jI, and if $[a_{k(r+1)}] = p$, then $I_j = L_p$.
2) If a_{kr} is of the form jD, and if $[a_{k(r+1)}] = p$, then $D_j = L_p$.
3) If $[a_{k1}] = p$, then $L_p \notin \mathcal{A}$; that is, no sequence starts in an 800X location.[8]

---

[7] We require, of course, that if $n \in N(\alpha)$ and $\beta \in K(\alpha)$, then $\beta + n \in \mathcal{D}$.

[8] This restriction could be lifted without difficulty by simply starting to apply the algorithm to the first non-800X address that occurs. The presentation of Figure 1 is somewhat simpler without this added complication.

These requirements imply not only that all instructions are executed according to the 650's normal operating sequence, but that each location address L_j is the actual location from where the instruction Q_j will be called into the program register at running time.

Notice that we do not necessarily expect L_j to be unequal to L_p when $j \neq p$. Two or more different instructions in the list may well have identical location addresses. In fact, the example sequence S_1 given above goes in one case from Q_1 to Q_2, and later from Q_1 to Q_4, while Q_1 is *not* a branch instruction; this situation *requires* that $L_2 = L_4$. It is important to allow such freedom, because many instructions are modified during running time or executed from locations different from where they are permanently stored; true optimization must take such things into account. (For example, if a certain instruction is placed into the exit line of a subroutine it must be optimized as it appears there.)

It is also important to make sure that the relevant parts of each instruction are given exactly as they will appear in the machine at running time. If an instruction has a data address that varies from 0222 to 0226 in steps of one, say, it should be listed five times, once with each possible D-address, if it is to be rigorously optimized.[9] Furthermore, if a certain loop in the coding is to be executed 100 times at running time, the operating sequence S_k should indicate this in order to give that loop the proper priority at running time. Abbreviative conventions could easily be devised to indicate such things. All of these features of the input language must be adhered to strictly if the true optimal solution is desired; but whenever approximations will be good enough, an approximate solution can be obtained by taking short-cuts and liberties with the input code.

Each sequence S_k describes one complete "run" of the program. Since given programs usually operate differently when given different inputs, each sequence describes the branches to be taken given a certain class of inputs. Associated with each sequence is an integer priority number, p_k, which indicates the relative number of times that the sequence S_k will be used when the program is running. As an example, consider our floating-point addition subroutine. We might make up five lists: S_1 for the case when $|\texttt{EXP2}| \le |\texttt{EXP1}| \le |\texttt{EXP2}| + 9$; S_2 for the case when $|\texttt{EXP2}| - 9 \le |\texttt{EXP1}| < |\texttt{EXP2}|$; S_3 when $|\texttt{EXP1}| > |\texttt{EXP2}| + 9$; S_4 when

[9] If a certain address is to occur randomly and is unknown to the programmer, we break the sequence at this instruction and form two sequences; this causes the preceding and succeeding parts of the program to be optimized independently of each other and of the random address, as they should be.

|EXP1| < |EXP2| − 9; and S_5 for the special case when the result is zero.[10] We might take $p_1 = 30$, $p_2 = 10$, $p_3 = p_4 = 1$, $p_5 = 2$, or whatever suits our fancy. The weightings in Table 4 were based on $p_1 = 3$, $p_2 = 1$, $p_3 = p_4 = p_5 = 0$ (in which case S_3, S_4, and S_5 may be discarded).

Optimizing Information Required

Without loss of generality we will assume that no read or punch commands are in any of the lists S_k; for if $a_{kr} = j$ and O_j is a read or punch, we can break S_k into two sequences

$$a_{k1}, a_{k2}, \ldots, a_{k(r-1)} \quad \text{and} \quad a_{k(r+1)}, \ldots, a_{km_k},$$

and relabel the subscripts. We also temporarily prohibit the table-lookup instruction; it is easy to handle, but does not fit into the general pattern so it will be considered later.

Table 5 gives precise optimizing information for the remaining basic opcodes. We classify the various operations into five types with respect to the optimization of both data and instruction addresses. (Here '$*$' is a symbolic address, either L or D.)

Type 1: $(1, *, n)$ Wait n word-times after base location $*$.

Type 2: $(2, *, n)$ Wait one word-time if $*$ is even, then n more word-times after base location $*$.

Type 3: $(3, *, n)$ Wait one word-time if $*$ is odd, then n more word-times after base location $*$.

Type 4: $(4, *, n, m)$ Wait one word-time if $*$ is even, then optimize by lower limit n, upper limit m.

Type 5: $(5, *, n, m)$ Wait one word-time if $*$ is odd, then optimize by lower limit n, upper limit m.

Some of the information in Table 5 conflicts with the timings previously published[11] and some of it has never appeared before; all of these entries are marked with '†', and they have been verified experimentally.

---

[10] These lists can be further subdivided if desired by considering how many shifts are to be executed at line 28. We are assuming nine or ten shifts in this paper.

[11] In this case they will probably work on all 650s, but may only be peculiar to the machines the author tested. In case of doubt they can be tried out by the reader, who is cautioned in this regard against using 8000 for the timing since address 8000 is accessible only when the arithmetic interlock is off. By the way (a hint for those who are teaching classes about the 650), the problem of discovering the timing of the shift-and-round instruction makes a very good student exercise!

TABLE 5. Timing chart for IBM 650 instructions

To optimize when D-address $\in \mathcal{D}$ and I-address $\in \mathcal{D}$

Instruction	D-address	I-address
NOP, HLT	—	$(1, L, 4)$
AUP, SUP, ALO, SLO, AML, SML, RAU, RSU, RAL, RSL, RAM, RSM	$(1, L, 3)$	$(2, D, 4)$
DIV, DVR	$(1, L, 3)$	$(4, D, 4, 60{+}2\Sigma q)$†
MPY	$(1, L, 3)$	$(4, D, 4, 20{+}2\Sigma m)$†
STL	$(2, L, 4)$	$(1, D, 3)$
STU	$(3, L, 4)$	$(1, D, 3)$
SDA, SIA	$(3, L, 3)$	$(1, D, 3)$
STD, LDD	$(1, L, 3)$	$(1, D, 3)$
SRT, SLT, SCT $(S = 0)$	—	$(2, L, 5)$
SRT, SLT, SCT $(S = 1, 2)$; SRD $(S = 1)$	—	$(2, L, 6)$
SRT, SLT, SCT $(S > 2)$	—	$(4, L, 6, 2S{+}2)$
SRD $(S > 1)$	—	$(4, L, 6, 2S{+}3)$†
NZU	$(3, L, 3)$	$(3, L, 4)$
NZE	$(2, L, 3)$	$(2, L, 4)$
BMI	$(1, L, 3)$	$(1, L, 4)$
BOV, BD1–BD8	$(1, L, 3)$	$(1, L, 5)$
BD0, BD9	$(1, L, 4)$	$(1, L, 5)$

To optimize an I-address $\in \mathcal{D}$ when D-address $\in \mathcal{A}$

	8000	8001	8002	8003
AUP, ..., RSM	$(3, L, 7)$	$(2, L, 6)$†	$(2, L, 8)$	$(3, L, 7)$
LDD	$(1, L, 6)$	$(1, L, 5)$†	$(2, L, 6)$	$(3, L, 6)$
MPY	$(5, L, 7, 23{+}2\Sigma m)$†	$(4, L, 6, 22{+}2\Sigma m)$†	$(4, L, 8, 24{+}2\Sigma m)$†	$(5, L, 7, 23{+}2\Sigma m)$†
DIV, DVR	$(5, L, 7, 63{+}2\Sigma q)$†	$(4, L, 6, 62{+}2\Sigma q)$†	$(4, L, 8, 64{+}2\Sigma q)$†	$(5, L, 7, 63{+}2\Sigma q)$†

To optimize a D-address $\in \mathcal{A}$

	8000, 8001	8002	8003
NZU	$(3, L, 3)$	$(3, L, 4)$	$(3, L, 3)$
NZE	$(2, L, 3)$	$(2, L, 3)$	$(2, L, 4)$
BMI, BOV, BD1–BD8	$(1, L, 3)$	$(2, L, 3)$	$(3, L, 3)$
BD0, BD9	$(1, L, 4)$	$(3, L, 4)$	$(2, L, 4)$

To optimize an I-address $\in \mathcal{A}$

	8000, 8001	8002	8003
SRT, SLT, SCT	$(2, L, 2S{+}5)$†	$(2, L, 2S{+}5)$†	$(2, L, 2S{+}6)$†
SRD	$(2, L, 2S{+}6)$†	$(2, L, 2S{+}7)$†	$(2, L, 2S{+}6)$†
MPY, DIV, DVR	Same as I-address $\in \mathcal{D}$, plus 3; plus one more if necessary to make the result even for 8002 or odd for 8003.†		
All others	Same as I-address $\in \mathcal{D}$; plus one more if necessary to make the result even for 8002 or odd for 8003; plus two more if a complement cycle occurs.		

Key: $\Sigma m \equiv$ sum of multiplier digits
$\Sigma q \equiv$ sum of quotient digits
$S \equiv$ number of shifts

General Description of the Method

The process to be described next takes the input pseudo-code in the form described and translates it into a set of equations and inequalities in nonnegative integer variables. Hence this is finally reduced to an integer programming problem. Algorithms that find optimum integer solutions to linear programming problems have been discovered by Gomory [3] and [4]; so the problem is solved from a mathematical standpoint.

The procedure below is a major modification of the process described recently by Chadwick and Loeb [2]. Some of their manipulations have been simplified, and the methods have been extended to remove the severe restrictions imposed in that paper. For example, no distinction is made here between instructions and data. Branch commands, even-odd timing considerations, repetition of instructions, and many other things are no longer prohibited.

The translation process has curious resemblances to a *compiler* and to a *semitracer*. We "scan" the instructions and asynchronously "compile" equations or portions of them. And we trace through the program according to the sequence S_k, compiling only the first time a quadruple is encountered while remaining mute all succeeding times through a loop.

The Algorithm: First Version

We introduce a large number of variables, $x_{\alpha\beta}$, for $\alpha = 1, \ldots, \ell$ and $\beta \in K(\alpha)$. We intend to have

$$x_{\alpha\beta} = \begin{cases} 1, & \text{if SS}\alpha \text{ is assigned to } \beta; \\ 0, & \text{if SS}\alpha \text{ is not assigned to } \beta. \end{cases}$$

We begin with ℓ equations that restrict $x_{\alpha\beta}$ to just these two values and ensure that each symbol is assigned to exactly one location:

$$\sum_{\beta \in K(\alpha)} x_{\alpha\beta} = 1, \qquad \text{for } \alpha = 1, \ldots, \ell. \tag{1}$$

To make sure that each available location is assigned to at most one symbol we also have[12]

$$\sum_{\alpha=1}^{\ell} \sum_{\substack{n \in N(\alpha) \\ \beta - n \in K(\alpha)}} x_{\alpha(\beta-n)} \le 1, \qquad \text{for } \beta \in \mathcal{D}. \tag{2}$$

---

[12] A modification of this procedure, where certain symbols are allowed to be placed into the same locations without harm, could be made by making several sets of inequalities of this type, leaving out certain α's from the summations in each set.

If the number of distinct symbols is equal to the total number of available locations, these may all be replaced by equalities. Other variables t_i, w_a, z_u will be introduced as they are needed.

The remaining constraints are generated by the flow charts in Figures 1, 2, and 3. The notation used there needs some elucidation:

λ indicates any integer.

τ indicates any linear combination of t-variables.

Ω indicates any absolute address L or any symbolic address of the form $\mathrm{SS}\alpha + n$.

These charts also use the "bar convention," meaning that whenever a symbol occurs with a bar *over* it, the symbol is to be incremented automatically by one *after* the operation in which it appears has been completed.[13] Thus, '$\overline{r}$' means $r \leftarrow r+1$; '$\tau \leftarrow t_{\overline{\imath}} + t_j$' means $\tau \leftarrow t_i + t_j$ followed by $i \leftarrow i+1$; and '$\overline{k} = w$?' means that we should test if $k = w$, then increment k, then branch.

The functions F, F^E, and F^O are defined as follows: Let $R(\beta)$ be the remainder of β after dividing by 50. Then

$$F(L) = R(L); \qquad F(\mathrm{SS}\alpha + n) = \sum_{\beta \in K(\alpha)} R(\beta + n) x_{\alpha\beta};$$

$$F^E(L) = \begin{cases} R(L)+1, & L \text{ even}; \\ R(L), & L \text{ odd}; \end{cases} \qquad F^O(L) = \begin{cases} R(L), & L \text{ even}; \\ R(L)+1, & L \text{ odd}; \end{cases}$$

$$F^E(\mathrm{SS}\alpha + n) = \sum_{\substack{\beta \in K(\alpha) \\ \beta+n \text{ even}}} \bigl(R(\beta+n)+1\bigr) x_{\alpha\beta} + \sum_{\substack{\beta \in K(\alpha) \\ \beta+n \text{ odd}}} R(\beta+n) x_{\alpha\beta};$$

$$F^O(\mathrm{SS}\alpha + n) = \sum_{\substack{\beta \in K(\alpha) \\ \beta+n \text{ even}}} R(\beta+n) x_{\alpha\beta} + \sum_{\substack{\beta \in K(\alpha) \\ \beta+n \text{ odd}}} \bigl(R(\beta+n)+1\bigr) x_{\alpha\beta}.$$

As we compile equations we might think of ourselves as using a *metacomputer*, some mechanical instrument of our imagination. We will be using temporary storage in which we store formulas and arithmetic expressions, and this is not to be confused with the storage of the program we are optimizing. We use the symbolism

$$F = \text{“}E\text{”}, \qquad \text{or} \qquad F \leftarrow \text{“}E\text{”},$$

[13] Similarly, although not used here, a bar under a symbol means to decrement the symbol. If the same symbol is barred more than once in an expression it is to be operated on from left to right; thus $a_{\overline{\imath},\overline{\imath}} \leftarrow 1$ means $a_{i,i+1} \leftarrow 1$ and $i \leftarrow i+2$.

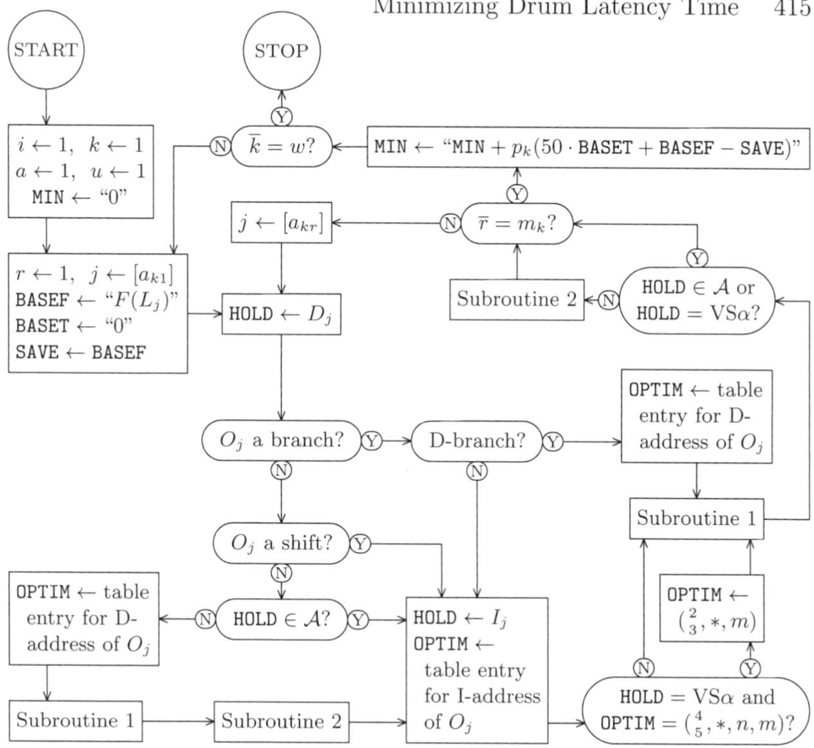

FIGURE 1. The equation-compiling algorithm.

to indicate that formula storage location F *contains*, or *is set to*, respectively, the linear expression E or the contents of formula storage location E, *after collecting terms*. Thus if storage F contains the expression t_1+t_0 we would have $F =$ "$t_0 + t_1$"; and the statement $F \leftarrow$ "$F + t_2 - t_1$" would mean that F is to be replaced by the expression $t_0 + t_2$.

The flow charts have not really been written for a machine, even though they might be programmed. They represent a procedure to be used by a human being who is trying to solve the optimizing problem, and the "machine language" is English and mathematics, plus a bit of common sense. This method of presenting mathematical procedures or proofs in terms of flow charts has broad applications to solving problems that would be very difficult to tackle in another way.

The main flow chart gives the basic order of processing, taking the operation sequences one by one, and going through them instruction by instruction. Subroutine 1 is used to count up the waiting time, and subroutine 2 compiles the constraints, which say essentially that so much time must elapse before the drum location can be used.

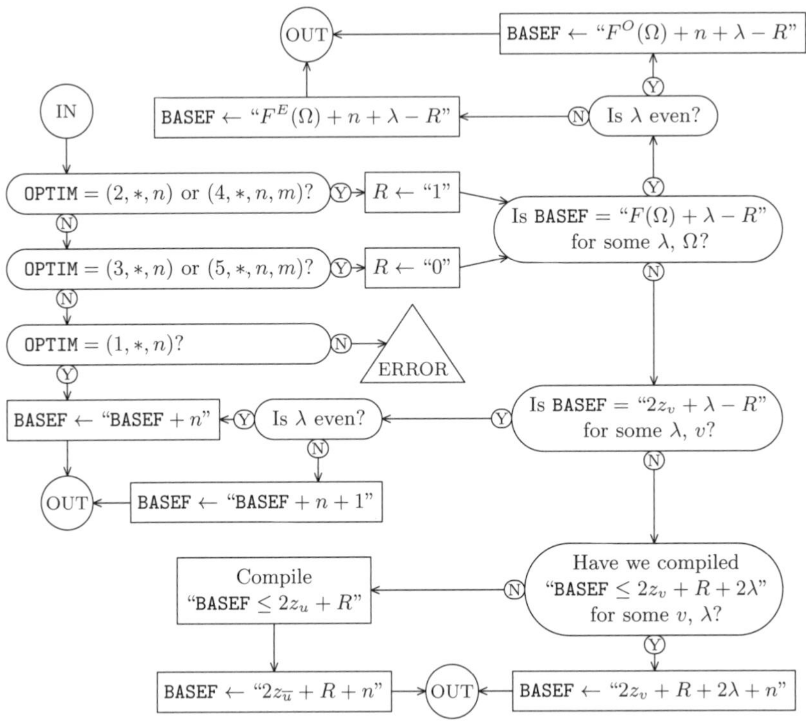

FIGURE 2. Subroutine 1.

The variables t_i represent the current drum-revolution number; and BASET and BASEF at compiling time represent the drum-revolution and dynamic-drumlevel, respectively, at running time. The z-variables are used to make certain drum levels *even* if they were *odd*, and w-variables are used to keep track of waiting time when optimizing the low-high-limit case in type 4 or 5 optimization. References to "VSα" are to be ignored for the moment; they are for version 2 of the algorithm.

No more explanation of the flow chart will be given here; the rest will become clear if the reader traces through some examples. After the STOP circle is encountered, the integer programming problem is to minimize "MIN" subject to (1), (2), and all inequalities compiled.

Example: Version 1

The number of x-variables is usually very great, and so we will choose a small problem for our example. The program chosen is a Stop-Codes routine that fits onto a single card and sets all 2000 drum locations to

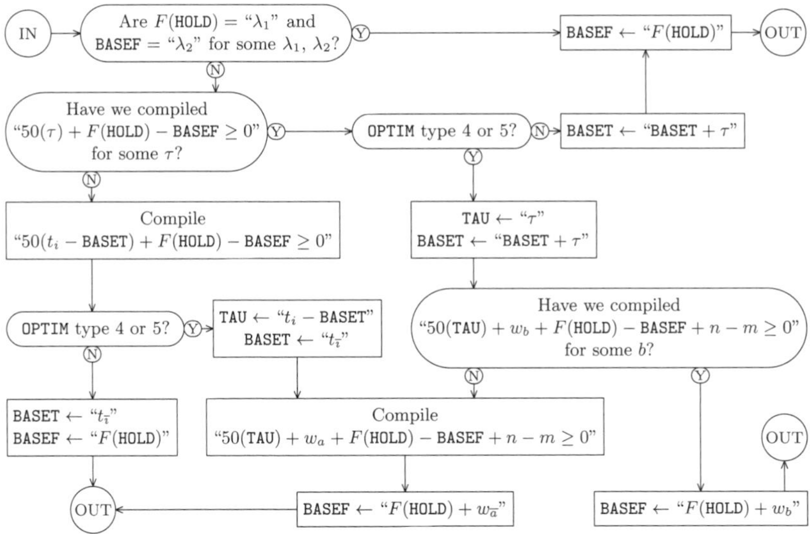

FIGURE 3. Subroutine 2.

stop codes. (More precisely, it sets location β to the 10-digit number '$\pm$ HLT β 8000' for $0 \le \beta < 2000$.) Here is the input pseudo-code:

j	L_j	O_j	D_j	I_j
1	SS1	60 (RAU)	0001	SS2
2	SS2	22 (SDA)	0000	SS3
3	SS3	65 (RAL)	SS4	SS5
4	SS5	10 (AUP)	SS6	8001
5	8001	17 (AML)	0000	8002
6	0002	18 (SML)	0001	8003
7	8003	17 (AML)	0000	8002
8	8002	22 (SDA)	1999	0002
⋮	⋮	⋮	⋮	⋮
$8+i$	8002	22 (SDA)	$1999-i$	0002
⋮	⋮	⋮	⋮	⋮
2005	8002	22 (SDA)	0002	0002

The single sequence S_1 is

$$1, 2, 3, 4, 5, 8, 6, 7, 9, 6, 7, 10, 6, 7, \ldots, 2004, 6, 7, 2005;$$

and we let $K(\alpha) = \{3, 4, 5, 6, 7, 8\}$ for every α.

Equations (1) are

$$x_{13} + x_{14} + x_{15} + x_{16} + x_{17} + x_{18} = 1,$$
$$x_{23} + x_{24} + x_{25} + x_{26} + x_{27} + x_{28} = 1,$$
$$\vdots$$
$$x_{63} + x_{64} + x_{65} + x_{66} + x_{67} + x_{68} = 1;$$

and equations (2), with equality in this case, are

$$x_{13} + x_{23} + x_{33} + x_{43} + x_{53} + x_{63} = 1,$$
$$x_{14} + x_{24} + x_{34} + x_{44} + x_{54} + x_{64} = 1,$$
$$\vdots$$
$$x_{18} + x_{28} + x_{38} + x_{48} + x_{58} + x_{68} = 1.$$

(The last equation can be dropped because it is redundant.) Note that the equations so far are just the constraints for a transportation problem. Now the algorithm produces:

$$\begin{aligned}
50t_1 - 3x_{13} - 4x_{14} - 5x_{15} - 6x_{16} - 7x_{17} - 8x_{18} - 2 &\geq 0;\\
50(t_2 - t_1) + 3x_{23} + 4x_{24} + 5x_{25} + 6x_{26} + 7x_{27} + 8x_{28} - 5 &\geq 0;\\
50(t_3 - t_2) - 4x_{23} - 4x_{24} - 6x_{25} - 6x_{26} - 8x_{27} - 8x_{28} - 3 &\geq 0;\\
50(t_4 - t_3) + 3x_{33} + 4x_{34} + 5x_{35} + 6x_{36} + 7x_{37} + 8x_{38} - 5 &\geq 0;\\
50(t_5 - t_4) + 3x_{43} + 4x_{44} + 5x_{45} + 6x_{46} + 7x_{47} + 8x_{48}&\\
- 3x_{33} - 4x_{34} - 5x_{35} - 6x_{36} - 7x_{37} - 8x_{38} - 3 &\geq 0;\\
50(t_6 - t_5) + 3x_{53} + 4x_{54} + 5x_{55} + 6x_{56} + 7x_{57} + 8x_{58}&\\
- 3x_{43} - 5x_{44} - 5x_{45} - 7x_{46} - 7x_{47} - 9x_{48} - 4 &\geq 0;\\
50(t_7 - t_6) + 3x_{63} + 4x_{64} + 5x_{65} + 6x_{66} + 7x_{67} + 8x_{68}&\\
- 3x_{53} - 4x_{54} - 5x_{55} - 6x_{56} - 7x_{57} - 8x_{58} - 3 &\geq 0;\\
50(t_8 - t_7) - 3x_{63} - 5x_{64} - 5x_{65} - 7x_{66} - 7x_{67} - 9x_{68} - 7 &\geq 0.
\end{aligned}$$

We want to minimize $50t_8 + 2 - 3x_{13} - 4x_{14} - 5x_{15} - 6x_{16} - 7x_{17} - 8x_{18}$.

By inspection, there are just two solutions:

$$t_1 = t_2 = 1, \quad t_3 = t_4 = t_5 = 2, \quad t_6 = t_7 = 3, \quad t_8 = 4; \quad x_{18} = x_{25} = 1;$$
$$\text{either } x_{33} = x_{46} = x_{54} = x_{67} = 1 \text{ or } x_{34} = x_{47} = x_{53} = x_{66} = 1;$$
$$\text{and all other } x\text{-variables are zero.}$$

The first solution gives us the Stop Codes card shown in Table 6; it is to be read in with 8000: 70 0008 xxxx. Actually only the first five instructions were optimized, because of the nature of the problem.

TABLE 6. "Optimized" Stop Codes card

word 1	01 0001 8000 ±	
word 2	18 0001 8003 +	
word 3	65 0006 0004 +	(SS3)
word 4	10 0007 8001 +	(SS5)
word 5	22 0000 0003 +	(SS2)
word 6	21 1998 2002 +	(SS4)
word 7	17 0000 8002 +	(SS6)
word 8	60 0001 0005 +	(SS1)

The Algorithm: Second Version

If we were to attempt to optimize the floating-addition subroutine by the preceding algorithm, we would need 2916 x-variables, 7 w-variables, 73 t-variables, and 7 z-variables, for a grand total of exactly 3003 variables. Obviously this is far too many to solve the problem, at least with methods known today. But when the problem is this large we can use an approximate method that reduces the number of variables considerably.

The approximation is to assume that the drum is essentially infinite; that is, on any given level, we may place any number of symbols. After we get this approximate solution, experience indicates that it is almost always possible to shuffle the locations around slightly and fit everything into the available locations without losing any running time; for there will be places where drum revolutions must go by without tight optimization, and a lot of freedom can be employed as we decide where to waste that time. When using this approximation we may consider those locations that are used in only one place of the program to be placed in immediate access storage, since we can always fit them into the proper place on an "infinite" drum.

With this groundwork we can give a formal presentation of the revised algorithm. The symbols SSα are broken into two groups: those that are very simple, in the sense that they occur at most once in the L-address and at most once in the D- and I-addresses in the instructions Q_j of any particular list S_k, and those that are not. Symbols that fall into this "very simple" category include the addresses that are blank in SOAP language, and the addresses of constants that are used only once. These single-reference symbols will be renamed VS1, VS2, etc. Locations of the form SS$\alpha + n$ are still allowed, but VS$\alpha + n$ is not legal for $n \neq 0$. There is one further restriction: The first location address in every operation sequence must not be a VS symbol or an 800X address.[14]

[14] We can remove this restriction as in footnote 8 as long as no branch command appears before the occurrence of the first acceptable address.

For this treatment we restrict the symbol Ω to refer to absolute addresses only. The only x-variables are the variables x_1, x_2, ... corresponding to SS1, SS2, ...; their meaning now is that the symbol SSα is to occur on the level x_α (mod 50). For this new interpretation we need a new definition of the function F:

$$F(L),\ F^E(L),\ F^O(L),\quad \text{as before;}$$
$$F(\mathrm{SS}\alpha + n) \;=\; x_\alpha + n.$$

We do not define $F^E(\mathrm{SS}\alpha + n)$, $F^O(\mathrm{SS}\alpha + n)$, or $F(\mathrm{VS}\alpha)$. Equations (1) and (2) are dropped.

No changes in the flow charts of the original algorithm are necessary.

Example: Version 2

Let us illustrate the revised algorithm for the floating-addition routine, our favorite example. Here we put

`FAD` = SS1,	`EXIT` = SS2,	`ARG1` = SS3,
`EXP1` = SS4,	`EXP2` = SS5,	`ARG2` = SS6,
`STORE` = SS7,	`SHIFT` = SS8,	`ADD` = SS9.

We have five lists S_k as described earlier. The equations compiled are:

$50t_1 + x_2 - x_1 - 3 \geq 0,$
$x_2 + 2 \leq 2z_1,$
$50(t_2 - t_1) + x_3 - 2z_1 - 24 \geq 0,$
$x_3 + 2 \leq 2z_2,$
$50(t_3 - t_2) + x_4 - 2z_2 - 5 \geq 0,$
$50(t_4 - t_3) - x_4 - 6 \geq 0,$
$50(t_5 - t_4) + x_5 - 15 \geq 0,$
$x_5 + 3 \leq 2z_3,$
$50(t_6 - t_5) + x_6 - 2z_3 - 4 \geq 0,$
$x_6 + 2 \leq 2z_4,$
$50(t_7 - t_6) + x_4 - 2z_4 - 12 \geq 0,$
$x_4 \leq 2z_5 + 1,$
$50(t_8 - t_7) + x_7 - 2z_5 - 35 \geq 0,$
$x_7 \leq 2z_6,$
$50(t_9 - t_8) + x_8 - 2z_6 - 3 \geq 0,$
$50(t_{10} - t_9) + x_6 - x_8 - 10 \geq 0,$
$x_6 \leq 2z_7 + 1,$
$50(t_{11} - t_{10}) + x_3 - 2z_7 - 8 \geq 0,$
$50(t_{12} - t_{11}) + x_8 - x_3 - 3 \geq 0,$
$x_8 \leq 2z_8 + 1,$
$50(t_{13} - t_{12}) + x_9 - 2z_8 - 7 \geq 0,$
$50(t_{13} - t_{12}) + x_9 + w_1 - 2z_8 - 21 \geq 0,$
$x_9 + w_1 \leq 2z_9 + 1,$
$50(t_{14} - t_{13}) + x_5 - 2z_9 - 35 \geq 0,$
$x_5 + 2 \leq 2z_{10},$
$50(t_{15} - t_{14}) + x_3 - 2z_{10} - 34 \geq 0,$
$50(t_{16} - t_{15}) + x_5 - x_3 - 6 \geq 0,$
$50(t_{17} - t_{16}) + x_9 - 2z_3 - 30 \geq 0,$
$50(t_{18} - t_{17}) + x_4 - 2z_9 - 8 \geq 0,$
$50(t_{19} - t_{18}) + x_3 - 2z_5 - 34 \geq 0,$
$x_3 \leq 2z_{11} + 1,$
$50(t_{20} - t_{19}) - 2z_{11} - 9 \geq 0,$
$50(t_{21} - t_{20}) + x_2 - 9 \geq 0,$
$50(t_{22} - t_9) + x_3 - x_8 - 9 \geq 0,$
$50(t_{23} - t_{22}) + x_5 - 2z_{11} - 8 \geq 0,$
$50(t_{24} - t_{23}) + x_4 - x_5 - 6 \geq 0,$
$50(t_{25} - t_{24}) + x_6 - x_4 - 6 \geq 0,$
$50(t_{26} - t_{25}) + x_8 - x_6 - 3 \geq 0,$
$50(t_{27} - t_7) + x_7 - 2z_5 - 21 \geq 0,$
$50(t_{13} - t_{12}) + x_9 + w_2 - 2z_8 - 24 \geq 0,$
$x_9 + w_2 \leq 2z_{12} + 1,$
$50(t_{28} - t_{27} + t_8 - t_{13}) + x_5 - 2z_{12} - 35 \geq 0,$
$50(t_{29} - t_{13}) - 2z_9 - 15 \geq 0,$
$50(t_{30} - t_{29}) + x_2 - 3 \geq 0.$

Subject to those inequalities, we want to minimize the quantity

$$p_1(50t_{21}+x_2-x_1)+p_2(50(t_{21}+t_{26}-t_{12})+x_2-x_1)+p_3(50(t_{21}+t_{28}-t_{14}) \\ +x_2-x_1)+p_4(50(t_{21}+t_{26}-t_{12}+t_{28}-t_{14})+x_2-x_1)+p_5(50t_{30}+x_2-x_1)$$

over all nonnegative integer values of the variables. As implied in footnote 14 we can drop x_1 and put $t_1 = 0$. This gives us 8 x-variables, 29 t-variables, 2 w-variables, and 12 z-variables, or 51 variables in all, compared with 3003 by the old method!

Refinements of the Algorithm

Now let us consider various things that the algorithm does not handle, and sketch ways to incorporate them.

(1) The TLU operation. In this command we simply treat the data address as 0000 when we do the D-address cycle, and as the location X of the argument found when we do the I-address cycle. With this convention, the optimization rules are $(1, L, 3)$ and $(3, X, 5)$.

(2) Indexed addresses. (The contents of the index register at running time must be specified somehow in the input code.) Simply add n_j (the number of tagged addresses in Q_j) to BASEF after setting HOLD $\leftarrow D_j$.

(3) Floating-point and index-register commands. These are done merely by adding optimizing information to the chart (Table 5).

(4) The interlock problem. This is the most difficult addition to make. It requires the careful introduction of more w-variables, to serve as waiting times for various interlocks (the read, punch, arithmetic, core, tape control, tape read, and disk storage interlocks, etc.), and it produces inequalities something like our treatment of low-high-limit commands. The treatment of BASEF will have to be altered (i.e., the boxes that say BASEF $\leftarrow$ "$F(\text{HOLD}) + w_a$" will have to be placed somewhere else), and significant problems arise when several interlocks are on at once.

(5) Immediate access storage. When the core and arithmetic interlocks are off, we treat this precisely as the address 8000 is treated now. But the problem of deciding *which* addresses to put in core and which to leave on the drum is too difficult to include in the algorithm without major changes. This latter problem falls in the category of "how do we include techniques (1), (4), and (5) of full optimizing into a mechanical optimizing algorithm?"

III. An "Artistic" Solution

In this section we will see how useful the mathematical solution is or could be, and will examine a way to facilitate the process of optimizing by hand.

Evaluation of Previous Results

The usefulness of the mathematical solution depends almost wholly upon developments in the new theory of integer programming. R. E. Gomory, as mentioned earlier, has developed algorithms that solve the integer programming problem; thus the time it will take is known to be finite. But of course in our application we have another finite algorithm (exhaustive trial); and while adaptations of the simplex method to integer problems will no doubt be more efficient then pure trial-and-error, we still have no assurance that the time required to solve the problem will be reasonable. For example, the 51-variable, 43-constraint problem developed in the previous section was put onto the 650 in an attempt to find the optimal solution. Using the most advanced integer programming techniques known at the time (April 1960), the program ran over 300 iterations and showed no sign whatever of being close to a solution; in fact, it seemed to be further away, if anything, than when it began.[15] In order to solve that set of equations we will apparently need to use a larger and much faster computer; we will find a very large machine necessary, just to optimize programs for the 650 system!

The major advantage of the mathematical solution is the fact that it is a precise equivalent formulation of the latency-minimizing problem, and such a problem in inequalities can be solved by techniques that need have nothing to do with the original concept of latency times. Thus any technique for solving integer programming problems, a class that encompasses a wide scope of combinatorial problems in general, will automatically facilitate 650 optimization. For example, if a way is found in which to start an integer programming problem with an initial near-optimum solution and to start iterating from there, we will have the answer in a few minutes. (The 300-plus iterations mentioned above were necessary solely because the initial guess required was of necessity a very poor one.) And such a device would also give us a most important additional advantage: We would be able to check a given solution obtained by hand to see if it is optimal or not. Such techniques are not yet known, but research is being conducted.[16]

---

[15] Each iteration took between two and five minutes, because the matrix was so large it had to be put into disk storage. According to Dr. Gomory, quite a few problems of this size have been known not to converge after many iterations, although some large problems have converged quickly.

[16] Dr. Gomory has recently informed the author in a private communication that no manner of plugging in arbitrary feasible solutions has yet been discovered, but that research leading to this end is currently in progress.

Let us return to the discussion of the definition of optimization, and evaluate our mathematical model from this viewpoint. How many of the additional techniques that are an intrinsic part of optimizing can be handled by the integer programming formulation?

Technique (3), giving priority to the most-used parts of the program, is already an essential part of our mathematical model, because of the construction of the operation sequences S_k and of the associated priority numbers p_k.

Technique (2), allowing a choice between several locations for temporary storage locations and for constants, can also be incorporated easily into the mathematical development. We introduce new symbolic names for these addresses. If we are using version one of the algorithm, we then leave out appropriate terms from the summations in the inequality (2), so that two symbols *can* be assigned to the same location; and if we are using the approximate-version two, we solve the problem with the new addresses and afterwards fit them into the drum levels available.

We cannot expect a mechanical device to perform the remaining techniques of full optimization, namely the permutation and alteration of instructions based on knowledge of the computer and of the program itself. But we could prepare a list of all possible variations to the program, and apply the mathematical model to each one, finally choosing the one that gives us the least running time.

All of this makes a fine theory, but it is a practical problem we are trying to solve. In our floating-add example, at least 1728 possible variations to the program are reasonable. Are we to set up the equations 1728 times and solve them, when a single setup has proved to be too time-consuming for the 650 to handle?[17] There is probably some algorithm that could significantly reduce the 1728 cases, but we still are lost — because problems to be optimized generally involve 200 or more lines of coding, not a mere 46 as in our floating-add routine. It is apparent that we cannot hope to achieve rigorously perfect optimization in a reasonable amount of time.

Approximations

Once we are resigned to accepting only a close approximation to rigorous optimizing, we wonder whether there are efficient mechanical methods for generating fairly decent results. In this field we already have SOAP,

---

[17] This setup did not include provision for incorporating "technique (2)," and we would have had at least two more variables in the program if we were to include that technique.

which gave us a time of 816 under the "tight" requirements, compared to 2326 by sequential programming and 428 by hand optimization.

The first question is whether to improve SOAP by giving it better optimizing rules, while retaining its policy of assigning locations on a first-come-first-served basis. Indeed, SOAP optimization can be improved in several respects: The `LDI` and `STI` commands often give bad results if you are not using a whole 50-word band; the index register commands always add a word time to prepare for a possible complement cycle, although certain instructions like `RAA 0005` never require one; etc.

But experience has shown that changes to SOAP's optimization rules make only negligible differences in program running times. In Case SOAP III [7], for example, the I-addresses of shift commands were picked starting at the lower limit, and the next instruction was optimized as if it appeared at the higher limit. The idea was that if locations were scarce, there would be more chance of finding an optimal available location, if the search began at the lower limit. Elaborate coding went into the SOAP processor to accommodate this feature; but after observing thousands of instructions assembled by Case SOAP III, the author has yet to find a single instance where this "improvement" saved even one drum revolution in the final output. Other optimizing inefficiencies covered up this advantage. We have found in general that it is not worth the effort to improve SOAP's rules.[18]

That brings us to a second question: Why not improve SOAP so that it does not decide on locations the first time it meets them, but iterates on the locations it picks, improving their relative relationships? For example, an approach like the following is plausible:

First systematically take out the "very simple" addresses that we called VSα in our earlier development; it will be sufficient to optimize the other addresses. As an initial location for the remaining addresses, take the ordinary SOAP equivalents. Now consider an address SSα: For the ith occurrence of SSα we can calculate the latency time $w_i(n)$ by which the machine must wait at this spot if SSα is assigned to drum level n, in a manner similar to our calculations for the mathematical model. And if p_i is a priority constant indicating the relative importance of the ith occurrence, let n' be a drum level that minimizes[19] $\sum p_i w_i(n)$. Call

---

[18] But it is occasionally useful to let programmers change the optimizing rules temporarily for their own use (see the pseudo-op `NXT` in [7] or [8]).

[19] For this purpose we can calculate m_i, the level that minimizes $w_i(n)$. Since $\sum p_i w_i(n)$ has discontinuities and changes in slope only at the points $\{m_i\}$, we can take n' to be one of those points.

n' the "optimum forward level." We now replace SS1 by its optimum forward level, based on the SOAP equivalents of SS2, ..., SSℓ; then replace SS2 by its optimum forward level, based on the new equivalent of SS1 and the SOAP equivalents of SS3, ..., SSℓ; and so on until we replace SSℓ with its optimum forward level, based on the new equivalents of SS1, ..., SS$(\ell - 1)$. This will complete the first iteration.

For the second iteration we use the "optimum backward level," which is obtained in the same manner except that $w_i(n)$ is replaced by $v_i(n)$, the latency time by which the machine must wait to get to the next address *after* the ith occurrence of SSα if the symbol is assigned to drum level n. The succeeding iterations alternate between forward and backward optimization; or perhaps experience may show that a random choice between forward and backward at each complete iteration step will give better results.

As an example, the floating-add coding was put through this procedure by hand, with the following results:

Iteration	EXIT	ARG1	EXP1	EXP2	ARG2	SHIFT	ADD	Time
0: SOAP	06	32	39	15	22	27	43	666
1: forward	09	36	00	44	06	39	28	566
2: backward	09	36	44	45	29	19	37	528
3: forward	09	36	44	23	30	39	06	566
4: backward	09	36	44	23	29	49	19	578
5: forward	09	36	44	15	22	39	48	528
6: backward	09	36	44	01	29	19	37	528

The execution times are given as weighted averages, as in Table 4; we have chopped 138 word times off SOAP optimization. But in order to save another 100 word times, to reach the hand-optimized level, we need much better methods than this, for we have only achieved optimization in a narrow sense. The final jump to minimum time must be one that incorporates techniques (1) through (5) of full optimization.

Therefore we find that improvements in SOAP are minor and generally not worth the labor involved to make them. The essential skewness of the floating-point program, as far as SOAP is concerned, lies in the choice of the particular sequence of instructions out of the 1728 equivalent reasonable formulations. The preparation of input cards for any improvement in SOAP requires more human labor anyway — the specification of priority numbers, etc., etc. — so we will be gaining little over the time it would take to sit down and optimize the thing by hand.

For all of these reasons, therefore, the following approach appears to be the most practical: Let the programmers do the optimizing, but let's make the job as easy as possible for them by doing all the routine calculations by machine. Such an approach enabled the author to hand optimize a large part of the SuperSoap assembly processor [8] in a single afternoon.

HAND SOAP

The comparison between previous methods of hand optimization and the method described here (called HAND SOAP) is the same as that between machine language coding and symbolic language coding. You can make a HAND SOAP deck from an ordinary SOAP deck by changing about fifty instructions and about two dozen plugboard wires.

To use HAND SOAP, you simply prepare a SOAP symbolic program as usual, except that everywhere an undefined address occurs, you specify a number between 00 and 49 that indicates the exact drum level on which this address is to appear. Among the advantages of such a process over the regular methods of hand optimizing are:

1. You get all the usual benefits of symbolic programming, including mnemonic opcodes, symbolic locations, and pseudo operations. You can debug the program from a symbolic deck before hand optimizing.

2. You need consider *only drum levels*, not machine locations, so you can concentrate your attention on the important aspects of optimizing.

3. You need not worry about overlaying two instructions into the same location, or using unavailable locations; HAND SOAP will do that bookkeeping for you.

4. After the work is done the final result is easily readable by others, because they can follow the symbolic language.

5. Corrections of errors is simple; just change a few cards and assemble again.

The SuperSoap assembly system [8] includes HAND SOAP as an optional feature and also allows the user to mix SOAP optimization with HAND SOAP optimization. The use of HAND SOAP techniques seems to increase a programmer's efficiency by a factor of about three, and to reduce debugging time by a factor of approximately five. Such quick techniques make it best to let human "artists" do the optimizing; automation is not yet capable of filling their shoes for this job.

Systematic Hand Optimizing ... by Hand

Now let us see how a person can do some of his or her creative hand optimization in a systematic way. We will consider again the floating-addition subroutine as an example, with the problem being to fit it into locations 0100–0153.

The equipment necessary will be an inexhaustible supply of yellow tablets, a felt pen ("Magic Marker" or some substitute), a 650 Optimum

Programming Chart,[20] a memory map form (easy to prepare, it contains 2000 boxes in a 40×50 array, one box for each 650 location), and several pencils with erasers.

0. First be absolutely sure your program is correct, for mistakes in the program are easier to correct before optimizing gets under way.

1. Prepare a tableau that contains the SOAP coding followed by three columns (for L-, D-, and I-addresses, respectively) in which you will put the drum levels as you assign them. Put a star in all places where the address will be *defined* when SOAP encounters it; put three stars where addresses will not be optimized. (See the starting tableau in Table 7.) Mark out all of the unavailable locations on your memory map.

2. Put in new dummy symbols wherever you have a choice of temporary storage from among several locations. For example, notice the

---

[20] Some comments should be appended on the use of the Optimum Programming Chart: (a) Core locations are immediate access only when the core interlock is off; it is set on tape and disk storage operations. (b) Addresses 8000–8007 are immediate access only when arithmetic interlock is off; after shift commands this means waiting several word times longer than you would expect, the common mistake being to optimize `SRT 0004 8002` incorrectly (see Table 5). (c) A "complement cycle" occurs whenever you add numbers with opposite signs (or subtract numbers with equal signs) and the result is negative. If `FOUR` contains the positive constant 0000000004 and `THREE` contains +0000000003, the commands `RSL FOUR` (+0 −4) and `ALO THREE` (−4 +3) both give a complement cycle. This extra cycle does not affect ordinary commands unless you have an I-address of 800X; but it profoundly affects the index register commands `RSA 0004`, `AXA 0003`, which need *six* word times, not five. (d) The `TLU` command always starts the table lookup at drum level 00. Even if you are doing a table lookup in core storage, the machine waits for the 00 home pulse before searching through the core. (e) The shift-and-round instruction takes one less word time than IBM's charts (to this date) say; see Table 5 for the correct information. (f) When doing continuous reading or punching you have about 55 drum revolutions between cards read in and about 110 drum revolutions for computing between cards punched out; these numbers vary from machine to machine. (g) Multiply and divide should be optimized as low-high limit instructions like the shift commands. The low limit is the same as for `AUP`. In particular, since `DIV` always takes at least 60 word times, the I-address of `DIV` can be placed anywhere and there will be no latency time. (h) Optimizing low-high limit instructions should always use the low limit only, when the next instruction does not interrogate the arithmetic interlock (e.g., `SET`, `LDI`, `STI`, `RCD`, `PCH`). The `NOP` command does interrogate the interlock.

TABLE 7. The starting tableau

01	FAD	STD	EXIT					
02		SLT	0008		*		* * *	
03		STU	ARG1		*			
04		STL	EXP1		*			
05		RAU	ACC		*	*	**00**	
06		SRT	0002		*		* * *	
07		STL	EXP2		*			
08		STU	ARG2		*			
09		RSM	8002		*		* * *	
10		AML	EXP1		*	*		
11		SLT	0001		*		* * *	
12		NZU	BIG		*			
13		SRT	0005		*		* * *	
14		LDD	SRT	STORE	*			
15	BIG	LDD	SRD	STORE	*			*
16	STORE	SDA	SHIFT		*			
17		BMI	SWICH		*			
18		RAU	ARG2		*	*		
19		LDD	ARG1	SHIFT	*	*		*
20	SWICH	RAU	ARG1		*	*		
21		LDD	EXP2		*	*		
22		STD	EXP1		*	*		
23		LDD	ARG2	SHIFT	*	*		*
24	SRT	SRT	0000	ADD	*		* * *	
25	SRD	SRD	0000	ADD	*		* * *	*
26	ADD	AUP	8001		*		* * *	
27		NZE		ZERO	*			
28		SCT	0000		*		* * *	
29		STL	TEMP1		*	*		
30		SRT	0002		*		* * *	
31		ALO	8002		*		* * *	
32		SLO	8002		*		* * *	
33		SCT	0004		*		* * *	
34		STU	TEMP2		*	*		
35		ALO	TEMP1		*	*		
36		SLT	0008		*		* * *	
37		RAM	8002		*		* * *	
38		SLO	ADD		*	*		
39		SML	EXP1		*	*		
40		BMI		LOW	*			
41		NZU	HIGH		*			
42		SRT	0008		*		* * *	
43		SML	TEMP2		*	*		
44		SIA	ACC		*	*	**00**	
45		RAL	8001	EXIT	*		* * *	*
46	ZERO	STL	ACC	EXIT	*	*	**00**	*

use of TEMP1 and TEMP2 in Table 7; they will eventually be identified with any two of the locations ARG1, ARG2, EXP2, ACC, SHIFT.

3. Fill in the known locations, in our case ACC; in general this may include read or punch band locations or TLU locations, etc. If no known

locations exist in your program, assign a level arbitrarily to the most frequently occurring address. In the FAD example you will now have filled in the starting tableau of Table 7.

4. Go once through the program quickly, lightly filling in arbitrary assignments. In this way you will get the feel of the problem and will see where tight optimization is possible without any trouble. See how well you come out — there is a chance that the program is easy to optimize, in which case you've saved a lot of trouble. Or you might find cases where the program has a lot of skewness.

In our example, we notice some unavoidable skewness in lines 16–23: We cannot put ARG2 just a little before ARG1 and also just a little after. There also seems to be a conflict between those lines and lines 07–10. Furthermore, how can we hope to optimize 'LDD ARG2 SHIFT', when ARG2 on line 18 is to come a little after SHIFT?

5. Experiment around the known locations. Write lightly until you are sure of yourself; you will be doing a lot of erasing!

In line 44 we can assign the I-address of SIA ACC to level 03; then EXIT can be made level 09. Looking around at the other appearances of EXIT shows that we can make FAD = 06, and then we'll have enough time to get to line 05 where ACC appears again. Therefore we decide that the best level for EXIT is 09; for FAD it is 06; and the I-address on line 44 should be 03. We had a choice of possibly permuting lines 01 and 02, but now we see that we're better off not doing so. Three locations are determined so far; we haven't enough information yet to choose any others.

6. Study the skewness problems, in this case lines 16–23. We could try moving lines 21–22 before line 20; but that's a lost cause, leading to the conflicting sequences 'RAU ARG2, LDD ARG1' and 'RAU ARG1, LDD ARG2'. We must resign ourselves to losing some time in those lines. Since the positive branch in line 17 is taken in the common case where the two exponents are equal, we decide to favor that branch by putting ARG2 a little before ARG1, while at the same time putting EXP2 before EXP1.[21]

7. Try to place symbols that occur relatively near themselves an integer multiple of drum revolutions apart. Consider the two occurrences of TEMP1. Inspection shows about one revolution between lines 29 and 35, so we do not want to interchange lines 34 and 35. But if we consider the two occurrences of ADD, we find that it occurs early in line 38, so we interchange lines 38 and 39.

---

[21] Many of the arguments in this example are not rigorous, of course; they are just a heuristic way of looking at the solution of this problem.

8. Work from the known locations. Looking toward the end, we see that `TEMP2` in line 43 should be placed on or before level 41; with this in mind we see that we can get from (the new) line 38 to line 43 in a drum revolution if `EXP1` is up in the forties. Hence instructions 03 and 04 should be kept in that order, and `ARG1` ought to be in level 36 while `EXP1` is in 43 or 44. This means that `TEMP2` (in the vicinity of 41) can only be identified with the temporary storage `SHIFT`—and we find this to be a fortuitous choice, because `SHIFT` should be approximately level 39 (according to line 19). Furthermore `TEMP1`, which lies a little after `TEMP2` by line 35, can be best identified with location `ACC` = 00. (When things start to click like this, you know you are in the right ballpark.)

We check to see that we can get from `EXP1` = 43 in (new) line 38 to `SHIFT` = 39 in line 43 in a little less than a drum revolution. There is some question about how to optimize the shifts in lines 24–25; how many shifts should we accommodate? Trying level 41 for `SHIFT` and level 00 for `TEMP1`, we find that we can safely allow time for the full shift-round of 10 in line 25 and a long shift-count in line 28.

9. Once a fair number of locations have been chosen, start at the beginning and write in the definite or approximate levels that are known so far. Indicate the chosen locations heavily, the approximate ones lightly. This step is shown in Table 8; when only a low or high limit is given it means that a few more or a few less are allowed. There still are areas where much arbitrariness is involved; we will do them last.

When you reach this step you may come to a spot where the timings are just terrible, because you forgot to consider one little point. Then you might decide to start all over, or to finish it this way now and another way later and compare the times. In our case we've had no such problems, except in places where we expected to take a beating.

It is important never to fill in a definite level at this stage unless there's no choice left. If there is any freedom (e.g., `EXP1` is either 43 or 44), leave it undecided at this point.

10. Begin to make the availability table. With the felt marker, place a dot in each of the cells of the drum map where you have definitely picked a location, simultaneously placing a dot on the tableau sheet next to the location you are reserving. In our case, levels 03, 06, 09, 36, and 39 have been reserved. Now we see that, for example, `SHIFT` cannot be placed in 39.

(You may have reserved locations that were not available to you—in such a case you must start over. But such trouble should not arise, because only a few locations where absolutely tight optimization occurs should be reserved at this point.)

TABLE 8. First assignments

01	FAD	STD	EXIT		**06**	**09**	12–13
02		SLT	0008		*	* * *	19–32
03		STU	ARG1		*	**36**	**39**
04		STL	EXP1		*	43–44	46–47
05		RAU	ACC		*	* **00**	05–
06		SRT	0002		*	* * *	11–
07		STL	EXP2		*	15–21	18–24
08		STU	ARG2		*	22–28	25–31
09		RSM	8002		*	* * *	33–40
10		AML	EXP1		*	* 43–44	47–
11		SLT	0001		*	* * *	03–
12		NZU	BIG		*	07–	08–
13		SRT	0005		*	* * *	15–
14		LDD	SRT	STORE	*	24–	27–37
15	BIG	LDD	SRD	STORE	* 07–		* 27–37
16	STORE	SDA	SHIFT		* 27–37	39–41	42–
17		BMI	SWICH		*	45–	46–
18		RAU	ARG2		*	* 22–28	–33
19		LDD	ARG1	SHIFT	*	* **36**	* 39–41
20	SWICH	RAU	ARG1		* 45–	* **36**	41–
21		LDD	EXP2		*	* 15–21	18–
22		STD	EXP1		*	* 43–44	47–
23		LDD	ARG2	SHIFT	*	* 22–28	* 39–41
24	SRT	SRT	0000	ADD	*	* * *	00–02
25	SRD	SRD	0000	ADD	*	* * *	* 00–02
26	ADD	AUP	8001		*	* * *	11–
27		NZE		ZERO	*	14–25	15–45
28		SCT	0000		*	* * *	–45
29		STL	ACC		*	* **00**	03–
30		SRT	0002		*	* * *	09–
31		ALO	8002		*	* * *	17–
32		SLO	8002		*	* * *	25–
33		SCT	0004		*	* * *	31–36
34		STU	SHIFT		*	* 39–41	44–47
35		ALO	ACC		*	* **00**	05–
36		SLT	0008		*	* * *	11–
37		RAM	8002		*	* * *	31–40
38		SML	EXP1		*	* 43–44	47–49
39		SLO	ADD		*	* 00–02	05–07
40		BMI		LOW	*	08–10	
41		NZU	HIGH		*		12–14
42		SRT	0008		*	* * *	19–36
43		SML	SHIFT		*	* 39–41	43–46
44		SIA	ACC		*	* **00**	**03**
45		RAL	8001	EXIT	*	* * *	* **09**
46	ZERO	STL	ACC	EXIT	* 15–45	* **00**	* **09**

11. The home stretch: The remaining task is to find locations that fit the approximate levels and that also remain available. This is not quite as easy as it sounds, because each location filled may restrict some other location. A systematic way is to fill locations first where only two

or three choices are possible, then branch out into the more free areas. The choices here are made by consulting the availability table and trying to maintain an even distribution among the available locations.

As each location is chosen, pencil it out on the availability table. Then after each group of 20 or so locations has been chosen, go back and dot all known locations that are not yet marked with the pen, in both the tableau and the availability table. (This double-checking of the availability table saves a lot of time, for it is amazing how often you will choose a location and forget to mark it, even if you try to be careful.) Then do 20 more in pencil, then double check with the marker, and so on until all locations are picked.

In line 01 we choose 13 for the I-address; in line 02 we have the choice of anything from 19–32, so we save it for later; in line 04 we set `EXP1` = 43 and choose 46 for the I-address; then we work on lines 38–43; etc. Hand optimization is usually not the *elimination* of latency times, but the *intelligent wasting* of time. With this procedure we continue to fill locations in, until finally only the I-addresses of the long shift commands remain together with the locations `ZERO`, `BIG`, `SRD`, `LOW`, `HIGH`. (`HIGH` and `LOW` have lowest priority because they are error indications and the computer will stop anyway.) These are fitted into the remaining spaces and the result of Table 1 appears.

12. Assembly: Punch all the levels you have defined onto the assembly input deck, and assemble with HAND SOAP. If you've made any errors by putting too many instructions on one level, or by failing to give a level for an undefined drum location, the assembler will point them out.

13. "Time Study." A computer program called Time Study has been developed to give you further help. Input your hand-optimized output cards to the Time Study program and it will calculate all the latency times, to check for errors. If the latency ever comes to 48 or 49, you have probably made a mistake reading the optimum timing chart. The output of Time Study is usually inspected manually or on a sorter, and it is a good way to double-check your work. With this program you need not be afraid to use tight optimization. Time Study will warn you whenever you get too stringent.

Conclusion

Two years of successful operation have demonstrated that the HAND SOAP and Time Study programs provide an efficient and practical approach to the problem of latency minimization. For such combinatorial tasks, the keen minds of men and women still outperform computer techniques, but machines are useful for the clerical details.

Although features of the IBM 650 computer have been used heavily for the exposition in this paper, the methods can be adapted for any other drum machine.

Acknowledgment

The flow charts in Figs. 1–3 were drawn by Jill Carter.

References

[1] R. V. Andree, *Programming the IBM 650 Magnetic Drum Computer and Data-Processing Machine* (New York: Holt, 1958), 56–57.

[2] F. M. Chadwick and H. L. Loeb, "On a mathematical model for the optimum programming of a drum computer," Convair-Astronautics Applied Mathematics Series #26 (1959), 14 pages.

[3] Ralph E. Gomory, "An algorithm for integer solutions to linear programs," Princeton–IBM Mathematics Research Project, Technical Report No. 1 (17 November 1958); *Recent Advances in Mathematical Programming*, edited by Robert L. Graves and Philip Wolfe (New York: McGraw–Hill, 1963), 269–302.

[4] Ralph E. Gomory, "All-integer integer programming algorithm," Research Report RC-189 (Yorktown Heights, New York: IBM Research Center, 29 January 1960), 18 + 4 pages.

[5] International Business Machines Corporation, "Optimum programming," 650 Data Processing System Bulletin G24-5002-0 (September 1958), 21–31.

[6] International Business Machines Corporation, "Notes on optimizing 650 programs," 650 Bulletin 13, Form 32-0313 (November 1957).

[7] Donald E. Knuth, "Case SOAP III," Case Institute of Technology Computer Center reports, series 4, vol. 1 (February 1958), 28 pages.

[8] Donald E. Knuth, "SuperSoap," Case Institute of Technology Computer Center reports, series 4, vol. 2 (August 1959), 55 pages.

[9] G. R. Trimble, Jr., and E. C. Kubie, "Principles of optimum programming the IBM Type 650," "An interpretive floating decimal system for the IBM Type 650," and "Floating decimal sub-routines for the IBM Type 650," in Technical Newsletter No. 8, IBM Applied Science Division (September 1954), 5–43.

[10] V. M. Wolontis, "A complete floating-decimal interpretive system for the IBM 650 magnetic drum calculator," Technical Newsletter No. 11, IBM Applied Science Division (March 1956), 63+xxi pages.

Historical Postscript

Most readers of this article in 1961 would have been familiar with the IBM 650, a computer that worked with signed 10-digit decimal numbers. Its instructions consisted of a 2-digit operation code, followed by a 4-digit data address, followed by the 4-digit address of the next instruction to be executed. There was a 20-digit "accumulator" for arithmetic operations, divided into an upper half (address 8003) and a lower half (address 8002). There also was a "distributor" register (address 8001).

To invoke the `FAD` subroutine that is described in this paper, a programmer who wanted to add the floating-point number in 8002 to the floating-point number in location `ACC` would give the instruction '`LDD NEXT FAD`', which meant, "Load the distributor with the contents of address `NEXT` and go to address `FAD`." Address `FAD` had the instruction '`STD EXIT`' (see line 01 of the program), which would store the distributor in the drum location whose address was `EXIT` and proceed to the instruction in line 02. That instruction, '`SLT 0008`', appeared in an address that needed no symbolic name, so its location has been left blank in the example; it means, "Shift the accumulator 8 digits to the left." Eventually control would pass to address `EXIT`, either in line 45 or line 46, and the user's program would be resumed — unless exponent overflow (`HIGH`) or exponent underflow (`LOW`) was detected.

All instructions and data on the 650 were stored on a rotating magnetic drum, in such a way that addresses (0000, 0050, 0100, ..., 1950) were accessible at the same time. Then addresses (0001, 0051, 0101, ..., 1951) would be accessible 96 microseconds later, and so on; you had to wait 4.8 milliseconds to get back to addresses (0000, 0050, 0100, ..., 1950) again. If the instructions and data were placed cleverly, several operations could be performed every time the drum revolved, but in the worst case a single instruction could last longer than two revolutions of the drum. Optimum placement could therefore speed things up significantly. Software programs called SOAP (the "Symbolic Optimum Assembly Program") were available to ameliorate this task. (For further information, see Chapter 13 of the author's book *Selected Papers on Computer Science*.)

Addendum

This paper was prompted by an "open request" published by Webb T. Comfort in *Communications of the ACM* **1**, 11 (November 1959), A11, where he proposed a meeting to exchange information about latency minimization. He stated that "some of the more recent applications of drum computers are in the area of high-speed real-time control systems. The latency problems in such applications are considerably more significant than in general applications (where slower programs can be offset by larger budgets). ... Discussion at the recent ACM meeting disclosed that there are at least a few people besides myself who are seriously interested in this problem."

However, significant improvements in the technology of magnetic core memory changed the picture completely during the early 1960s, rapidly making machines such as the 650 obsolete. As a result, I cannot recall ever meeting anyone who found this paper actually useful in practice, although I still believe that my approach to latency minimization was the best way to proceed in the days when drum computers were still important and dominant.

The integer programming problem presented above as "Example: Version 2" remains interesting as a benchmark for IP-solvers. (Its coefficients appear in the file `latency.data` on my website.) Many years after this paper was published, I learned that my inability to solve the problem in 1960 was not at all surprising, since it presented lots of trouble even to state-of-the-art integer programming systems in 1995! The first optimum solution was found at that time by Dimitris Alevras at Zuse-Institut Berlin, who reported that the current CPLEX system was unable to find a good upper bound in a reasonable amount of time. He got around this problem by replacing the nonnegative variables t_3 through t_{30} by unrestricted integer variables $u_k = t_k - t_{k-1}$ for $3 \le k \le 30$; then he was able to find a solution of cost 29300 rather quickly, but only after modifying CPLEX's default parameters for branching, sos, clique, and cover constraints. Adding the further constraint "objective function $\le$ 29300" led to a solution of cost 22996, after 8880 nodes of a branch-and-bound tree had been generated. Finally, the problem of verifying that 22996 is actually optimum caused CPLEX to investigate 732,200 further B&B branches(!).

Alevras remarked that this problem is "amazingly sensitive from a numerical point of view." However, its constraints have a peculiar form that is rather different from other problems in the literature; one can imagine that new techniques specifically intended for such constraints might work substantially better.

Here's the optimum solution that CPLEX eventually found:

$$t_2=0,\ t_3=0,\ t_4=1,\ t_5=1,\ t_6=1,\ t_7=1,\ t_8=3,\ t_9=2,\ t_{10}=3,\ t_{11}=3,$$
$$t_{12}=3,\ t_{13}=4,\ t_{14}=5,\ t_{15}=6,\ t_{16}=7,\ t_{17}=8,\ t_{18}=8,\ t_{19}=9,\ t_{20}=10,$$
$$t_{21}=10,\ t_{22}=3,\ t_{23}=4,\ t_{24}=4,\ t_{25}=5,\ t_{26}=5,\ t_{27}=3,\ t_{28}=5,\ t_{29}=5,\ t_{30}=5;$$
$$x_2=9,\ x_3=36,\ x_4=43,\ x_5=21,\ x_6=28,\ x_7=0,\ x_8=63,\ x_9=35;$$

$$w_1=0,\ w_2=2;\quad z_1=6,\ z_2=19,\ z_3=12,\ z_4=15,\ z_5=21,$$
$$z_6=0,\ z_7=14,\ z_8=31,\ z_9=17,\ z_{10}=12,\ z_{11}=18,\ z_{12}=18.$$

(As stated in the paper, we let $t_1 = 0$ and $x_1 = 0$.) The value $x_8 = 63$ is a surprise, since it represents a drum level; however, the transformation

$$x_8 \leftarrow x_8 - 50\delta,\ \ z_8 \leftarrow z_8 - 25\delta,\ \ t_9 \leftarrow t_9 + \delta,\ \ t_{12} \leftarrow t_{12} + \delta,\ \ t_{26} \leftarrow t_{26} + \delta$$

leaves the inequalities and objective function invariant, so we get another optimum solution with $x_8 = 13$, $z_8 = 6$, $t_9 = 3$, $t_{12} = 4$, $t_{26} = 5$. This solution is essentially equivalent to the one found by the heuristic "SOAP-forward-backward" iteration in part 3 of the paper.

A modern reader will immediately notice the fact that I gave no proof of correctness for the algorithm presented in Figures 1–3. Indeed, the very idea of correctness proofs for algorithms was nowhere visible on my intellectual horizon when I submitted this paper for publication in June 1960. Such radical notions never crossed my mind until I met Bob Floyd in September 1962. Even today I know of no easy way to demonstrate that the stated procedure is correct, although I haven't been able to come up with a counterexample. I must admit being somewhat relieved to learn in 1995 that the CPLEX solution agreed with my intuition about the `FAD`-latency problem — because CPLEX had been presented only with a purely mathematical system of inequalities, not with the concrete interpretation that I had placed on the abstract variables involved. I secretly feared that a gap in my reasoning might have meant that some very strange solutions to those inequalities could exist.

While preparing this chapter for republication I noticed that I had made an error when originally constructing those inequalities by hand: Instead of '$50(t_{18} - t_{17}) + x_4 - 2z_9 - 8 \geq 0$' I should have introduced a new z-variable and given two inequalities,

$$50(t_{18} - t_{17}) + x_4 - 2z_{13} - 8 \ \geq\ 0, \qquad x_9 \ \leq\ 2z_{13} + 1.$$

Egon Balas, Matteo Fischetti, and Arrigo Zanette ["A hard integer program made easy by lexicography," *Mathematical Programming* **A135** (1012), 509–514] have used dramatically improved integer programming techniques to show that the optimum cost remains 22996 when this correction is made.

See George R. Trimble, Jr., "A brief history of computing: Memoirs of living on the edge," *IEEE Annals of the History of Computing* **23**, 3 (July–September 2001), 44–59, especially page 49, for his account of the hand-optimized programs in [9] that inspired those of [10].

Index